D0903570

Product Integration
with Applications to
Differential Equations

GIAN-CARLO ROTA, *Editor*
ENCYCLOPEDIA OF MATHEMATICS AND ITS APPLICATIONS

Volume		Section
1	LUIS A. SANTALÓ **Integral Geometry and Geometric Probability,** 1976	Probability
2	GEORGE E. ANDREWS **The Theory of Partitions,** 1976	Number Theory
3	ROBERT J. McELIECE **The Theory of Information and Coding** A Mathematical Framework for Communication, 1977	Probability
4	WILLARD MILLER, Jr. **Symmetry and Separation of Variables,** 1977	Special Functions
5	DAVID RUELLE **Thermodynamic Formalism** The Mathematical Structures of Classical Equilibrium Statistical Mechanics, 1978	Statistical Mechanics
6	HENRYK MINC **Permanents,** 1978	Linear Algebra
7	FRED S. ROBERTS **Measurement Theory** with Applications to Decisionmaking, Utility, and the Social Sciences, 1979	Mathematics and the Social Sciences
8	L. C. BIEDENHARN and J. D. LOUCK **Angular Momentum in Quantum Physics:** Theory and Application In preparation	Mathematics of Physics
9	L. C. BIEDENHARN and J. D. LOUCK **The Racah-Wigner Algebra in Quantum Theory** In preparation	Mathematics of Physics
10	JOHN D. DOLLARD and CHARLES N. FRIEDMAN **Product Integration** with Application to Differential Equations, 1979	Analysis

Other volumes in preparation

ENCYCLOPEDIA OF MATHEMATICS and Its Applications

GIAN-CARLO ROTA, *Editor*
ENCYCLOPEDIA OF MATHEMATICS AND ITS APPLICATIONS
Volume 10

Section: Analysis
Felix E. Browder, *Section Editor*

Product Integration with Applications to Differential Equations

John D. Dollard and **Charles N. Friedman**
Department of Mathematics
University of Texas, Austin, Texas

Foreword by
Felix E. Browder
University of Chicago

Appendix by
P. R. Masani
University of Pittsburgh

1979

Addison-Wesley Publishing Company
Advanced Book Program
Reading, Massachusetts

London•Amsterdam•Don Mills, Ontario•Sydney•Tokyo

Library of Congress Cataloging in Publication Data

Dollard, John Day.
Product integration with applications to differential equations.

(Encyclopedia of mathematics and its applications ; v. 10)
Bibliography: p.
Includes index.
1. Differential equations. 2. Differential equations, Partial. 3. Integral equations.
I. Friedman, Charles N., joint author. II. Title.
III. Series.
QA371.D64 515′.35 79-20454
ISBN 0-201-13509-4

American Mathematical Society (MOS) Subject Classification Scheme (1970):
28-02, 34-01, 34-02, 34B05, 34D05, 34G05, 35-02, 35F05, 35F10, 35G05, 35G10, 35J15, 45-02, 45A05, 45N05

ABCDEFGHIJ-HA-79

Printed in the United States of America

Contents

Editor's Statement

A large body of mathematics consists of facts that can be presented and described much like any other natural phenomenon. These facts, at times explicitly brought out as theorems, at other times concealed within a proof, make up most of the applications of mathematics, and are the most likely to survive changes of style and of interest.

This ENCYCLOPEDIA will attempt to present the factual body of all mathematics. Clarity of exposition, accessibility to the non-specialist, and a thorough bibliography are required of each author. Volumes will appear in no particular order, but will be organized into sections, each one comprising a recognizable branch of present-day mathematics. Numbers of volumes and sections will be reconsidered as times and needs change.

It is hoped that this enterprise will make mathematics more widely used where it is needed, and more accessible in fields in which it can be applied but where it has not yet penetrated because of insufficient information.

This is the first survey of the product integral since the turn of the century. Product integration, an idea going back to Volterra, has been and is being used today in several disparate circumstances; the present definitive treatment will contribute to make better known this useful technique.

P. R. Masani, in his Appendix, gives a different overview of the history of the subject. Professor Masani has adopted the colorful notation of the first workers. His subject is largely complementary to the main text. His different notation for the product integral will be useful to those who wish to gain access to the early literature.

GIAN-CARLO ROTA

Foreword

An editor's preface to a mathematics book does not have a clearly defined role in contemporary usage. If some past precedents were followed, I would merely remark that the present work by John Dollard and Charles Friedman is a completely self-contained treatment of the product integral on a simple and elementary basis. As such, I believe it to be unique as far as this topic is concerned. The applications that are presented fall mainly within the domain of ordinary differential equations. Some amplifications of the generality of the theme with applications to a wider circle of mathematical topics are described in the accompanying Appendix by P. R. Masani.

For the benefit of some readers at least, I shall go beyond this conventional restriction of the editor's function. Though in the last analysis, mathematical topics must be treated in full technical detail and with logical completeness (as they are indeed treated in the body of the present work), it is often useful to preface such a detailed development with a more discursive and less technical discussion.

What is the product integral? As the text tells us, it is an analytic process or class of processes first put forward by Volterra in the last decades of the nineteenth century for the study of various questions relating to the theory of ordinary differential equations. As Professor Masani reminds us in his Appendix, it was extensively developed by Schlesinger in the early part of the twentieth century, particularly in connection with differential equations in the complex domain. In later decades, it was investigated in connection with topics in functional analysis, stochastic processes, and the theory of analytic functions with operator values. It is based on the same heuristic idea which in the domain of quantum field theory was used by Feynman with such remarkable impact in his celebrated path-integral approach.

What is this heuristic idea? To describe it, let us form a simple model of the definition of a product integral. Consider a metric space M and a class $T(M)$ of self-mappings of M. Assume that this class is closed under compositions. Suppose that we are given a mapping $S(t, h)$ in this class $T(M)$ for each real parameter t in a given interval $[a, b]$ of the real line and for each real h with $0 < h < d$. For each partition of the interval $[a, b]$ with

$$a = t_0 < t_1 < t_2 < \cdots < t_n = b$$

such that each successive difference $t_{j+1} - t_j < d$, we can form the iterated composition of the mappings

$$S(t_{n-1}, t_n - t_{n-1}) \cdots S(t_1, t_2 - t_1)S(t_0, t_1 - t_0)$$

(where we note that the factors have to be composed in the prescribed *time-ordered* fashion). If, as the mesh of the partition approaches zero, these iterated composition mappings converge to a limit mapping T in a prescribed topology of convergence in $T(M)$, then we call this limit T the product integral, written in the form

$$T = \prod_a^b S(t, dt).$$

We have thus defined a Riemann product integral, with possible extensions to corresponding more-general product integrals of the Lebesgue type.

Where do such iterated compositions and their limits arise in a natural way? First and foremost, in the theory of ordinary differential equations in both the finite- and infinite-dimensional cases. Consider a differential equation of first order on a manifold M,

$$\frac{du}{dt} = A(t, u(t)).$$

Suppose that for a solution $u(t)$ of this equation on an interval $[a, b]$ under suitable conditions on the equation, for $s < t$, $u(s)$ determines $u(t)$. Then set

$$T(s, t)(u(s)) = u(t)$$

thereby defining the two-parameter family of mappings $T(s, t)$ called the *propagator* mappings corresponding to the given equation. Suppose that in addition we are given in some natural way a nice class of *pseudo-propagators* $S(t, h)$, i.e., mappings of M into M which satisfy the differential equation

$$\frac{d}{dh} S(t, h)(u)|_{h=0} = A(t, u).$$

One can form the product integral

$$W_{a,b} = \prod_a^b S(t, dt)$$

and can hope that $w_{a,b} = T(a, b)$ for all such pairs (a, b). In a wide variety of cases, this procedure yields useful machinery for constructing and studying the solutions of differential equations.

The simplest such case is that in which M reduces to a vector space V (as it always does locally) and we set

$$S(t, h)(u) = u + hA(t, u).$$

In this particularly simple case, the iterated composition of these particular pseudo-propagators corresponds simply to the application of the classical Euler polygon method for solving an initial-value problem for the ordinary differential equation. In other words, we approximate the differential equation by the system of difference equations

$$u(t_{j+1}) - u(t_j) = (t_{j+1} - t_j)A(t_j, u(t_j)),$$

and take the limit as the mesh of the partition goes to zero. If this approach works, we can write

$$T(s, r) = \prod_s^r (I + dt\, A(t)).$$

A more sophisticated case appears when the operators $A(t)$ are not everywhere defined and therefore cannot be applied to an arbitrary element of V, a situation characteristic of ordinary differential equations in infinite-dimensional spaces V obtained from initial-value problems for partial differential equations. In this case, we modify the difference equation of the preceding paragraph to obtain instead

$$u(t_{j+1}) - u(t_j) = (t_{j+1} - t_j)A(t_j, u(t_{j+1})).$$

If we solve this latter system, we find that

$$u(t_{j+1}) = (I - (t_{j+1} - t_j)A(t_j))^{-1}(u(t_j)).$$

The successive application of this formula yields an iterated composition of mappings defining the approximation for the product integral corresponding to $S(t, h) = (I - hA(t))^{-1}$. Where this approximation works, we find that

$$T(s, r) = \prod_s^r (I - dtA(t))^{-1}.$$

A classical case for linear operators in a Banach space is that treated in the Hille-Yosida theorem in which $A(t)$ is a fixed closed linear operator A satisfying the condition that

$$\|(I - hA)^{-1}\| \leq 1.$$

Here the product integral yields the semigroup generated by A.

A third illuminating special case involves solving the differential equation

$$\frac{du}{dt}(t) = A(t, u(t)) + B(t, u(t))$$

by what is called the fractional-step method in numerical analysis. Suppose that for each j we define

$$v_j - u(t_j) = (t_{j+1} - t_j)A(t_j, v_j)$$
$$u(t_{j+1}) - v_j = (t_{j+1} - t_j)B(t_j, u(t_{j+1})).$$

If we add the two formulas, we find a reasonable looking approximation to our differential equation, namely,

$$u(t_{j+1}) - u(t_j) = (t_{j+1} - t_j)[A(t_j, v_j) + B(t_j, u_{j+1})].$$

This suggests using the product integral for

$$S(t, h) = (I - hB(t))^{-1}(I - hA(t))^{-1}.$$

The success of this process in various cases leads to the representation of propagators for such equations by Trotter formulas.

The reader should bear in mind some general facts about product integrals.

Product integration is in fact a generalization in principle of the ordinary additive process of integration if one interprets the latter as applying to commuting operator-valued functions. If V is a normed linear space and if $A(t)$ is a continuous family of bounded linear operators on V with $A(t)$ and $A(s)$ commuting for each pair s and t, then

$$\prod_r^s \exp(dt\, A(t)) = \exp\left(\int_r^s A(t)\, dt\right)$$

where the exponential function is defined in the simplest way by the infinite power series, $\exp(T) = \Sigma_{j=0}^{\infty} T^j/j!$. However, product integration is fundamentally *noncommutative*. Since transformations (even linear ones) generally do not commute, the definition of the product integrals depends essentially on the *time-ordered* structure of the integrand.

Product integration is not a linear operation, nor in general is it restricted to linear operators or transformations. The linear case is often simpler in its structure, but the basic machinery is also applicable to the nonlinear case with appropriate modifications.

Finite dimensionality is also not a basic requirement for the application of product integrals. Indeed, the second illustrative case mentioned above is significant mainly in the context of operators defined in infinite-dimensional spaces. Even if we begin with a finite-dimensional context and then introduce such basic kinds of additional structure as random processes, we are led immediately and naturally into an infinite-dimensional situation. (For a formal account, we refer to Professor Masani's Appendix.)

When Volterra first defined product integrals in the late 1880s, the definition was part of the same movement to formulate analytical processes in explicitly self-conscious and general terms that led to the development of the beginnings of twentieth-century functional analysis—a process in which Volterra was one of the most explicit public spokesmen as shown by his address to the Paris International Congress of Mathematicians in 1900. Though overshadowed by more fashionable parts of analysis and functional analysis, the theory of the product integral in its various forms continues to be a focus of important activity. This is particularly the case in the theory of nonlinear evolution equations and their extension to nonstationary evolution processes. In the past decade and a half, iteration processes of the product integral type have been adapted to the study of a general theory of nonlinear semigroups, initiated in Hilbert space by Komura and extensively developed in a more general Banach space context as a nonlinear analogue of the older Hille-Yosida theory of linear semigroups.

Professors Dollard and Friedman have worked primarily on mathematical questions originating in mathematical physics. In the present book they have given the reader a concrete, simply written presentation of a basic tool in present-day analysis which should be of value to applied mathematicians and mathematical physicists as well as to mathematical analysts.

FELIX E. BROWDER
General Editor, Section on Analysis

Preface

This monograph is intended as an introduction to product integration for the general scientific audience and a reference book for workers in differential equations. Chapter 1 can be understood by a reader with only a knowledge of matrix algebra and elementary calculus. Later chapters assume additional knowledge on the part of the reader.

In a nutshell, the product integral is to the product what the ordinary integral is to the sum. The product integral arises in connection with *equations of evolution* and the associated initial-value problems. These have the form

$$y'(x) = A(x)y(x) \qquad y(x_0) = I \tag{1}$$

where y and A are operator-valued functions (in the simplest case they are nxn matrix-valued functions), and I is the identity operator. Many important scientific equations can be analyzed by solving initial-value problems of the type given in (1). Examples are the heat equation, the Schrödinger equation, and any linear ordinary differential equation. The product integral is a construction which solves the initial-value problem (1). Analogously, the ordinary integral $\int_{x_0}^{x} A(s)ds$ is a construction which solves the initial-value problem

$$y'(x) = A(x) \qquad y(x_0) = 0 \tag{2}$$

It is well known that there are many advantages to be gained by studying the ordinary integral as an object in its own right, rather than restricting oneself to the terminology "the solution of (2)." There are similar advantages to be gained by studying the product integral as an object in its own right. Strangely, this is usually not done, and many scientists are unfamiliar with the concept of product integration. The present authors attribute this to the fact that no comprehensive modern treatise on product integration has been available. V. Volterra, who invented product integration, wrote a monograph with B. Hostinski in 1938 [VV4], which deals with the fundamentals of the subject. However, their account is limited in generality by modern standards, and is difficult to read because of unnatural notational conventions and an approach which obscures the simplicity of the subject. As a consequence, those wishing to exploit product integral ideas have tended to reinvent the subject for themselves or rely on scanty accounts in research papers. It is a testimony to the usefulness of product integral concepts that advanced workers in differential equations have learned and exploited these concepts despite the unavailability of a standard text on the subject. The present authors think that the time is ripe for a monograph which will introduce product integrals to the general scientific audience.

In this monograph we deal almost exclusively with "linear" product integration, which is the theory obtained when $A(x)$ of eq. (1) is a linear operator for each x. This is the only case for which a systematic theory can be said to exist. The literature contains a number of interesting and important applications to cases in which $A(x)$ is nonlinear. However, the existence theory for such cases typically depends on very specialized hypotheses and cannot be viewed as resulting from a general theory in a natural way. For this reason, an account of this work would consist in little more than a recital of the contents of the relevant papers, and we chose to give only some brief remarks on this work along with references to the literature. We believe that our account of the linear case will serve as a sufficient introduction to the literature on nonlinear product integration.

Our account of the theory of linear product integration is quite extensive. Beginning with the case in which $A(x)$ is an $n \times n$ matrix, we point out the beautiful simplifications brought to the theory of linear ordinary differential equations by viewing them from the product integral viewpoint. Building on the foundation laid in the discussion of $n \times n$ matrices, we arrive by successive generalizations at an advanced theory suitable for the study of partial differential equations. (In this theory $A(x)$ could be, say, an unbounded operator on a Hilbert space.) Enough applications are given to illustrate the usefulness and flexibility of the theory.

JOHN D. DOLLARD
CHARLES N. FRIEDMAN

Introduction

We indicate here, briefly, the content of the chapters. Chapter 1 is an elementary introduction to the product integral of a continuous matrix-valued function and its properties. (The generalization to "Lebesgue product integrals" is sketched in Section 8.) This chapter should be accessible to readers with quite minimal mathematical background. It is prerequisite for understanding of the other chapters. Chapter 2 deals with contour product integrals; the development is parallel to that of the theory of ordinary contour integrals in complex variable theory. Contour product integrals are not used in later chapters. In Chapter 3 we present a theory of product integration in a much more general setting. More mathematical background is required here, for example, familiarity with functional analysis in Banach space and some integration theory. Included are results on the equation of evolution (1) with unbounded $A(x)$. The other chapters (except Chapter 4, Section 5) are independent of this chapter. Chapter 4 presents applications of product integration to the theory of differential equations; some new results concerning solutions of the Schrödinger equation with rather singular potentials are included. In Chapter 5 we discuss product integration of (matrix-valued) measures. Some familiarity with measure theory is assumed here, but nothing very sophisticated is required. Chapter 6 contains a discussion of work on product integration by various other authors and some remarks on generalizations of the theory. Following Chapter 6 is an appendix on matrix theory containing elementary definitions and results and a few special results which may not be familiar to all readers. We have included, finally, a fairly extensive and hopefully complete list of references together with notes. This includes all papers, articles, books, etc., known to us, which develop, discuss, or make use of product integration.

In each chapter, theorems, definitions, and formulas are numbered consecutively within each section. Thus within Chapter 1, definition 1.2 refers to the second definition of Section 1, the notation (2.3) refers to the third formula of Section 2, etc. References in a chapter to theorems (definitions, etc.), in a previous chapter are given by stating the theorem number and chapter, e.g., "Theorem 5.1 of Chapter 2" or by prefixing an Arabic numeral indicating the chapter; e.g., the theorem just cited would be referred to as "Theorem 2.5.1."

The authors would like to thank Professors M. Crandall, J. R. Dorroh,

W. E. Fitzgibbon, J. Goldstein, J. Helton, J. S. MacNerney, P. R. Masani, J. Neuberger, R. Showalter, B. Simon, M. K. Smith, and G. Webb for suggestions and help in compiling the list of references. Thanks are also due to B. Simon for several interesting discussions on product integration and applications of the theory.

The authors are also pleased to acknowledge a grant of one thousand dollars from the University Research Institute of the University of Texas at Austin.

Product Integration
with Applications to
Differential Equations

CHAPTER 1

Product Integration of Matrix-Valued Functions

1.0 Introduction

In this chapter we define product integration for $n \times n$ matrix-valued functions. As will be seen, the concept of product integration can be used as a central unifying idea in the study of systems of linear differential or integral equations. A large portion of the present chapter is devoted to an elaboration of this point. We begin with a brief explanation of the intuitive connection between linear differential equations and product integrals. (For the definition of various standard terms in matrix theory, see Appendix 1.)

Consider a system of n linear differential equations in n unknowns, having the form

$$\begin{aligned} y_1'(s) &= a_{11}(s)y_1(s) + a_{12}(s)y_2(s) + \cdots + a_{1n}(s)y_n(s) \\ &\vdots \\ y_n'(s) &= a_{n1}(s)y_1(s) + a_{n2}(s)y_2(s) + \cdots + a_{nn}(s)y_n(s). \end{aligned} \tag{0.1}$$

The coefficients $a_{ij}(s)$ are assumed to be continuous on an interval $[a,b]$. At the endpoints of the interval, the derivatives are one-sided. Suppose that the values $y_1(a), y_2(a), \ldots, y_n(a)$ are given. Consider the problem of finding $y_1(b), y_2(b), \ldots, y_n(b)$. We first convert to matrix notation: writing

$$y(s) = \begin{bmatrix} y_1(s) \\ \vdots \\ y_n(s) \end{bmatrix} \tag{0.2}$$

ENCYCLOPEDIA OF MATHEMATICS and Its Applications, Gian-Carlo Rota (ed.). Vol. 10: John D. Dollard and Charles N. Friedman, Product Integration with Applications to Differential Equations

ISBN 0-201-13509-4

and

$$A(s)=\begin{bmatrix} a_{11}(s) & \cdots & a_{1n}(s) \\ \vdots & & \vdots \\ a_{n1}(s) & \cdots & a_{nn}(s) \end{bmatrix}, \tag{0.3}$$

Eq. (0.1) take on the form

$$Y'(s)=A(s)Y(s), \tag{0.4}$$

and the problem is to calculate $Y(b)$ given $Y(a)$. An approximate value for $Y(b)$ can be found by a variant of the Euler tangent-line method: let $P=\{s_0,s_1,\ldots,s_n\}$ be a partition of the interval $[a,b]$. Let $\Delta s_k=s_k-s_{k-1}$ for $k=1,\ldots,n$. On the interval $[s_0,s_1]$ we approximate $A(s)$ by the constant value $A(s_1)$. We solve the differential equation (0.4) on the interval $[s_0,s_1]$ with $A(s)$ replaced by $A(s_1)$ and with initial value $Y(a)$. The solution is $e^{A(s_1)(s-a)}Y(a)$. This leads to the following approximate value for the exact solution Y at the point s_1:

$$Y(s_1)\cong e^{A(s_1)(s_1-a)}Y(a)=e^{A(s_1)\Delta s_1}Y(a). \tag{0.5}$$

Now proceeding on the interval $[s_1,s_2]$ from the approximate initial value (0.5) and replacing $A(s)$ by the constant value $A(s_2)$, we find

$$Y(s_2)\cong e^{A(s_2)\Delta s_2}e^{A(s_1)\Delta s_1}Y(a). \tag{0.6}$$

Proceeding in this manner, the following approximate value is obtained for $Y(b)$:

$$Y(b)=Y(s_n)\cong e^{A(s_n)\Delta s_n}\cdots e^{A(s_1)\Delta s_1}Y(a)=\prod_{k=1}^{n} e^{A(s_k)\Delta s_k}Y(a). \tag{0.7}$$

We remark that on the right-hand side of Eq. (0.7) the order of the exponentials is important: they may not commute, because in general the values $A(s_1),A(s_2),\ldots,A(s_n)$ will not commute. The ordered product of exponentials in (0.7) will be denoted $\Pi_P(A)$:

$$\Pi_P(A)=\prod_{k=1}^{n} e^{A(s_k)\Delta s_k}. \tag{0.8}$$

Let $\mu(P)$ denote the mesh of the partition P (length of the longest subinterval). The function A is continuous on $[a,b]$, hence uniformly continuous. Thus if $\mu(P)$ is small, for all $k=1,\ldots,n$, the value $A(s_k)$ will be close to the values taken by A on the interval $[s_{k-1},s_k]$. It is thus

ISBN 0-201-13509-4

reasonable to suppose that when $\mu(P)$ is small the calculation given above produces a value for $Y(b)$ quite close to the true value. We may then expect to find $Y(b)$ exactly by the prescription

$$Y(b) = \lim_{\mu(P)\to 0} \Pi_P(A)\,Y(a). \tag{0.9}$$

This is indeed correct, as will be seen. In fact, the limit of the matrix $\Pi_P(A)$ exists independent of the initial value $Y(a)$, and it is the limit of $\Pi_P(A)$ that defines the product integral of A over $[a,b]$.

We will in the sequel be mainly concerned with *analytical* properties of differential equations and product integrals. However, it is important to note that the construction of the product integral, as outlined above, is very close in spirit to procedures for finding *numerical* solutions to differential equations. In fact, the method suggested above for finding $Y(b)$ is a possible (although by no means the most efficient) method for beginning a numerical calculation of $Y(b)$. By taking a sufficiently fine partition and making sufficiently good approximations of the exponentials occurring, the value of $Y(b)$ can be calculated to any desired degree of accuracy. It is this connection with the ideas of numerical analysis that gives the subject of product integration its very *constructive* flavor. In this subject, we seek not only to prove *existence* of solutions of differential equations, but to represent these solutions concretely as limits of certain approximate solutions. Many of the theorems of this chapter concerning existence of the product integral can be reinterpreted as theorems concerning the accuracy of certain numerical approximations to the solutions of differential equations. The reader who bears in mind the connection with numerical analysis will have a deeper understanding of the subject.

We now turn to the study of the product integral. In the course of our study, we will justify the analysis of Eq. (0.4) given above.

1.1 Product Integration

The ingredients of the approximation procedure discussed above are (i) a step-function, namely, the step-function taking the value $A(s_1)$ on $[s_0,s_1]$, $A(s_2)$ on $(s_1,s_2]$, etc., and (ii) the construction of the corresponding ordered product (0.8) used in approximating $Y(b)$.

We now formalize our construction, extending it to any step-function and any $x\in[a,b]$.

DEFINITION 1.1. *Let B be a function defined on the interval $[a,b]$ of real numbers and taking values in the space $\mathbb{C}_{n\times n}$ of $n\times n$ matrices with complex entries. B is called a* step-function *if and only if there is a partition $P=\{s_0,s_1,\ldots,s_n\}$ of $[a,b]$ such that B is constant on each open subinterval (s_{k-1},s_k) for $k=1,\ldots,n$. The value of B on (s_{k-1},s_k) is then denoted B_k.*

ISBN 0-201-13509-4

DEFINITION 1.2. *Let $A:[a,b]\to\mathbb{C}_{n\times n}$ be a function, and let $P=\{s_0,s_1,\dots,s_n\}$ be a partition of $[a,b]$. The* point-value approximant A_P *corresponding to A and P is the step-function taking value $A(s_1)$ on $[s_0,s_1]$, $A(s_2)$ on $(s_1,s_2],\dots,$ and $A(s_n)$ on $(s_{n-1},s_n]$.*

DEFINITION 1.3. *Let $B:[a,b]\to\mathbb{C}_{n\times n}$ be a step-function. Using notation as in Definition 1.1, and letting $\Delta s_k=s_k-s_{k-1}$, the function $E_B:[a,b]\to\mathbb{C}_{n\times n}$ is defined as follows:*

$$\begin{aligned} E_B(x)&=e^{B_1(x-s_0)}, && x\in[s_0,s_1]\\ &=e^{B_2(x-s_1)}e^{B_1\Delta s_1}, && x\in[s_1,s_2]\\ &\ \vdots\\ &=e^{B_n(x-s_{n-1})}\cdots e^{B_2\Delta s_2}e^{B_1\Delta s_1}, && x\in[s_{n-1},s_n]. \end{aligned} \tag{1.1}$$

We can now describe the approximation procedure of the Introduction as follows: Given the continuous function A and a partition P of $[a,b]$, we formed the point-value approximant A_P and constructed the ordered product $\Pi_P(A)$ of Eq. (0.8), which is precisely $E_{A_P}(b)$. The conclusion was that $E_{A_P}(b)Y(a)$ should be close to $Y(b)$ if the mesh of P is small. In view of the definition of $E_{A_P}(x)$, it is equally reasonable to conclude that $E_{A_P}(x)Y(a)$ should be close to $Y(x)$ when the mesh of P is small. This will follow from our analysis.

Let A be continuous (hence uniformly continuous) on $[a,b]$. Taking $\varepsilon>0$, choose $\delta>0$ such that $x,y\in[a,b]$ and $|x-y|<\delta$ imply $\|A(x)-A(y)\|<\varepsilon$. It follows immediately that for $\mu(P)<\delta$ we have

$$\|A_P(x)-A(x)\|<\varepsilon \qquad \text{for all} \qquad x\in[a,b]. \tag{1.2}$$

Thus we have uniformly on $[a,b]$

$$\lim_{\mu(P)\to 0} A_P(x)=A(x). \tag{1.3}$$

It follows immediately that the step-function A_P converges *in the L^1 sense* to A as $\mu(P)\to 0$. This means the following: If B is an $n\times n$ matrix-valued function on $[a,b]$, all of whose entries are integrable over $[a,b]$, we set

$$\|B\|_1=\int_a^b\|B(s)\|\,ds. \tag{1.4}$$

To say that A_P converges to A in the L^1 sense means that the following equation holds:

$$\lim_{\mu(P)\to 0}\|A_P-A\|_1=0. \tag{1.5}$$

ISBN 0-201-13509-4

This will be the crucial property of the point-value approximants in justifying our approximation procedure for the differential equation (0.4). The fact that A_P converges to A uniformly is of interest only in deducing Eq. (1.5). In fact, our approximation procedure could equally well be carried out using *any* sequence of step-functions converging to A in the L^1 sense. This fact will be reflected in Theorems 1.1 and 1.2, which justify the approximation procedure. Before proceeding to Theorem 1.1, we prove two simple lemmas on the properties of the functions $E_B(x)$ of Definition 1.3.

LEMMA 1.1. *Let* $B:[a,b]\to\mathbb{C}_{n\times n}$ *be a step-function and let* E_B *be as in Definition* 1.3. *Then*

(*i*) $E_B(a)=I$ (*I denotes the* $n\times n$ *identity matrix*).
(*ii*) $E_B(x)$ *is nonsingular for each* $x\in[a,b]$.
(*iii*) E_B *and* E_B^{-1} *are continuous on* $[a,b]$ *and satisfy the integral equations*

$$E_B(x)=I+\int_a^x B(s)E_B(s)\,ds \tag{1.6}$$

and

$$E_B^{-1}(x)=I-\int_a^x E_B^{-1}(s)B(s)\,ds. \tag{1.7}$$

(*iv*) E_B *and* E_B^{-1} *satisfy the bounds*

$$\|E_B(x)\|\leqslant e^{\int_a^x\|B(s)\|\,ds},$$
$$\|E_B^{-1}(x)\|\leqslant e^{\int_a^x\|B(s)\|\,ds}. \tag{1.8}$$

Proof. Parts (i) and (ii) are obvious. Note that $E_B^{-1}(x)$ is obtained by reversing the order of the factors in $E_B(x)$ and inserting minus signs in the exponents. E_B and E_B^{-1} are continuous by inspection. Also, if P is the partition associated with B, then inspection of the definition (1.1) shows that E_B is differentiable except at the division points of P, and

$$E_B'(x)=B(x)E_B(x) \tag{1.9}$$

except at the division points. Since $E_B(a)=I$, the integral equation (1.6) is essentially just the fundamental theorem of calculus applied to E_B. Some additional justification is necessary because E_B is not differentiable at all points of $[a,b]$; however, E_B is continuous on $[a,b]$ and is continuously differentiable on each open subinterval (s_{k-1},s_k) determined by P. Using the fundamental theorem of calculus on each open subinterval and piecing the results together, one easily obtains Eq. (1.6). Equation (1.7) is obtained in a similar manner. Thus (iii) is proved. To prove part (iv), we adopt the

ISBN 0-201-13509-4

notation of Definition 1.3. Letting $x \in [s_{k-1}, s_k]$, we have

$$\begin{aligned}\|E_B(x)\| &= \|e^{B_k(x-s_{k-1})} \cdots e^{B_1 \Delta s_1}\| \\ &\leqslant \|e^{B_k(x-s_{k-1})}\| \cdots \|e^{B_1\Delta s_1}\| \\ &\leqslant e^{\|B_k\|(x-s_{k-1})} \cdots e^{\|B_1\|\Delta s_1} \\ &= e^{\|B_k\|(x-s_{k-1}) + \cdots + \|B_1\|\Delta s_1} = e^{\int_a^x \|B(s)\|\,ds}. \end{aligned} \tag{1.10}$$

The corresponding bound for E_B^{-1} is proved in an analogous manner. Thus the lemma is proved. ■

Lemma 1.2. *Let $B, C: [a,b] \to \mathbb{C}_{n\times n}$ be step-functions. Then*

$$E_B(x) - E_C(x) = E_C(x) \int_a^x E_C^{-1}(s)(B(s) - C(s)) E_B(s)\,ds. \tag{1.11}$$

Proof. The proof of (1.11) is very similar to the proof of Eq. (1.6) above. Let

$$G(x) = E_C^{-1}(x) E_B(x). \tag{1.12}$$

Then $G(a)$ is the identity, G is continuous, and, except at the division points of the partitions associated with B and C, G is differentiable. The product rule for differentiation and the definitions of E_B and E_C^{-1} show that except at the division points we have

$$G'(x) = E_C^{-1}(x)(B(x) - C(x)) E_B(x). \tag{1.13}$$

Thus, arguing as in the proof of the integral equation (1.6), we have

$$G(x) = I + \int_a^x E_C^{-1}(s)(B(s) - C(s)) E_B(s)\,ds. \tag{1.14}$$

Equation (1.11) is obtained from (1.14) by multiplying on the left with $E_C(x)$. This proves the lemma. ■

We are now in a position to prove the following:

Theorem 1.1. *Let $A: [a,b] \to \mathbb{C}_{n\times n}$ be continuous and let $\{A_n\}$ be any sequence of step-functions converging to A in the L^1 sense. Then the sequence $\{E_{A_n}(x)\}$ converges uniformly on $[a,b]$ to a matrix denoted $\prod_a^x e^{A(s)\,ds}$, called the* product integral of A over $[a,x]$.

ISBN 0-201-13509-4

Proof. By Lemma 1.2, we have

$$E_{A_n}(x)-E_{A_m}(x)=E_{A_m}^{-1}(x)\int_a^x E_{A_m}(s)(A_n(s)-A_m(s))E_{A_n}(s)\,ds. \tag{1.15}$$

Thus

$$\|E_{A_n}(x)-E_{A_m}(x)\|\leq\|E_{A_m}^{-1}(x)\|\int_a^x\|E_{A_m}(s)\|\,\|A_n(s)-A_m(s)\|\,\|E_{A_n}(s)\|\,ds. \tag{1.16}$$

The norms of the functions $E_{A_m}^{-1}$, E_{A_m}, and E_{A_n} can be estimated using (1.8), and we may estimate the integrals in (1.8) by integrals from a to b. Using also the notation (1.4), we then obtain from (1.16) the estimate

$$\begin{aligned}\|E_{A_n}(x)-E_{A_m}(x)\|&\leq e^{2\|A_m\|_1}e^{\|A_n\|_1}\int_a^x\|A_n(s)-A_m(s)\|\,ds\\&\leq e^{2\|A_m\|_1}e^{\|A_n\|_1}\|A_n-A_m\|_1.\end{aligned} \tag{1.17}$$

Now because $\{A_n\}$ converges to A in the L^1 sense, the norms $\|A_n\|_1$ are bounded and the sequence $\{A_n\}$ is Cauchy in the L^1 sense. Thus the right-hand side of (1.17) approaches zero as $n,m\to\infty$. Furthermore, the right-hand side of (1.17) is independent of x. This shows that the sequence $\{E_{A_n}(x)\}$ is uniformly Cauchy, hence uniformly convergent. If $\{B_n\}$ and $\{C_n\}$ are two sequences converging to A in the L^1 sense, then estimating $\|E_{B_n}(x)-E_{C_n}(x)\|$ as above, one sees that $\{E_{B_n}\}$ and $\{E_{C_n}\}$ have the same limit. Thus the theorem is proved. We note for later reference that since $E_{A_n}(a)=I$ for all n, we have $\prod_a^a e^{A(s)ds}=I$. ■

It should be noted that in the proof of Theorem 1.1 no use was made of the fact that A was continuous. All that was needed was a sequence of step-functions converging to A in the L^1 sense. This fact permits the definition of the product integral for any matrix function A with Lebesgue integrable entries, as we shall remark later. However, for the present we continue to develop the theory under the hypothesis that A is continuous. This is the most natural assumption when studying the differential equation (0.4).

Since we know in particular that for continuous A there is a sequence of point-value approximants converging to A in the L^1 sense, Theorem 1.1 implies that in our previous notation

$$\prod_a^b e^{A(s)ds}=\lim_{\mu(P)\to 0}\prod_{k=1}^n e^{A(s_k)\Delta s_k}, \tag{1.18}$$

ISBN 0-201-13509-4

where $P=\{s_0,s_1,\ldots,s_n\}$ denotes a partition of $[a,b]$. Naturally the product integral $\prod_a^x e^{A(s)\,ds}$ can be evaluated in an analogous manner using partitions P of the interval $[a,x]$.

We can make a slight generalization of (1.18), which is occasionally convenient. Namely, given a partition P, instead of considering the point-value approximant to A we can consider a step-function that, on the subinterval $(s_{k-1},s_k]$ of P, takes the value $A(s_k')$, where s_k' is any number in $[s_{k-1},s_k]$. When $\mu(P)\to 0$ such a step-function will still approach A in the L^1 sense (even uniformly) and we therefore have

$$\prod_a^b e^{A(s)\,ds}=\lim_{\mu(P)\to 0}\prod_{k=1}^{n} e^{A(s_k')\Delta s_k}. \tag{1.18$'$}$$

We now establish some basic properties of the product integral.

THEOREM 1.2. *Let $A:[a,b]\to\mathbb{C}_{n\times n}$ be continuous, and for $x\in[a,b]$ let*

$$F(x,a)=\prod_a^x e^{A(s)\,ds}. \tag{1.19}$$

Then F satisfies the integral equation

$$F(x,a)=I+\int_a^x A(s)F(s,a)\,ds. \tag{1.20}$$

The function F is also a solution of the initial value problem

$$\frac{dF(x,a)}{dx}=A(x)F(x,a),\qquad F(a,a)=I \tag{1.21}$$

for $x\in[a,b]$. (One-sided derivatives are meant at a and b.)

Proof. Let $\{A_n\}$ be a sequence of step-functions converging to A in the L^1 sense. The functions E_{A_n} are continuous on $[a,b]$ and converge uniformly to F as $n\to\infty$. Thus $F(x,a)$ is continuous. Now by Lemma 1.1 we have

$$E_{A_n}(x)=I+\int_a^x A_n(s)E_{A_n}(s)\,ds. \tag{1.22}$$

Taking the limit of (1.22) as $n\to\infty$, and using convergence of A_n in the L^1 sense and uniform convergence of E_{A_n}, we easily obtain Eq. (1.20). Then because the integrand in (1.20) is continuous, the differential equation in (1.21) follows immediately from (1.20). That $F(a,a)=I$ was remarked earlier. This completes the proof. ∎

ISBN 0-201-13509-4

We note that according to this proof the product integral of a continuous function is continuously differentiable.

Theorem 1.2 provides the justification for the approximation procedure discussed earlier for Eq. (0.4), because it is now clear that if $F(x,a)$ is defined as in (1.19), then the function $F(s,a)Y(a)$ satisfies (0.4) and reduces to $Y(a)$ when $x=a$. As is well known and as we shall soon prove, there is only one function satisfying these conditions, hence $Y(s)=F(s,a)Y(a)$ for all s, and in particular for $s=b$, the case studied in the approximation procedure.

It is useful to remark that the product integral of a commutative family can be found explicitly:

THEOREM 1.3. *Let $A:[a,b]\to\mathbb{C}_{n\times n}$ be continuous, and suppose that the family $\{A(s)|s\in[a,b]\}$ is* commutative, *i.e.*,

$$A(s)A(s')=A(s')A(s) \qquad \textit{for all } s,s' \textit{ in } [a,b]. \tag{1.23}$$

Then

$$\prod_a^x e^{A(s)\,ds}=e^{\int_a^x A(s)\,ds}. \tag{1.24}$$

Proof. Letting P denote a partition of $[a,x]$, we have

$$\begin{aligned}\prod_a^x e^{A(s)\,ds} &= \lim_{\mu(P)\to 0}\prod_{k=1}^n e^{A(s_k)\Delta s_k}\\ &= \lim_{\mu(P)\to 0} e^{\sum_{k=1}^n A(s_k)\Delta s_k},\end{aligned} \tag{1.25}$$

where we have used commutativity to obtain the second equality. But $\sum_{k=1}^n A(s_k)\Delta s_k$ is just a Riemann sum for the integral $\int_a^x A(s)\,ds$, so the right-hand side of (1.25) agrees with that of (1.24). Thus the theorem is proved. ■

We now return to the development of the general theory. The next theorem establishes a fundamental property of product integrals.

THEOREM 1.4. *Let $A:[a,b]\to\mathbb{C}_{n\times n}$ be continuous. Then for each $x\in[a,b]$ the product integral $\prod_a^x e^{A(s)\,ds}$ is nonsingular. The following formula holds*:

$$\det\left(\prod_a^x e^{A(s)\,ds}\right)=e^{\int_a^x \operatorname{tr}A(s)\,ds}\neq 0. \tag{1.26}$$

Proof. Clearly if (1.26) is established, then nonsingularity of the product integral follows. To establish (1.26), note that if P denotes a partition of

ISBN 0-201-13509-4

$[a,x]$, then

$$\det\left(\prod_a^x e^{A(s)\,ds}\right)=\det\left(\lim_{\mu(P)\to 0}\prod_{k=1}^n e^{A(s_k)\Delta s_k}\right)$$
$$=\lim_{\mu(P)\to 0}\det\left(\prod_{k=1}^n e^{A(s_k)\Delta s_k}\right)=\lim_{\mu(P)\to 0}\prod_{k=1}^n \det\left(e^{A(s_k)\Delta s_k}\right)$$
$$=\lim_{\mu(P)\to 0}\prod_{k=1}^n e^{\operatorname{tr}A(s_k)\Delta s_k}=\lim_{\mu(P)\to 0} e^{\sum_{k=1}^n \operatorname{tr}A(s_k)\Delta s_k}$$
$$=e^{\int_a^x \operatorname{tr}A(s)\,ds}, \tag{1.27}$$

where we have used some elementary manipulations and the fact that the Riemann sum $\sum_{k=1}^n \operatorname{tr}A(s_k)\Delta s_k$ approaches $\int_a^x \operatorname{tr}A(s)\,ds$ as $\mu(P)\to 0$. Thus the theorem is proved. ■

If y is a point of $[a,b]$, then by applying the above analysis to the interval $[y,b]$ instead of $[a,b]$, we can say that we have defined $\prod_y^x e^{A(s)\,ds}$ for all $x\in[y,b]$, that is to say for all $x\geqslant y$. It will be convenient also to define $\prod_y^x e^{A(s)\,ds}$ when $x<y$. We do this by analogy with the definition

$$\int_y^x f(s)\,ds=-\int_x^y f(s)\,ds \tag{1.28}$$

in the usual theory of integration. Equation (1.28) states that $\int_y^x f(s)\,ds$ is the additive inverse of $\int_x^y f(s)\,ds$. For product integrals, we merely replace "additive" by "multiplicative":

Definition 1.4. *Let $A:[a,b]\to\mathbb{C}_{n\times n}$ be continuous and let $x,y\in[a,b]$ with $x<y$. Then $\prod_y^x e^{A(s)\,ds}$ is defined by the formula*

$$\prod_y^x e^{A(s)\,ds}=\left(\prod_x^y e^{A(s)\,ds}\right)^{-1}. \tag{1.29}$$

Note that combining Definition 1.4 with Theorem 1.4 and Eq. (1.28), we obtain for any $x,y\in[a,b]$ the formula

$$\det\prod_y^x e^{A(s)\,ds}=e^{\int_y^x \operatorname{tr}A(s)\,ds}. \tag{1.30}$$

In the ordinary theory of integration one proves the following additive property of the integral:

$$\int_z^x f(s)\,ds=\int_y^x f(s)\,ds+\int_z^y f(s)\,ds. \tag{1.31}$$

ISBN 0-201-13509-4

Using Definition 1.4, it can be shown that product integrals have an analogous *multiplicative* property:

THEOREM 1.5. *Let* $A:[a,b]\rightarrow \mathbb{C}_{n\times n}$ *be continuous, and let* $x,y,z\in[a,b]$. *Then*

$$\prod_{z}^{x} e^{A(s)\,ds} = \prod_{y}^{x} e^{A(s)\,ds} \prod_{z}^{y} e^{A(s)\,ds}. \tag{1.32}$$

Proof. First suppose that $z \leqslant y \leqslant x$. The left-hand side of (1.32) can be written as

$$\prod_{z}^{x} e^{A(s)\,ds} = \lim_{\mu(P)\rightarrow 0} \prod_{k=1}^{n} e^{A(s_k)\Delta s_k}, \tag{1.33}$$

where P denotes a partition of $[z,x]$. If we always choose P so that the point y belongs to P, then the product on the right-hand side of (1.33) factors in an obvious way into the product of two terms that respectively converge to the two terms on the right-hand side of (1.32). Thus (1.32) is established for $z \leqslant y \leqslant x$. This result together with Definition 1.4 can now be used to establish (1.32) for any $x,y,z\in[a,b]$, proving the theorem. ■

Definition 1.4 and Theorem 1.5 point up the analogy between the theory of product integration and the usual theory of integration. This analogy is very far-reaching, and when trying to intuit what properties should be possessed by product integrals, it is often useful to remember properties of ordinary integrals and attempt to convert statements about additivity into corresponding statements about multiplicativity. If this is done with care, the resulting conjectures about properties of product integrals usually turn out to be true.

Before proceeding, we make a further remark on Theorem 1.5: Recall that a *propagator* for a differential equation is an operator that, acting on the value of a solution at one point, produces the value at another point. Thus if $P(x,y)$ is the propagator and $u(x)$ is the solution, we have $P(x,y)u(y)=u(x)$. The propagator $P(x,y)$ must satisfy the equation $P(x,z)=P(x,y)P(y,z)$. Theorem 2.1 will show that the product integral is just such a propagator, and Eq. (1.32) reflects this fact.

We now summarize the differentiability properties of product integrals in

THEOREM 1.6. *Let* $A:[a,b]\rightarrow \mathbb{C}_{n\times n}$ *be continuous. For* $x,y\in[a,b]$ *define*

$$F(x,y) = \prod_{y}^{x} e^{A(s)\,ds}. \tag{1.34}$$

ISBN 0-201-13509-4

Then

$$\frac{\partial F(x,y)}{\partial x} = A(x)F(x,y) \tag{1.35}$$

and

$$\frac{\partial F(x,y)}{\partial y} = -F(x,y)A(y). \tag{1.36}$$

Proof. Equation (1.35) is already known to hold for $y=a$ (Theorem 1.2). Theorem 1.5 allows us to write

$$F(x,y) = F(x,a)F(a,y), \tag{1.37}$$

and Eq. (1.35) follows immediately. Having established differentiability in the first variable, we can verify (1.36) by the rule for differentiating inverses:

$$\frac{\partial F(x,y)}{\partial y} = \frac{\partial F^{-1}(y,x)}{\partial y} = -F^{-1}(y,x)\left(\frac{\partial F(y,x)}{\partial y}\right)F^{-1}(y,x)$$

$$= -F(x,y)(A(y)F(y,x))F(x,y) = -F(x,y)A(y). \tag{1.38}$$

This completes the proof. ∎

1.2 Product Integral Analysis of Linear Ordinary Differential Equations

We can now prove the fundamental existence and uniqueness theorem for equations of the type (0.4). This proof is especially simple because with the use of product integration it can be modeled directly on the usual proof for scalar equations of the type (0.4).

THEOREM 2.1. *Let $A:[a,b]\rightarrow \mathbb{C}_{n\times n}$ be continuous. Fix $c\in[a,b]$ and consider the initial-value problem*

$$Y'(x) = A(x)Y(x), \qquad Y(c) = Y_0, \tag{2.1}$$

where Y_0 is a given $n\times 1$ or $n\times n$ matrix. This problem has the unique solution

$$Y(x) = \prod_c^x e^{A(s)\,ds}\,Y_0. \tag{2.2}$$

Proof. Y as given by (2.2) is certainly a solution of (2.1) by Theorem 1.6 and the fact that $\prod_c^c e^{A(s)\,ds} = I$. To prove uniqueness, let Z be any solution

ISBN 0-201-13509-4

of (2.1) and consider the function H defined by

$$H(x) = \prod_{x}^{c} e^{A(s)\,ds} Z(x). \tag{2.3}$$

Then

$$H(c) = Z(c) = Y_0. \tag{2.4}$$

Also, using Theorem 1.6 and the product rule for differentiation, the derivative $H'(x)$ is found to be identically zero. Thus

$$H(x) = H(c) = Y_0, \qquad x \in [a,b]. \tag{2.5}$$

Multiplying (2.5) on the left by $\prod_c^x e^{A(s)\,ds}$, we find that Z agrees with Y of (2.2). This completes the proof. ■

If $Y_1, Y_2, \ldots, Y_m$ are $n \times 1$ matrix functions that are solutions of the differential equation in (2.1) (not necessarily satisfying the initial condition), then the set $\{Y_1, Y_2, \ldots, Y_m\}$ is called a *fundamental* set of solutions of this differential equation on $[a,b]$ if and only if any $n \times 1$ solution Y of the differential equation on $[a,b]$ can be written

$$Y(x) = c_1 Y_1(x) + c_2 Y_2(x) + \cdots + c_m Y_m(x) \tag{2.6}$$

with uniquely determined constants $c_1, c_2, \ldots, c_m$. This amounts to saying that $\{Y_1, Y_2, \ldots, Y_m\}$ is a basis for the vector space S of all solutions to the equation. It follows that any fundamental set has a number of elements equal to the dimension of S. Our next theorem shows that the product integral defines a fundamental set of solutions in a natural way.

THEOREM 2.2. *Let $A : [a,b] \to \mathbb{C}_{n \times n}$ be continuous. Let $C_1(x)$, $C_2(x), \ldots, C_n(x)$ denote the columns of the matrix $\prod_a^x e^{A(s)\,ds}$. Then $\{C_1, C_2, \ldots, C_n\}$ is a fundamental set of solutions of the equation*

$$Y'(x) = A(x)Y(x) \tag{2.7}$$

on the interval $[a,b]$.

Proof. Let E_j denote the $n \times 1$ matrix having a 1 in the jth row and zeros in all other rows. Then by the rules of matrix multiplication we have

$$C_j(x) = \prod_{a}^{x} e^{A(s)\,ds} E_j. \tag{2.8}$$

By Theorem 1.6, C_j is a solution of (2.7). To see that the set $\{C_1, C_2, \ldots, C_n\}$

ISBN 0-201-13509-4

is fundamental, let Z be a solution of (2.7) and write

$$Z(a)=\begin{bmatrix} z_1 \\ \vdots \\ z_n \end{bmatrix}. \tag{2.9}$$

By Theorem 2.1, Z is given uniquely by

$$Z(x)=\prod_a^x e^{A(s)\,ds}Z(a). \tag{2.10}$$

Using on the right-hand side of (2.10) the familiar rules for multiplying partitioned matrices, we have

$$Z(x)=z_1C_1(x)+z_2C_2(x)+\cdots+z_nC_n(x). \tag{2.11}$$

This shows that any solution of (2.8) can be written as a linear combination of $C_1, C_2, \ldots, C_n$. This can be done in only one way, because if an equation of the form (2.11) holds, then setting $x=a$ and noting that $C_j(a)=E_j$ it is seen that $z_1, z_2, \ldots, z_n$ must be the entries of $Z(a)$. Thus the theorem is proved. ■

We note that according to Theorem 2.2, the vector space S of all solutions of (2.7) has dimension n when A is an $n \times n$ matrix.

COROLLARY *Let* $\{D_1, D_2, \ldots, D_n\}$ *be any fundamental set of solutions of Eq.* (2.7) *on* $[a,b]$. *Let* D *be the matrix with columns* $D_1, D_2, \ldots, D_n$. *Then the determinant of* D *has the form*

$$\det D(x)=\beta e^{\int_a^x \operatorname{tr} A(s)\,ds} \tag{2.12}$$

with $\beta \neq 0$.

Proof. Since $D_1, D_2, \ldots, D_n$ can be expressed in terms of the solutions $C_1, C_2, \ldots, C_n$ studied in Theorem 2.2 and conversely, the matrix $\prod_a^x e^{A(s)\,ds}$ with columns $C_1, C_2, \ldots, C_n$ must be related to D in the following manner:

$$D(x)=\prod_a^x e^{A(s)\,ds}B, \tag{2.13}$$

where B is a nonsingular matrix. Writing β for the determinant of B, Eq. (2.12) is then an immediate consequence of Eq. (1.26) for the determinant of a product integral. ■

ISBN 0-201-13509-4

The proofs of the last two theorems consist of a sequence of obvious remarks. Yet these theorems are very powerful. The reader should remark that they cover in particular the existence and uniqueness theory and the theory of fundamental sets of solutions for scalar equations of the type

$$y^{(n)}(x)+a_{n-1}(x)y^{(n-1)}(x)+\cdots+a_0(x)y(x)=0, \tag{2.14}$$

where $a_k(x)$ is continuous for each $k=0,1,\ldots,n-1$. Equation (2.14) can be converted to the form (2.7) in the following standard way: Let

$$Y(x)=\begin{bmatrix} y(x) \\ y'(x) \\ \vdots \\ y^{(n-1)}(x) \end{bmatrix} \tag{2.15}$$

and let

$$A(x)=\begin{bmatrix} 0 & 1 & 0 & \cdots & 0 \\ 0 & 0 & 1 & \cdots & 0 \\ \vdots & & & & \vdots \\ & & & & 1 \\ -a_0(x) & -a_1(x) & -a_2(x) & \cdots & -a_{n-1}(x) \end{bmatrix}. \tag{2.16}$$

Then Eq. (2.14) is equivalent to (2.7) with this choice of Y and A. Note that the trace of A in (2.7) is just $-a_{n-1}(x)$. The Corollary of the last theorem, interpreted for this case, is just the familiar result that the Wronskian of a fundamental set of solutions of (2.14) has the form $\beta e^{-\int_a^x a_{n-1}(s)\,ds}$. If the coefficients $a_k(x)$ in (2.14) are constants, then the corresponding $A(x)$ of (2.7) is constant, so that the family $\{A(s)|s\in[a,b]\}$ is commutative. By Theorem 1.3 we then have

$$\prod_a^x e^{A(s)\,ds}=e^{\int_a^x A(s)\,ds}=e^{A(x-a)}, \tag{2.17}$$

where A is the constant value of $A(x)$. A fundamental set of solutions of (2.5) can then be found using Theorem 2.2 and the fact that $e^{A(x-a)}$ can be computed explicitly.

Example. We give here an example that could easily be analyzed without product integrals, but which is instructive for that very reason. The formulas to be obtained will be useful to us later. Consider the equation

$$y''+Ey=0, \tag{2.18}$$

ISBN 0-201-13509-4

where E is a positive real number. Setting

$$Y=\begin{pmatrix} y \\ y' \end{pmatrix} \tag{2.19}$$

we obtain as above the equation

$$Y'=A_0 Y \tag{2.20}$$

with

$$A_0=\begin{pmatrix} 0 & 1 \\ -E & 0 \end{pmatrix}. \tag{2.21}$$

The solution is

$$Y(x)=e^{A_0(x-a)}Y(a). \tag{2.22}$$

To evaluate $e^{A_0(x-a)}$, note that the characteristic equation

$$\det(\lambda I-A_0)=0 \tag{2.23}$$

yields the eigenvalues $\lambda=\pm ik$, where k is the positive square root of E. Thus A_0 is similar to the diagonal matrix with entries $\pm ik$. Explicitly, setting

$$M=\begin{pmatrix} 1 & 1 \\ ik & -ik \end{pmatrix} \tag{2.24}$$

and

$$\Lambda=\begin{pmatrix} ik & 0 \\ 0 & -ik \end{pmatrix}, \tag{2.25}$$

we have

$$A_0=M\Lambda M^{-1}. \tag{2.26}$$

Thus

$$\begin{aligned} e^{A_0(x-a)}&=Me^{\Lambda(x-a)}M^{-1} \\ &=M\begin{pmatrix} e^{ik(x-a)} & 0 \\ 0 & e^{-ik(x-a)} \end{pmatrix}M^{-1}. \end{aligned} \tag{2.27}$$

Comparing (2.22) and (2.27), we see that we have proved the expected result: the general solution of (2.18) is a linear combination of the functions e^{ikx} and e^{-ikx}.

Product integration can be used to give an account of the familiar technique of *variation of parameters*. The setting is as follows: We consider the initial-value problem for an inhomogeneous equation,

$$Y'(x)=A(x)Y(x)+F(x), \qquad Y(a)=Y_0. \tag{2.28}$$

ISBN 0-201-13509-4

In (2.28) we assume that A and F are given continuous functions: $A(x)$ is as usual an $n\times n$ matrix, and $F(x)$ has the same matrix dimensions as the initial condition Y_0 (either $n\times n$ or $n\times 1$). To solve (2.28), let $R(x,y)$ be given by

$$R(x,y)=\prod_{y}^{x} e^{A(s)\,ds}. \tag{2.29}$$

Consider the function

$$Z(x)=R(a,x)\,Y(x). \tag{2.30}$$

Using (2.28) and the product rule for differentiation, we find

$$Z'(x)=R(a,x)F(x), \qquad Z(a)=R(a,a)\,Y(a)=Y_0. \tag{2.31}$$

Upon integrating (2.31) we find

$$Z(x)=Y_0+\int_a^x R(a,s)F(s)\,ds. \tag{2.32}$$

Multiplying (2.32) on the left by $R(x,a)$ yields

$$\begin{aligned} Y(x)&=R(x,a)\,Y_0+R(x,a)\int_a^x R(a,s)F(s)\,ds \\ &=R(x,a)\,Y_0+\int_a^x R(x,s)F(s)\,ds. \end{aligned} \tag{2.33}$$

This solves (2.28) provided that the solution $R(x,a)$ of the homogeneous equation is known. To see how this works out in a familiar case, consider the situation to which the technique of variation of parameters is usually applied: we have a scalar equation of the form

$$y''+py'+qy=f. \tag{2.34}$$

This equation takes the form (2.28) if we put

$$\begin{aligned} Y(x)&=\begin{pmatrix} y(x) \\ y'(x) \end{pmatrix}, \\ A(x)&=\begin{pmatrix} 0 & 1 \\ -q(x) & -p(x) \end{pmatrix}, \\ F(x)&=\begin{pmatrix} 0 \\ f(x) \end{pmatrix}. \end{aligned} \tag{2.35}$$

ISBN 0-201-13509-4

Now by Theorem 2.2 the columns of $R(x,a)$ are solutions of the equation $Y'=AY$, and because of the form of A any such solution has the form $Y = \binom{y}{y'}$. Thus $R(x,a)$ has the form

$$R(x,a) = \begin{pmatrix} y_1(x) & y_2(x) \\ y_1'(x) & y_2'(x) \end{pmatrix}, \tag{2.36}$$

where y_1 and y_2 are solutions of the scalar equation obtained from $Y' = AY$:

$$y'' + py' + qy = 0, \tag{2.37}$$

and y_1 and y_2 are so chosen that $R(x,a)$ reduces to the identity when $x = a$. Thus

$$R(a,s) = R^{-1}(s,a) = \frac{1}{W(s)} \begin{pmatrix} y_2'(s) & -y_2(s) \\ -y_1'(s) & y_1(s) \end{pmatrix}, \tag{2.38}$$

where

$$W(s) = y_1(s)y_2'(s) - y_1'(s)y_2(s) \tag{2.39}$$

is the Wronskian of y_1 and y_2. Thus

$$R(a,s)F(s) = \frac{1}{W(s)} \begin{pmatrix} -y_2(s)f(s) \\ y_1(s)f(s) \end{pmatrix} \tag{2.40}$$

and

$$R(x,s)F(s) = R(x,a)R(a,s)F(s) = \frac{1}{W(s)} \begin{pmatrix} \{y_2(x)y_1(s) - y_1(x)y_2(s)\} f(s) \\ \{y_2'(x)y_1(s) - y_1'(x)y_2(s)\} f(s) \end{pmatrix}. \tag{2.41}$$

Then setting $Y_0 = \binom{c_1}{c_2}$, the top entry of the matrix equation (2.33) reads

$$y(x) = c_1 y_1(x) + c_2 y_2(x) + \int_a^x \frac{y_2(x)y_1(s) - y_1(x)y_2(s)}{W(s)} f(s)\, ds. \tag{2.42}$$

This is of course the result usually obtained. It is of some interest that (2.42) follows directly via product integration without the imposition of any subsidiary conditions on the "varied parameters," as is necessary in the usual treatment. Naturally the product integral formula (2.33) holds for matrices of any dimension, and thus allows solution of inhomogeneous scalar equations of any order provided that the corresponding homoge-

ISBN 0-201-13509-4

neous equations can be solved. As above, it is unnecessary to discuss subsidiary conditions on the "varied parameters."

After the developments of Section 3, we shall be able to give a product integral account of the method of "reduction of order."

In closing this section, we remark that Theorem 2.1 can be simply generalized to equations of the type

$$R'(x) = A(x)R(x) - R(x)B(x), \qquad R(a) = R_0, \qquad x \in [a,b], \tag{2.43}$$

where R_0 is a given $n \times n$ matrix. Writing for $x,y \in [a,b]$

$$P(x,y) = \prod_y^x e^{A(s)\,ds} \tag{2.44}$$

and

$$Q(x,y) = \prod_y^x e^{B(s)\,ds}, \tag{2.45}$$

the unique solution of (2.43) is

$$R(x) = P(x,a)R_0Q(a,x). \tag{2.46}$$

The proof is the same as for Theorem 2.1: by Theorem 1.6, R of (2.46) satisfies (2.43). On the other hand, if S is any solution of (2.43) one immediately shows that $P(a,x)S(x)Q(x,a)$ is constantly equal to R_0, so $S = R$.

Similarly, one can easily show that the unique solution of

$$R'(x) = A(x)R(x) - R(x)B(x) + L(x), \qquad R(a) = R_0 \tag{2.47}$$

is

$$R(x) = P(x,a)R_0Q(a,x) + \int_a^x P(x,s)L(s)Q(s,x)\,ds. \tag{2.48}$$

1.3 Further Properties of Product Integrals

DEFINITION 3.1. *Let* $P:[a,b] \to \mathbb{C}_{n\times n}$ *be a nonsingular function* (*i.e.*, $P(x)$ *is nonsingular for each* $x \in [a,b]$). *We will say that* P *is an* indefinite product integral *if there is a continuous function* $A:[a,b] \to \mathbb{C}_{n\times n}$ *such that*

$$\prod_a^x e^{A(s)\,ds} = P(x)P^{-1}(a). \tag{3.1}$$

We will now show that a nonsingular function P is an indefinite product integral if and only if P is a C^1 (continuously differentiable) function.

ISBN 0-201-13509-4

From Theorem 1.2 this condition is clearly necessary. To establish the sufficiency, we first define the operation L, which is a kind of logarithmic derivative of a nonsingular function and which plays approximately the same role in product integration as does the operation of differentiation in the usual theory of integration.

DEFINITION 3.2. *Let $P:[a,b]\to\mathbb{C}_{n\times n}$ be nonsingular and differentiable. We define*

$$(LP)(x)=P'(x)P^{-1}(x). \tag{3.2}$$

We remark that if (3.1) holds, then

$$P(x)=\prod_a^x e^{A(s)\,ds}P(a) \tag{3.3}$$

so that

$$P'(x)=A(x)P(x) \tag{3.4}$$

and

$$(LP)(x)=A(x). \tag{3.5}$$

Conversely, we can prove

THEOREM 3.1. *Suppose that $P:[a,b]\to\mathbb{C}_{n\times n}$ is nonsingular and C^1 on $[a,b]$. Then*

$$\prod_a^x e^{(LP)(s)\,ds}=P(x)P^{-1}(a). \tag{3.6}$$

Remark. Theorem 3.1 is of course analogous to the fundamental theorem of calculus

$$\int_a^x f'(s)\,ds=f(x)-f(a) \tag{3.7}$$

in the usual theory of integration. We shall refer to Theorem 3.1 as "the fundamental theorem of product integration."

Proof of Theorem 3.1. By the hypotheses of the theorem, LP is a continuous function, so that the product integral of LP has a meaning. The function P is by inspection a solution of the initial-value problem

$$P'(x)=(LP)(x)P(x), \qquad P(a)=P(a). \tag{3.8}$$

ISBN 0-201-13509-4

By Theorem 2.1, the unique solution of (3.8) is given by

$$P(x)=\prod_a^x e^{(LP)(s)\,ds}P(a) \tag{3.9}$$

so that (3.6) is established. This proves the theorem.

Theorem 3.1 can be used as a starting point for the explicit evaluation of product integrals in much the same way as (3.7) is used as a starting point for the evaluation of ordinary integrals. To apply (3.7), one begins from a known function f and learns to evaluate the integral of f'. To apply Theorem 3.1, one starts from a known function P and learns to evaluate the product integral of LP. An interesting problem in the theory of product integration is this: find a convenient class of explicitly product-integrable continuous functions that can play a role in product integration analogous to the role of the polynomials in ordinary integration theory.

We note that the operation L has the following properties:

$$LP^{-1}=(P^{-1})'(P^{-1})^{-1}=(-P^{-1}P'P^{-1})P=-P^{-1}P' \tag{3.10}$$

and

$$L(PQ)=(P'Q+PQ')Q^{-1}P^{-1}=LP+P(LQ)P^{-1}. \tag{3.11}$$

Using (3.11) we can obtain two very useful rules for manipulating with product integrals:

THEOREM 3.2 (The sum rule). *Let* $A,B:[a,b]\to\mathbb{C}_{n\times n}$ *be continuous. Write*

$$P(x)=\prod_a^x e^{A(s)\,ds}. \tag{3.12}$$

Then

$$\prod_a^x e^{\{A(s)+B(s)\}\,ds}=P(x)\prod_a^x e^{P^{-1}(s)B(s)P(s)\,ds}. \tag{3.13}$$

Proof. Let

$$Q(x)=\prod_a^x e^{P^{-1}(s)B(s)P(s)\,ds}. \tag{3.14}$$

Then

$$LQ=P^{-1}BP. \tag{3.15}$$

ISBN 0-201-13509-4

Thus (3.11) yields

$$L(PQ)=A+B \tag{3.16}$$

so that by the fundamental theorem

$$\prod_a^x e^{\{A(s)+B(s)\}ds}=P(x)Q(x)(P(a)Q(a))^{-1}=P(x)Q(x). \tag{3.17}$$

Thus Theorem 3.2 is proved.

THEOREM 3.3 (The similarity rule). *Let* $B:[a,b]\to\mathbb{C}_{n\times n}$ *be continuous and let* $P:[a,b]\to\mathbb{C}_{n\times n}$ *be* C^1 *and nonsingular. Then*

$$P(x)\prod_a^x e^{B(s)ds}P^{-1}(a)=\prod_a^x e^{\{(LP)(s)+P(s)B(s)P^{-1}(s)\}ds}. \tag{3.18}$$

Proof. Put

$$Q(x)=\prod_a^x e^{B(s)ds}. \tag{3.19}$$

Then by the fundamental theorem we have

$$P(x)Q(x)(P(a)Q(a))^{-1}=\prod_a^x e^{L(PQ)(s)ds}. \tag{3.20}$$

By (3.11) and the fact that $Q(a)=I$, we see that (3.20) is the same as (3.18), completing the proof. ∎

The typical application of (3.18) is as follows: $B(s)$ is given explicitly and it is desired to analyze $\prod_a^x e^{B(s)ds}$. The matrix $P(s)$ is chosen so that $P(s)B(s)P^{-1}(s)$ is diagonal. The right-hand side of (3.18) is then analyzed (by the sum rule!). Then (3.18) is used to make statements about $\prod_a^x e^{B(s)ds}$. We note for reference that if in (3.18) the function $P(x)$ is constantly equal to the value P, then $LP=0$ and (3.18) becomes

$$P\prod_a^x e^{B(s)ds}P^{-1}=\prod_a^x e^{PB(s)P^{-1}ds}, \tag{3.21}$$

an equation that can also easily be deduced directly from the definition of the product integral.

It is sometimes handy to use (3.18) with P replaced by P^{-1}. Using (3.10) we find

$$P^{-1}(x)\prod_a^x e^{B(s)ds}P(a)=\prod_a^x e^{\{P^{-1}(s)B(s)P(s)-P^{-1}(s)P'(s)\}ds}. \tag{3.22}$$

ISBN 0-201-13509-4

We are now in a position to give an account of the method of "reduction of order" for second-order scalar differential equations. Our analysis will be based on the following theorem, which is of some interest in itself and will also be used in Section 6:

THEOREM 3.4. *Let* $A:[a,b]\to\mathbb{C}_{2\times 2}$ *be continuous, and suppose that either one column or one row of* A *is zero. Then the product integral of* A *can be evaluated explicitly. For example, if*

$$A(x)=\begin{pmatrix} 0 & f(x) \\ 0 & g(x) \end{pmatrix}, \tag{3.23}$$

then writing

$$G(x)=\int_a^x g(s)\,ds \tag{3.24}$$

we have

$$\prod_a^x e^{A(s)\,ds}=\begin{pmatrix} 1 & 0 \\ 0 & e^{G(x)} \end{pmatrix}\begin{bmatrix} 1 & \int_a^x f(s)e^{G(s)}\,ds \\ 0 & 1 \end{bmatrix}. \tag{3.25}$$

Proof. We will prove (3.25), which will illustrate the method for proving the theorem in all cases. The idea is to write A as a sum of two terms and use the sum rule to separate off the term with a nonzero diagonal entry. With A as in (3.23) we write

$$A(x)=\begin{pmatrix} 0 & 0 \\ 0 & g(x) \end{pmatrix}+\begin{pmatrix} 0 & f(x) \\ 0 & 0 \end{pmatrix}\equiv A_1(x)+A_2(x). \tag{3.26}$$

We have using Theorem 1.3

$$P_1(x)=\prod_a^x e^{A_1(s)\,ds}=e^{\int_a^x A_1(s)\,ds}=e^{\begin{pmatrix} 0 & 0 \\ 0 & G(x) \end{pmatrix}}=\begin{pmatrix} 1 & 0 \\ 0 & e^{G(x)} \end{pmatrix}. \tag{3.27}$$

The sum rule gives

$$\prod_a^x e^{A(s)\,ds}=P_1(x)\prod_a^x e^{\hat{A}_2(s)\,ds} \tag{3.28}$$

with

$$\hat{A}_2(s)=P_1^{-1}(s)A_2(s)P_1(s)=\begin{pmatrix} 0 & f(s)e^{G(s)} \\ 0 & 0 \end{pmatrix}. \tag{3.29}$$

ISBN 0-201-13509-4

Now using Theorem 1.3 we have

$$\prod_a^x e^{\hat{A}_2(s)\,ds} = e^{\int_a^x \hat{A}_2(s)\,ds} = e^{\begin{pmatrix} 0 & \int_a^x f(s) e^{G(s)} ds \\ 0 & 0 \end{pmatrix}}$$

$$= I + \begin{bmatrix} 0 & \int_a^x f(s) e^{G(s)}\, ds \\ 0 & 0 \end{bmatrix}. \tag{3.30}$$

Combining (3.27), (3.28), and (3.30), the result is proved. ■

To study "reduction of order," let A be a 2×2 matrix continuous on $[a,b]$ and consider the equation

$$Y'(x) = A(x)Y(x). \tag{3.31}$$

We assume that a nontrivial solution of (3.31) is known, say

$$Y(x) = \begin{pmatrix} z_1(x) \\ z_2(x) \end{pmatrix}. \tag{3.32}$$

We also assume, for illustrative purposes, that $z_1(x) \neq 0$ on $[a,b]$. (If this is false, we can use a similar analysis on any subinterval on which z_1 or z_2 is nonzero.) Using $Y(x)$, we will show how to evaluate the product integral $\prod_a^x e^{A(s)ds}$ and hence find a second solution of (3.31), linearly independent from $Y(x)$. This is the method of reduction of order. Let

$$Z(x) = \begin{pmatrix} z_1(x) & 0 \\ z_2(x) & 1 \end{pmatrix}. \tag{3.33}$$

[Note that Z is obtained by replacing the first column of the identity matrix with $Y(x)$.] Using (3.22) with P replaced by Z and B by A, we have

$$Z^{-1}(x) \prod_a^x e^{A(s)\,ds} Z(a) = \prod_a^x e^{C(s)\,ds} \tag{3.34}$$

with

$$C(s) = Z^{-1}(s)A(s)Z(s) - Z^{-1}(s)Z'(s). \tag{3.35}$$

Now by our choice of $Z(s)$ we have

$$Z'(s) = A(s)\begin{pmatrix} z_1(s) & 0 \\ z_2(s) & 0 \end{pmatrix}. \tag{3.36}$$

ISBN 0-201-13509-4

Thus (3.35) becomes

$$C(s) = Z^{-1}(s)A(s)\begin{pmatrix} 0 & 0 \\ 0 & 1 \end{pmatrix}. \tag{3.37}$$

Equation (3.37) shows that the first column of $C(s)$ is zero, hence the product integral of C can be evaluated by Theorem 3.4. Then $\prod_a^x e^{A(s)\,ds}$ can be found from (3.34). In detail: writing

$$A = \begin{pmatrix} a_{11} & a_{12} \\ a_{21} & a_{22} \end{pmatrix} \tag{3.38}$$

we have by an elementary computation

$$C(s) = \frac{1}{z_1(s)}\begin{pmatrix} 0 & a_{12}(s) \\ 0 & a_{22}(s)z_1(s) - a_{12}(s)z_2(s) \end{pmatrix}. \tag{3.39}$$

For simplicity, let us apply our analysis to the situation in which reduction of order is usually used: study of the equation

$$y'' + p(x)y' + q(x)y = 0. \tag{3.40}$$

This equation can be put in the form (3.31) with $Y = \binom{y}{y'}$ and

$$A(x) = \begin{pmatrix} 0 & 1 \\ -q(x) & -p(x) \end{pmatrix}. \tag{3.41}$$

Equation (3.39) then becomes (note $z_2 = z_1'$ by the form of Y)

$$C(s) = \frac{1}{z_1(s)}\begin{pmatrix} 0 & 1 \\ 0 & -z_1'(s) - p(s)z_1(s) \end{pmatrix}. \tag{3.42}$$

Now letting $P(x)$ denote $\int_a^x p(s)\,ds$, we have

$$\int_a^x \left\{ \frac{-z_1'(s)}{z_1(s)} - p(s) \right\} ds = -\log\frac{z_1(x)}{z_1(a)} - P(x). \tag{3.43}$$

Then (3.25) yields

$$\prod_a^x e^{C(s)\,ds} = \begin{bmatrix} 1 & 0 \\ 0 & \dfrac{z_1(a)}{z_1(x)}e^{-P(x)} \end{bmatrix} \begin{bmatrix} 1 & \displaystyle\int_a^x \frac{z_1(a)}{z_1(s)^2}e^{-P(s)}\,ds \\ 0 & 1 \end{bmatrix} \tag{3.44}$$

ISBN 0-201-13509-4

and (3.34) shows that

$$\prod_a^x e^{A(s)\,ds} = Z(x)\prod_a^x e^{C(s)\,ds}Z^{-1}(a)$$

$$= \frac{1}{z_1(a)}\begin{pmatrix} z_1(x) & z_3(x) \\ z_1'(x) & z_3'(x) \end{pmatrix}\begin{pmatrix} 1 & 0 \\ -z_1'(a) & z_1(a) \end{pmatrix}, \tag{3.45}$$

where

$$z_3(x) = z_1(x)\int_a^x \frac{z_1(a)}{z_1(s)^2}e^{-P(s)}\,ds. \tag{3.46}$$

Equation (3.46) is [up to the constant factor $z_1(a)$] the result usually obtained for a second solution of (3.40) if the solution $z_1(x)$ is known.

Theorem 3.4 and the technique of reduction of order can be generalized to $n\times n$ matrices. Namely, if A is a continuous $n\times n$ matrix-valued function having exactly one nonzero row (or column), then the product integral of A can be evaluated explicitly. Suppose, for instance, that only the kth column of A is nonzero. Then $a_{kk}(s)$ is the trace of A. Let $\hat{A}(s)$ denote the matrix obtained from A by setting $a_{kk}(s)=0$. Let E_k denote the matrix having 1 for its (k,k) entry, with all other entries zero. Then

$$A(s) = (\operatorname{tr} A(s))E_k + \hat{A}(s). \tag{3.47}$$

Now let

$$T(x) = \int_a^x \operatorname{tr} A(s)\,ds \tag{3.48}$$

and

$$Q(x) = \prod_a^x e^{(\operatorname{tr} A(s))E_k\,ds}$$

$$= \begin{bmatrix} 1 & & & & & \\ & \ddots & & & & \\ & & 1 & & & \\ & & & e^{T(x)} & & \\ & & & & \ddots & \\ & & & & & 1 \end{bmatrix} \leftarrow k\text{th row}. \tag{3.49}$$

Then by the sum rule

$$\prod_a^x e^{A(s)\,ds} = Q(x)\prod_a^x e^{Q^{-1}(s)\hat{A}(s)Q(s)\,ds} \equiv Q(x)\prod_a^x e^{\hat{B}(s)\,ds}. \tag{3.50}$$

ISBN 0-201-13509-4

A simple computation shows that

$$\hat{B}(s) = e^{T(s)}\hat{A}(s) \tag{3.51}$$

and that the family $\{\hat{B}(s) | s \in [a,b]\}$ is commutative. Thus

$$\prod_a^x e^{A(s)\,ds} = Q(x)e^{\int_a^x e^{T(s)}\hat{A}(s)\,ds} = Q(x)(I + R(x)) = Q(x) + R(x), \tag{3.52}$$

where $R(x)$ is given by

$$\begin{aligned} R_{ij} &= 0 \qquad (j \neq k), \\ R_{kk} &= 0, \\ R_{ik} &= \int_a^x e^{T(s)} a_{ik}(s)\,ds \qquad (i \neq k). \end{aligned} \tag{3.53}$$

Using the preceding observations, it is easy to show that if A is any continuous $n \times n$ matrix-valued function and $n-1$ linearly independent solutions of $Y' = AY$ are known, then $\prod_a^x e^{A(s)\,ds}$ can be evaluated. One merely follows the approach given above for 2×2 matrices, using for $Z(x)$ the matrix obtained by replacing the first $n-1$ columns of the identity matrix by the $n-1$ linearly independent solutions already found. We leave the details to the reader.

Just as in the ordinary theory of Riemann integration, there is a change of variable formula for product integrals:

THEOREM 3.5. *Let $\phi : [a,b] \to \mathbb{R}$ be a function that is continuously differentiable. Let A be an $n \times n$ matrix-valued function that is continuous on the range of ϕ. Then for $x, y \in [a,b]$ we have*

$$\prod_y^x e^{A(\phi(s))\phi'(s)\,ds} = \prod_{\phi(y)}^{\phi(x)} e^{A(s)\,ds}. \tag{3.54}$$

Proof. Fix y. The two sides of (3.43) agree when $x = y$. Furthermore, it is easy to see via the chain rule that considered as functions of x the two sides satisfy the same differential equation. Thus by Theorem 2.1 the two sides agree for $x \in [a,b]$ and we are done. ■

In particular, let us replace y by $-d$ and x by $-c$ where $d > c$. Taking $\phi(s) = -s$, (3.54) then yields

$$\prod_d^c e^{A(s)\,ds} = \prod_{-d}^{-c} e^{-A(-s)\,ds}. \tag{3.55}$$

Since $d > c$, the left-hand side of (3.55) has previously been defined only as

ISBN 0-201-13509-4

the inverse of $\prod_c^d e^{A(s)ds}$. Formula (3.55) shows how to write $\prod_d^c e^{A(s)ds}$ as a product integral in our original sense (limit of Riemann products) because the right-hand side of (3.55) is a product integral from a smaller to a larger number and thus falls under our original definition. The representation (3.55) will occasionally be useful to us later on.

1.4 Estimates of Size, and the Product Integral as a Time-Ordered Exponential

In this section we derive some bounds on product integrals that will be useful to us later. We shall also see how to "expand" a product integral as an infinite series of ordinary integrals, usually referred to as a "time-ordered exponential."

The two bounds that will be most directly useful to us are given in

THEOREM 4.1. *Let $A:[a,b]\to\mathbb{C}_{n\times n}$ be continuous. Let $x,y\in[a,b]$. Then*

$$\left\|\prod_y^x e^{A(s)ds}\right\| \leqslant e^{\left|\int_y^x\|A(s)\|ds\right|} \tag{4.1}$$

and

$$\left\|\prod_y^x e^{A(s)ds}-I\right\| \leqslant e^{\left|\int_y^x\|A(s)\|ds\right|}-1. \tag{4.2}$$

Remark. If $y>x$, the integral $\int_y^x\|A(s)\|\,ds$ is nonpositive. This is the reason for the absolute value signs in (4.1) and (4.2).

Proof of Theorem 4.1. We will prove the theorem for $y\leqslant x$ (hence without the absolute value signs). The result for $y>x$ can then be obtained using (3.55). Now if $y=x$, then (4.1) and (4.2) are obvious, so we may assume $y<x$. We prove (4.1) directly from the definition of the product integral: let $P=\{s_0,s_1,\ldots,s_n\}$ be a partition of $[y,x]$. Then

$$\begin{aligned}\left\|\prod_y^x e^{A(s)ds}\right\| &= \lim_{\mu(P)\to 0}\left\|\prod_{k=1}^n e^{A(s_k)\Delta s_k}\right\| \\ &\leqslant \lim_{\mu(P)\to 0}\prod_{k=1}^n\|e^{A(s_k)\Delta s_k}\| \leqslant \lim_{\mu(P)\to 0}\prod_{k=1}^n e^{\|A(s_k)\|\Delta s_k} \\ &= \lim_{\mu(P)\to 0} e^{\sum_{k=1}^n\|A(s_k)\|\Delta s_k} = e^{\int_y^x\|A(s)\|ds}. \end{aligned} \tag{4.3}$$

ISBN 0-201-13509-4

This proves (4.1). To prove (4.2) let

$$F(x,y)=\prod_{y}^{x} e^{A(s)\,ds}. \tag{4.4}$$

Using Eq. (1.20) of Theorem 1.2 (with the integration variable replaced by t) and the bound (4.1) we have

$$\begin{aligned}\|F(x,y)-I\| &= \left\|\int_y^x A(t)F(t,y)\,dt\right\|\\ &\leqslant \int_y^x \|A(t)\|e^{\int_y^t\|A(s)\|\,ds}\,dt=\int_y^x \frac{d}{dt}e^{\int_y^t\|A(s)\|\,ds}\,dt\\ &= e^{\int_y^x\|A(s)\|\,ds}-1.\end{aligned} \tag{4.5}$$

Thus Theorem 4.1 is proved. ■

Theorem 4.1 is actually a special case of Theorem 4.2. We have isolated the results of Theorem 4.1 for clarity and ease of reference. The inequality (4.2) estimates the difference between the product integral $F(x,y)$ and its "leading term," the identity matrix. Of course, I is not a very good approximation to $F(x,y)$. A better approximation can be found by iterating the integral equation (1.20) once to find

$$F(x,y)=I+\int_y^x A(t_1)\,dt_1+\int_y^x A(t_1)\left\{\int_y^{t_1}A(t_2)F(t_2,y)\,dt_2\right\}dt_1. \tag{4.6}$$

Using the bound (4.1) in the integral on the right-hand side of (4.6) yields for $y\leqslant x$

$$\begin{aligned}\left\|F(x,y)-I-\int_y^x A(t_1)\,dt_1\right\| &\leqslant \int_y^x \|A(t_1)\|\left\{\int_y^{t_1}\|A(t_2)\|e^{\int_y^{t_2}\|A(s)\|\,ds}\,dt_2\right\}dt_1\\ &= \int_y^x \|A(t_1)\|\left[e^{\int_y^{t_1}\|A(s)\|\,ds}-1\right]dt_1=e^{\int_y^x\|A(s)\|\,ds}-1-\int_y^x\|A(s)\|\,ds,\end{aligned} \tag{4.7}$$

where we have calculated the t_1 integral as in (4.5). The inequality (4.7) shows that the quantity $I+\int_y^x A(t_1)\,dt_1$ is a better approximation to $F(x,y)$ than is I alone. (Technically, it is the estimate of the difference that is better.) Note that the error estimate consists of the exponential $e^{\int_y^x\|A(s)\|\,ds}$ minus the first two terms of the power series expansion of this exponential. Our next theorem is a direct generalization of this result.

THEOREM 4.2. *Let* $A:[a,b]\to\mathbb{C}_{n\times n}$ *be continuous. For* $x,y\in[a,b]$ *let*

$$J_0(x,y)=I \tag{4.8}$$

ISBN 0-201-13509-4

and for $n \geqslant 1$ define $J_n(x,y)$ inductively by

$$J_n(x,y)=\int_y^x A(s)J_{n-1}(s,y)\,ds. \tag{4.9}$$

Let

$$L(x,y)=\int_y^x \|A(s)\|\,ds. \tag{4.10}$$

[*Note that $L(x,y)$ is positive or negative according as $y<x$ or $y>x$.*] *Then for $n\geqslant 0$ we have*

$$\begin{aligned}\left\|\prod_y^x e^{A(s)ds}-\sum_{k=0}^{n} J_n(x,y)\right\| &\leqslant e^{|L(x,y)|}-\sum_{k=0}^{n}\frac{|L(x,y)|^k}{k!}\\ &\leqslant \frac{|L(x,y)|^{n+1}}{(n+1)!}e^{L(b,a)}.\end{aligned} \tag{4.11}$$

Remark. Note that (4.2) and (4.7) are respectively the cases $n=0$ and $n=1$ in (4.11). Equation (4.1) can be formally obtained from (4.11) by setting $n=-1$. We shall prove (4.11) for the case $y\leqslant x$. If $x<y$, the general result can then be obtained using (3.55).

Proof of Theorem 4.2 with $y\leqslant x$. The proof is by induction on n. The result is known for $n=0$. Assume that the result holds for $n=l$. Note that

$$\sum_{k=0}^{l+1} J_k(x,y)=I+\int_y^x A(s)\sum_{k=0}^{l} J_k(s,y)\,ds. \tag{4.12}$$

Using the notation (4.4) and the integral equation for $F(x,y)$, we have

$$F(x,y)-\sum_{k=0}^{l+1} J_k(x,y)=\int_y^x A(s)\left(F(s,y)-\sum_{k=0}^{l} J_k(s,y)\right)ds. \tag{4.13}$$

Thus by the induction hypothesis

$$\begin{aligned}\left\|F(x,y)-\sum_{k=0}^{l+1} J_k(x,y)\right\| &\leqslant \int_y^x \|A(s)\|\left(e^{L(s,y)}-\sum_{k=0}^{l}\frac{L(s,y)^k}{k!}\right)ds\\ &=\int_y^x\left\{\frac{d}{ds}e^{L(s,y)}-\frac{d}{ds}\left(\sum_{k=0}^{l+1}\frac{L(s,y)^k}{k!}\right)\right\}ds\\ &=e^{L(x,y)}-\sum_{k=0}^{l+1}\frac{L(x,y)^k}{k!}.\end{aligned} \tag{4.14}$$

ISBN 0-201-13509-4

This proves the first inequality in (4.11). To obtain the second, we use the elementary estimate

$$e^{\mu} - \sum_{k=0}^{n} \frac{\mu^n}{n!} \leqslant \frac{\mu^{n+1}}{(n+1)!} e^{\mu} \qquad (\mu > 0) \tag{4.15}$$

and the fact that $|L(x,y)|$ is bounded by the constant $L(b,a)$. We also note that since $|L(x,y)| \leqslant L(b,a)$, we have

$$\left\| \prod_y^x e^{A(s)ds} - \sum_{k=0}^{n} J_n(x,y) \right\| \leqslant e^{L(b,a)} - \sum_{k=0}^{n} \frac{L(b,a)^k}{k!} \leqslant \frac{L(b,a)^{n+1}}{(n+1)!} e^{L(b,a)}. \tag{4.16}$$

Equation (4.16) gives an estimate that is uniform for x,y in the interval $[a,b]$. As an immediate application of this estimate, we obtain

THEOREM 4.3. *Let* $A:[a,b] \to \mathbb{C}_{n \times n}$ *be continuous. Then with notation as in Theorem* 4.2 *we have*

$$\prod_y^x e^{A(s)ds} = \sum_{k=0}^{\infty} J_k(x,y), \tag{4.17}$$

the series on the right converging uniformly for $x,y \in [a,b]$.

Proof. According to (4.16),

$$\left\| \prod_y^x e^{A(s)ds} - \sum_{k=0}^{n} J_k(x,y) \right\| \leqslant M \frac{L(b,a)^{n+1}}{(n+1)!} \to 0, \qquad n \to \infty \tag{4.18}$$

and we are done. ■

The series in (4.17) is often known as the "time-ordered exponential" of A. The reason for this terminology is as follows: the nth term of the series is

$$\begin{aligned} J_n(x,y) &= \int_y^x A(s_1) \int_y^{s_1} A(s_2) \cdots \int_y^{s_{n-1}} A(s_n)\, ds_n \cdots ds_2\, ds_1 \\ &= \int_{x \geqslant s_1 \geqslant s_2 \geqslant \cdots \geqslant y} \cdots \int A(s_1)A(s_2) \cdots A(s_n)\, ds_n \cdots ds_2\, ds_1. \end{aligned} \tag{4.19}$$

In the integrals occurring in (4.19) the arguments in the product of A's are ordered: the A with the smallest argument stands furthest to the right, etc. If the parameters s_k are thought of as representing time, then in the product of A's earlier terms occur further to the right. Thus the product is

ISBN 0-201-13509-4

"time-ordered." If the family $\{A(s)|s\in[a,b]\}$ is commutative, then the ordering is irrelevant. In this case we can write

$$J_n(x,y)=\frac{1}{n!}\int_y^x\cdots\int_y^x A(s_1)A(s_2)\cdots A(s_n)\,ds_n\cdots ds_2\,ds_1. \qquad (4.20)$$

The factor of $1/n!$ occurs because the integral on the right in (4.20) can be split into $n!$ integrals each corresponding to a certain ordering of $\{s_1,s_2,\ldots,s_n\}$, and all (because of commutativity) having the value $J_n(x,y)$. From (4.20) we have

$$J_n(x,y)=\frac{1}{n!}\left(\int_y^x A(s)\,ds\right)^n \qquad (4.21)$$

and thus

$$\sum_{k=0}^{\infty} J_k(x,y)=e^{\int_y^x A(s)\,ds}. \qquad (4.22)$$

Thus in the commutative case the "time-ordered exponential" really is an exponential. Of course, (4.22) agrees with the result of Theorem 1.3 on the product integral of a function whose values commute.

As a final remark on bounds for product integrals, we note that if $A:[a,b]\to\mathbb{C}_{n\times n}$ is *skew-adjoint*, i.e.,

$$A^*(s)=-A(s), \qquad s\in[a,b], \qquad (4.23)$$

then for $x,y\in[a,b]$ we have

$$\left\|\prod_y^x e^{A(s)\,ds}\right\|=1. \qquad (4.24)$$

The reason is simple: if $y\leqslant x$, then $\prod_y^x e^{A(s)ds}$ is a limit of products of terms having the form $e^{A(s_k)\Delta s_k}$. Each such term is unitary since $A(s_k)$ is skew-adjoint. Hence the product of the terms is unitary and so is the limit of the product, proving (4.24) for $y\leqslant x$. The inequality for $y>x$ is then immediate since the inverse of a unitary operator is unitary.

1.5 Dependence on a Parameter

In this section we consider a function $A:[a,b]\times[c,d]\to\mathbb{C}_{n\times n}$, where $[a,b]$ and $[c,d]$ are subintervals of $\mathbb{R}$. Writing $A=A(s,\lambda)$, we consider the product integral of A with respect to s. Assuming various regularity properties (continuity, differentiability) for A as a function of λ, we ask whether these properties are inherited by the product integral of A. We shall need the following result (compare Lemma 1.2):

ISBN 0-201-13509-4

THEOREM 5.1 (Duhamel's Formula): *Let $A,B:[a,b]\to\mathbb{C}_{n\times n}$ be continuous. For $x,y\in[a,b]$, let*

$$P(x,y)=\prod_y^x e^{A(s)\,ds}, \qquad Q(x,y)=\prod_y^x e^{B(s)\,ds}, \tag{5.1}$$

then

$$P(x,y)-Q(x,y)=\int_y^x Q(x,s)(A(s)-B(s))P(s,y)\,ds. \tag{5.2}$$

Proof. Fix y and let

$$H(x,y)=Q(y,x)P(x,y). \tag{5.3}$$

Using the product rule for differentiation and the differentiation formulas of Theorem 1.6, we find

$$\frac{dH(x,y)}{dx}=Q(y,x)(A(x)-B(x))P(x,y). \tag{5.4}$$

Equation (5.4) combined with the fact that that $H(y,y)=I$ yields

$$H(x,y)=I+\int_y^x Q(y,s)(A(s)-B(s))P(s,y)\,ds. \tag{5.5}$$

Equation (5.2) is now obtained by multiplying (5.5) on the left with $Q(x,y)$. Thus theorem 5.1 is proved.

COROLLARY. *Let A and B satisfy the hypotheses of Theorem 5.1. Then*

$$\left\|\prod_y^x e^{A(s)\,ds}-\prod_y^x e^{B(s)\,ds}\right\|\leqslant e^{\|A\|_1+\|B\|_1}\left|\int_y^x \|A(s)-B(s)\|\,ds\right|, \tag{5.6}$$

where $\|A\|_1$ and $\|B\|_1$ denote the L^1 norms of A and B. (The absolute value sign is necessary if $y>x$.)

Proof. The inequality (5.6) follows directly from Duhamel's formula (5.2): we estimate the integral on the right-hand side of (5.2) by $|\int_y^x \|Q(x,s)\|\,\|A(s)-B(s)\|\,\|P(s,y)\|\,ds|$, and use the first bound of Theorem 4.1 to estimate $\|Q(x,s)\|$ by $e^{\|B\|_1}$ and $\|P(s,y)\|$ by $e^{\|A\|_1}$.

We can now prove ■

THEOREM 5.2. *Let $[a,b]$ and $[c,d]$ be intervals in $\mathbb{R}$, and let $A:[a,b]\times[c,d]\to\mathbb{C}_{n\times n}$ be a function such that $A(s,\lambda)$ is continuous in s for each fixed*

ISBN 0-201-13509-4

$\lambda \in [c,d]$. *Suppose that* $A(s,\lambda)$ *is continuous in* λ in the L^1 sense, *i.e.*,

$$\lim_{\lambda' \to \lambda} \int_a^b \|A(s,\lambda') - A(s,\lambda)\| \, ds = 0, \qquad \lambda', \lambda \in [c,d]. \tag{5.7}$$

Let $x,y \in [a,b]$ *and let*

$$P(x,y;\lambda) = \prod_y^x e^{A(s,\lambda)\,ds}. \tag{5.8}$$

Then $P(x,y;\lambda)$ is jointly continuous in $(x,y;\lambda)$ for $x,y \in [a,b]$, $\lambda \in [c,d]$.

Proof. First note that by (5.7) the function

$$M_1(\lambda) = \int_a^b \|A(s,\lambda)\| \, ds$$

is continuous and hence bounded on $[c,d]$ by some number M. By Theorem 4.1 we then have

$$\|P(x,y;\lambda)\| \leqslant e^M, \qquad x,y \in [a,b], \qquad \lambda \in [c,d]. \tag{5.9}$$

Now

$$\|P(x',y';\lambda') - P(x,y;\lambda)\| \leqslant \|P(x',y';\lambda') - P(x',y';\lambda)\| \\ + \|P(x',y';\lambda) - P(x,y';\lambda)\| + \|P(x,y';\lambda) - P(x,y;\lambda)\|. \tag{5.10}$$

Using the Corollary to Theorem 5.1, we find

$$\|P(x',y';\lambda') - P(x',y';\lambda)\| \leqslant e^{2M} \left| \int_{y'}^{x'} \|A(s,\lambda') - A(s,\lambda)\| \, ds \right| \\ \leqslant e^{2M} \int_a^b \|A(s,\lambda') - A(s,\lambda)\| \, ds. \tag{5.11}$$

Using Theorem 4.1 and the multiplicative property (Theorem 1.5), we have

$$\|P(x',y';\lambda) - P(x,y';\lambda)\| = \|(P(x',x;\lambda) - I)P(x,y';\lambda)\| \\ \leqslant \left[e^{|\int_x^{x'} \|A(s,\lambda)\| \, ds|} - 1 \right] e^M. \tag{5.12}$$

Treating the third term on the right-hand side of (5.9) in the same way, we find

$$\lim_{\substack{x' \to x \\ y' \to y \\ \lambda' \to \lambda}} P(x',y';\lambda') = P(x,y;\lambda), \tag{5.13}$$

ISBN 0-201-13509-4

as desired. We note that the hypotheses of Theorem 5.2 are satisfied if $A(s,\lambda)$ is jointly continuous in s and λ on $[a,b]\times[c,d]$, or (by Lebesgue's dominated convergence theorem) if $A(s,\lambda)$ is continuous in λ for almost every $s\in[a,b]$ and bounded in norm by a fixed integrable function $g(s)$. The hypotheses of Theorem 5.2 have been stated in such a way as to allow the theorem to be generalized easily to the case in which A is merely integrable in s for each fixed λ. We make this generalization later. We now prove a theorem on differentiability. ■

THEOREM 5.3. *Let* $A:[a,b]\times[c,d]\to\mathbb{C}_{n\times n}$ *be a function such that* $A(s,\lambda)$ *is continuous in s for each fixed* $\lambda\in[c,d]$. *Suppose that A is differentiable in* λ in the L^1 sense, *i.e., suppose there is a function* $A_2:[a,b]\times[c,d]\to\mathbb{C}_{n\times n}$ *such that* $A_2(s,\lambda)$ *is integrable in s for each* $\lambda\in[c,d]$ *and*

$$\lim_{\lambda'\to\lambda}\int_a^b\left\|\frac{A(s,\lambda')-A(s,\lambda)}{\lambda'-\lambda}-A_2(s,\lambda)\right\|ds=0,\qquad \lambda\in[c,d]. \tag{5.14}$$

Then the function $P(x,y;\lambda)$ *of* (5.8) *is differentiable with respect to* λ, *and*

$$\frac{\partial}{\partial\lambda}P(x,y;\lambda)=\int_y^x P(x,s;\lambda)A_2(s,\lambda)P(s,y;\lambda)\,ds. \tag{5.15}$$

Remark. The notation $A_2(s,\lambda)$ has been used to indicate differentiation of $A(s,\lambda)$ with respect to the second variable. The reader can observe that formula (5.15) is the continuous analogue of the formula for differentiating a product of n factors.

Proof of Theorem 5.3. First note that the integral on the right-hand side of (5.15) is well-defined because A_2 is integrable in s, and $P(x,s;\lambda)$ and $P(s,y;\lambda)$ are continuous in s. Temporarily denoting this integral by $I(x,y;\lambda)$ we have, using Duhamel's formula (or really its negative),

$$\left\|\frac{P(x,y;\lambda')-P(x,y;\lambda)}{\lambda'-\lambda}-I(x,y;\lambda)\right\|=$$

$$\left\|\int_y^x P(x,s;\lambda')\left\{\frac{A(s,\lambda')-A(s,\lambda)}{\lambda'-\lambda}-A_2(s,\lambda)\right\}P(s,y;\lambda)\,ds\right.$$

$$\left.+\int_y^x\{P(x,s;\lambda')-P(x,s;\lambda)\}A_2(s,\lambda)P(s,y;\lambda)\,ds\right\|. \tag{5.16}$$

The hypothesis (5.14) of L^1-differentiability implies L^1-continuity [Eq. (5.7)] in an obvious way, so that the bound (5.9) holds for $P(x,y;\lambda)$. This together with the hypothesis (5.14) shows that the first integral on the right in (5.16) approaches zero when $\lambda'\to\lambda$. Because A satisfies (5.7) we also

ISBN 0-201-13509-4

have from Theorem 5.2 that $P(x,y;\lambda)$ is jointly continuous for $x,y \in [a,b], \lambda \in [c,d]$, and thus uniformly continuous, since the domain of the variables is compact. It follows easily that the second integral on the right in (5.16) approaches zero when $\lambda' \to \lambda$, and the theorem is proved. ∎

We remark that if λ is allowed to vary in $\mathbb{C}$ and the derivatives with respect to λ are interpreted as complex derivatives, then the proof given is still valid and Theorem 5.3 deduces analyticity in λ of $P(x,y;\lambda)$ from "L^1-analyticity" of $A(s,\lambda)$.

Under the hypotheses of Theorem 5.3, let us further assume $A_2 \in L^1([a,b]\times[c,d])$. Defining $P(x,y;\lambda)$ as above, we have that

(1) $P(x,y;\lambda)$ is differentiable with respect to λ for every $\lambda \in [c,d]$ and

(2) $\partial P(x,y;\lambda)/\partial\lambda$ is integrable with respect to λ over $[c,d]$.

Statement (2) follows from (5.15), the bound (5.9) for $P(x,y;\lambda)$, and our new assumption on A_2. In fact, the function $P(x,s;\lambda)A_2(s,\lambda)P(s,y;\lambda)$ is clearly in $L^1([a,b]\times[c,d])$, and Fubini's theorem on integrals [KY2] then shows that statement (2) holds. Statements (1) and (2) imply (though it is somewhat subtle to prove [WR, p. 168 Theorem 8.21,]) that for $\lambda,\mu \in [c,d]$ we have

$$P(x,y;\lambda) - P(x,y;\mu) = \int_\mu^\lambda \frac{\partial P(x,y;\sigma)}{\partial\sigma}\, d\sigma. \tag{5.17}$$

Incorporating (5.15) we have proved

THEOREM 5.4. *Assume the hypotheses of Theorem* 5.3 *and further assume* $A_2 \in L^1([a,b]\times[c,d])$. *Then for* $x,y \in [a,b], \lambda,\mu \in [c,d]$ *we have*

$$P(x,y;\lambda) - P(x,y;\mu) = \int_\mu^\lambda \int_y^x P(x,s;\sigma)A_2(s,\sigma)P(s,y;\sigma)\, ds\, d\sigma. \tag{5.18}$$

Example. Consider the equation

$$y''(x) = \left(\frac{\lambda}{x^2} - k^2\right)y(x), \qquad x \in (0,\infty), \tag{5.19}$$

where λ and k are real numbers. This equation results from Bessel's equation via some elementary transformations. Putting $Y = \binom{y}{y'}$, Eq. (5.19) becomes

$$\frac{d}{dx} Y(x,\lambda) = A(x,\lambda)\, Y(x,\lambda) \tag{5.20}$$

with

$$A(x,\lambda) = \begin{pmatrix} 0 & 1 \\ \lambda/x^2 - k^2 & 0 \end{pmatrix}. \tag{5.21}$$

ISBN 0-201-13509-4

If $a \in (0,\infty)$, we have

$$Y(x,\lambda) = \prod_a^x e^{A(s,\lambda)\,ds}\, Y(a,\lambda) \equiv P(x,y;\lambda)\, Y(a,\lambda). \tag{5.22}$$

Assuming that $Y(a,\lambda) \equiv Y_0$ does not vary with λ, we can use (5.18) to compare solutions of (5.18) with different values of λ but the same initial conditions. We have

$$A_2(x,\lambda) = \begin{pmatrix} 0 & 0 \\ 1/x^2 & 0 \end{pmatrix}. \tag{5.23}$$

It will follow from our work in Section 6 that $\|P(x,y;\sigma)\|$ is bounded by some number M for $x,y \in [a,\infty)$ and σ belonging to some finite interval $[c,d]$. [In Section 6 we discuss the case $\lambda = l(l+1)$, l a nonnegative integer, but this plays no role in the proof.] Equation (5.18) then yields for $\lambda, \mu \in [c,d], x,y \in (0,\infty)$;

$$\|P(x,a;\lambda)\, Y_0 - P(x,a;\mu)\, Y_0\| \leqslant M^2 \int_\mu^\lambda \int_a^x \frac{1}{s^2}\, ds\, d\sigma = M^2(\lambda - \mu)\left(\frac{1}{a} - \frac{1}{x}\right). \tag{5.24}$$

This bound is clearly uniform in x on $[a,\infty)$. Thus for Eq. (5.19) a small change in the parameter λ produces a change in the solution y, which is uniformly small for $x \geqslant a$. The same is true for the derivative y'.

1.6 Improper Product Integration

In the usual theory of Riemann integration, one defines "improper" integrals in two types of situations. In the first type, the integration region is a finite interval $[a,b]$ but there is difficulty at an endpoint. For instance, the function f to be integrated might be Riemann integrable over $[a',b]$ for each $a' > a$, but not Riemann integrable over $[a,b]$, as in the integral

$$\int_0^1 \frac{1}{x^{1/2}}\, dx = \lim_{\varepsilon \to 0} \int_\varepsilon^1 \frac{1}{x^{1/2}}\, dx. \tag{6.1}$$

In the second type, the integration region is infinite, as in the integral

$$\int_1^\infty \frac{1}{x^2}\, dx = \lim_{M \to \infty} \int_1^M \frac{1}{x^2}\, dx. \tag{6.2}$$

As indicated in (6.1) and (6.2), both types of improper integrals are defined by taking limits of ordinary integrals. The theory of the two types is so similar that they can be discussed simultaneously.

ISBN 0-201-13509-4

The situation is much the same for product integrals. In this section we discuss improper product integration. Rather than strive for maximum generality, we shall discuss only improper product integrals of the second type. This will render the material somewhat more readable. The reader should be aware, however, that for every theorem we shall prove there is a corresponding theorem for improper product integrals of the first type. The modifications required should be obvious in all cases.

DEFINITION 6.1. *Suppose that* $a \in \mathbb{R}$ *and that* $A:[a,\infty)\to \mathbb{C}_{n\times n}$ *is continuous. We define the* improper product integral $\prod_a^{\infty} e^{A(s)ds}$ *by*

$$\prod_a^{\infty} e^{A(s)\,ds} = \lim_{x\to\infty} \prod_a^{x} e^{A(s)\,ds} \tag{6.3}$$

provided that the indicated limit exists.

Naturally, we define similarly improper product integrals of the type $\prod_{-\infty}^{b} e^{A(s)ds}$ and $\prod_{-\infty}^{\infty} e^{A(s)ds}$, and the improper product integral $\prod_a^b e^{A(s)ds}$ for the case in which A is continuous on (a,b) but not on $[a,b]$.

The study of the improper integral in (6.3) is particularly interesting because it yields information on the asymptotic behavior (for large x) of solutions of differential equations. For example, if the improper product integral of (6.3) exists, then every solution of $Y'(x)=A(x)Y(x)$ approaches a constant as $x\to\infty$. In Chapter IV (Applications) we shall use the theory of improper product integration to study the asymptotic behavior of solutions of the radial Schrödinger equation.

Unlike ordinary product integrals, improper product integrals can be singular. For example, let $A(x) = -I$ on $[a,\infty)$. Then

$$\prod_a^{x} e^{A(s)\,ds} = e^{-I(x-a)} \underset{x\to\infty}{\to} 0. \tag{6.4}$$

We shall be interested in establishing conditions under which $\prod_a^{\infty} e^{A(s)ds}$ exists and is nonsingular. Our first result is a well-known theorem in the theory of differential equations.

THEOREM 6.1. *Suppose that* $A:[a,\infty)\to\mathbb{C}_{n\times n}$ *is continuous. Further suppose that* $A\in L^1(a,\infty)$, *i.e.,*

$$\|A\|_1 \equiv \int_a^{\infty} \|A(s)\|\,ds < \infty. \tag{6.5}$$

Then $\prod_a^{\infty} e^{A(s)ds}$ *exists and is nonsingular.*

Proof. For $x,y\in[a,\infty)$, let

$$F(x,y) = \prod_y^{x} e^{A(s)\,ds}. \tag{6.6}$$

ISBN 0-201-13509-4

We wish to show that $F(x,a)$ has a nonsingular limit as $x \to \infty$. Assuming $y \leqslant x$ and using Theorem 4.1 and the multiplicative property of product integrals, we have

$$\|F(x,a)-F(y,a)\| = \|(F(x,y)-I)F(y,a)\|$$
$$\leqslant \|F(x,y)-I\|\,\|F(y,a)\| \leqslant \left[e^{\int_y^x \|A(s)\|\,ds}-1\right]e^{\int_a^y \|A(s)\|\,ds}. \tag{6.7}$$

Now the second factor on the right of (6.7) is bounded by $e^{\int_a^\infty \|A(s)\|\,ds}$. Also, because of (6.5), the integral $\int_y^x \|A(s)\|\,ds$ can be made as small as desired by taking x and y large enough. The same is then true for $(e^{\int_y^x \|A(s)\|\,ds}-1)$. Thus (6.7) shows that $F(x,a)$ has the Cauchy property and hence converges as $x \to \infty$. Also the limit of $F(x,a)$ is nonsingular because using Theorem 1.4 we have

$$\det \lim_{x\to\infty} F(x,a) = \lim_{x\to\infty} \det F(x,a)$$
$$= \lim_{x\to\infty} e^{\int_a^x \operatorname{tr} A(s)\,ds} = e^{\int_a^\infty \operatorname{tr} A(s)\,ds} \neq 0 \tag{6.8}$$

where $\int_a^\infty \operatorname{tr} A(s)\,ds$ exists because of (6.5). This proves the theorem. ■

We note that taking the limit as $x \to \infty$ in (6.7) gives the estimate

$$\left\| \prod_a^\infty A(s)\,ds - \prod_a^y e^{A(s)\,ds} \right\| \leqslant \left[e^{\int_y^\infty \|A(s)\|\,ds}-1\right]e^{\int_a^y \|A(s)\|\,ds}$$
$$\leqslant \left[e^{\int_y^\infty \|A(s)\|\,ds}-1\right]e^{\int_a^\infty \|A(s)\|\,ds}. \tag{6.9}$$

Using the elementary inequality

$$|e^\lambda - 1| \leqslant |\lambda| e^{|\lambda|} \qquad (\lambda \in \mathbb{C}) \tag{6.10}$$

we obtain from (6.9) the bound

$$\left\| \prod_a^\infty e^{A(s)\,ds} - \prod_a^y e^{A(s)\,ds} \right\| \leqslant \int_y^\infty \|A(s)\|\,ds e^{2\|A\|_1} \qquad (y \geqslant a), \tag{6.11}$$

which provides an estimate of the speed with which the improper product integral of A converges.

Corollary to Theorem 6.1. *Suppose that $A,B:[a,\infty)\to\mathbb{C}_{n\times n}$ are continuous, that $B \in L^1(a,\infty)$, and that $\prod_a^\infty e^{A(s)ds}$ exists and is nonsingular. Then $\prod_a^\infty e^{\{A(s)+B(s)\}ds}$ exists and is nonsingular.*

ISBN 0-201-13509-4

Proof. Let

$$P(x)=\prod_a^x e^{A(s)\,ds}. \tag{6.12}$$

Then

$$\prod_a^x e^{\{A(s)+B(s)\}\,ds}=P(x)\prod_a^x e^{P^{-1}(s)B(s)P(s)\,ds}. \tag{6.13}$$

Now by hypothesis $P(x)$ has a *nonsingular* limit as $x\to\infty$. It follows that $P^{-1}(x)$ also tends to a limit as $x\to\infty$. Since P and P^{-1} are continuous, they are bounded in norm on any finite interval $[a,b]$, and because they have limits at ∞ they are bounded in norm on $[a,\infty)$. Since $B\in L^1(a,\infty)$ it follows that $P^{-1}BP\in L^1(a,\infty)$. Thus both the first and second factors on the right-hand side of (6.13) have nonsingular limits as $x\to\infty$, and we are done. ■

As an example of the use of the preceding theorem, we reconsider Eq. (5.19) with λ replaced by $l(l+1)$, where l is a nonnegative integer:

$$y''(x)=\left(\frac{l(l+1)}{x^2}-k^2\right)y(x),\qquad x\in(0,\infty). \tag{6.14}$$

[*Note*: we choose the value $l(l+1)$ for λ not because this value plays any role in our analysis but because Bessel function asymptotics are most familiar in this case and because this is the value that occurs in the radial Schrödinger equation, which will be studied later.]

If J_p denotes the Bessel function of order p, then it is well known [LIS] that two linearly independent solutions of (6.14) are given by

$$y_\pm(x)=\left(\frac{\pi kx}{2}\right)^{1/2}J_{\pm(l+1/2)}(kx). \tag{6.15}$$

It is also known that

$$y_+(x)\approx\cos\left(kx-\frac{(l+1)\pi}{2}\right)\qquad(x\to\infty) \tag{6.16}$$

and

$$y_-(x)\approx(-1)^{l+1}\sin\left(kx-\frac{(l+1)\pi}{2}\right)\qquad(x\to\infty), \tag{6.17}$$

where the notation $\approx$ means that the difference between the two sides approaches zero as $x\to\infty$.

ISBN 0-201-13509-4

Any solution of (6.14) must asymptotically behave like a linear combination of the asymptotic forms given in (6.16) and (6.17), or equivalently like a linear combination of the functions e^{ikx} and e^{-ikx}. That is, as might be expected from the form of (6.14), asymptotically any solution of this equation behaves like a solution of $y''=-k^2y$. Without making use of Bessel functions, we now show how to deduce the stated asymptotic form via the theory of product integration, complete with an estimate of the rate of convergence.

THEOREM 6.2. *Let $a>0$ and let y be any solution of* (6.14). *Then there exist constants d_+, d_-, and N such that*

$$|y(x)-d_+e^{ikx}-d_-e^{-ikx}|\leqslant N/x \qquad (x\geqslant a). \tag{6.18}$$

The constants d_+ and d_- are both zero if and only if $y(x)$ is identically equal to zero.

Proof. Let

$$Y(x)=\begin{pmatrix} y(x) \\ y'(x) \end{pmatrix}. \tag{6.19}$$

Then (6.14) is equivalent to

$$Y'(x)=A(x)Y(x) \tag{6.20}$$

with

$$A(x)=\begin{bmatrix} 0 & 1 \\ \dfrac{l(l+1)}{x^2}-k^2 & 0 \end{bmatrix}. \tag{6.21}$$

By Theorem 2.1, the unique solution of (6.20) is

$$Y(x)=\prod_a^x e^{A(s)ds}Y(a). \tag{6.22}$$

We now analyze the behavior of the product integral in (6.22). Write

$$A(x)=A_0+A_1(x) \tag{6.23}$$

with

$$A_0=\begin{pmatrix} 0 & 1 \\ -k^2 & 0 \end{pmatrix} \tag{6.24}$$

ISBN 0-201-13509-4

and

$$A_1(x)=\begin{bmatrix} 0 & 0 \\ \dfrac{l(l+1)}{x^2} & 0 \end{bmatrix}. \tag{6.25}$$

We know that A_0 is diagonalized by the matrix M of (2.24) and that the product integral of A_0 simplifies upon transformation by M. [See (2.27).] Since for large x, A_0 is the leading term of $A(x)$, we can hope that the product integral of A will also simplify upon transformation by M. Using (3.21), we write

$$\prod_a^x e^{A(s)\,ds}=M\prod_a^x e^{B(s)\,ds}M^{-1} \tag{6.26}$$

with

$$\begin{aligned} B(s)&=M^{-1}A(s)M=M^{-1}A_0M+M^{-1}A_1(s)M \\ &=\begin{pmatrix} ik & 0 \\ 0 & -ik \end{pmatrix}+\frac{il(l+1)}{2ks^2}\begin{pmatrix} -1 & -1 \\ 1 & 1 \end{pmatrix}\equiv B_0+B_1(s). \end{aligned} \tag{6.27}$$

The sum rule yields

$$\prod_a^x e^{B(s)\,ds}=Q(x)\prod_a^x e^{Q^{-1}(s)B_1(s)Q(s)\,ds} \tag{6.28}$$

with

$$Q(s)=\prod_a^x e^{B_0\,ds}=e^{B_0(x-a)}=\begin{pmatrix} e^{ik(x-a)} & 0 \\ 0 & e^{-ik(x-a)} \end{pmatrix}. \tag{6.29}$$

Since $Q(s)$ is unitary, we have

$$\|Q^{-1}(s)B_1(s)Q(s)\|=\|B_1(s)\|=\frac{l(l+1)}{2ks^2}\left\|\begin{pmatrix} -1 & -1 \\ 1 & 1 \end{pmatrix}\right\|=\frac{l(l+1)}{ks^2}. \tag{6.30}$$

Thus $Q^{-1}B_1Q$ belongs to $L^1(a,\infty)$ so that, writing $R=Q^{-1}B_1Q$, the product integral $\Pi_+=\prod_a^\infty e^{R(s)\,ds}$ exists and is nonsingular. We have by (6.11) the estimate

$$\left\|\prod_a^x e^{R(s)\,ds}-\Pi_+\right\|\leqslant\int_x^\infty\|R(s)\|\,ds\,e^{2\|R\|_1}=\frac{l(l+1)}{kx}e^{2\|R\|_1}\qquad(x\geqslant a). \tag{6.31}$$

ISBN 0-201-13509-4

Combining the above results we have

$$\prod_a^x e^{A(s)\,ds} = MQ(x)\Pi_+ M^{-1} + \Delta(x) \tag{6.32}$$

with

$$\|\Delta(x)\| = \left\| MQ(x)\left(\prod_a^x e^{R(s)\,ds} - \Pi_+\right)M^{-1}\right\|$$

$$\leqslant \|M\|\,\|M^{-1}\|\frac{l(l+1)}{kx}e^{2\|R\|_1} \equiv \frac{C}{x}. \tag{6.33}$$

Now examining equation (6.22) for $Y(x)$ we see that y is not identically zero if and only if $Y(a)$ is not zero. Since Π_+ and M^{-1} are nonsingular, $\Pi_+M^{-1}Y(a)$ is then not zero. Writing this last matrix as $\begin{pmatrix} d_+ \\ d_- \end{pmatrix}$, Eq. (6.32) then implies that

$$Y(x) = \begin{pmatrix} d_+e^{ikx} + d_-e^{-ikx} \\ ikd_+e^{ikx} - ikd_-e^{-ikx} \end{pmatrix} + \Delta(x)\,Y(a) \tag{6.34}$$

and (6.33) shows that (6.18) holds so that the proof of Theorem 6.2 is finished. ■

It should be clear from the above proof that an estimate similar to (6.18) can be obtained if the function $l(l+1)/x^2$ of (6.14) is replaced by any function W in $L^1(a,\infty)$.* Instead of N/x on the right-hand side of (6.18), one will then have $N\int_x^\infty |W(s)|\,ds$. In particular, if

$$W(x) = \frac{l(l+1)}{x^2} + V(x), \tag{6.35}$$

then (6.14) becomes the radial Schrödinger equation with angular momentum l, and if $V \in L^1(a,\infty)$ one obtains the result described above, which is usually expressed by saying that asymptotically the solution $y(x)$ behaves like a linear combination of the "plane waves" e^{ikx} and e^{-ikx}.

In some applications one requires a convergence theorem more delicate than Theorem 6.1. Instead of assuming $A \in L^1(a,\infty)$, suppose we merely assume that the improper integral (we use "imp" for emphasis)

$$\operatorname{imp}\int_a^\infty A(s)\,ds \equiv \lim_{x\to\infty}\int_a^x A(s)\,ds \tag{6.36}$$

* W is of course a scalar function and $W \in L^1(a,\infty)$ means $\int_a^\infty |W(s)|\,ds < \infty$.

ISBN 0-201-13509-4

exists. [This can occur through cancellations even though $\|A(s)\|$ may not be integrable, as is shown by the scalar example $A(s)=(\sin s)/s$.] We can ask whether the existence of the integral in (6.36) guarantees existence of $\prod_a^\infty e^{A(s)\,ds}$. If the family $\{A(s)|s\in[a,\infty)\}$ is *commutative*, then the answer is affirmative because by Theorem 1.3 we have in this case

$$\prod_a^x e^{A(s)\,ds}=e^{\int_a^x A(s)\,ds}, \tag{6.37}$$

so that, by (6.36), $\prod_a^\infty e^{A(s)\,ds}$ exists and is nonsingular. If $\{A(s)|s\in[a,\infty)\}$ is *not* assumed to be commutative, then the improper integral of (6.36) may exist while $\prod_a^\infty e^{A(s)\,ds}$ does not. We give the following example:

$$A(x)=\begin{bmatrix} 0 & \dfrac{\sin x}{\sqrt{x}} \\ 0 & \dfrac{d}{dx}\left(\dfrac{\sin x}{\sqrt{x}}\right) \end{bmatrix}. \tag{6.38}$$

An elementary calculation shows that $\operatorname{imp}\int_a^\infty A(s)\,ds$ exists. However, Eq. (3.25) of Theorem 3.4 shows that the (1,2) entry of $\prod_a^x e^{A(s)\,ds}$ is given by

$$h(x)=\int_a^x \frac{\sin s}{\sqrt{s}}\, e^{\frac{\sin s}{\sqrt{s}}-\frac{\sin a}{\sqrt{a}}}\,ds \tag{6.39}$$

Expanding $e^{(\sin s)/\sqrt{s}}$ to two terms, we find

$$h(x)=e^{-(\sin a)/\sqrt{a}}\left(\int_a^x \frac{\sin s}{\sqrt{s}}\,ds+\int_a^x \frac{\sin^2 s}{s}\,ds+\int_a^x g(s)\,ds\right), \tag{6.40}$$

where

$$|g(s)|=\left|\frac{\sin s}{\sqrt{s}}\left(e^{(\sin s)/\sqrt{s}}-1-\frac{\sin s}{\sqrt{s}}\right)\right|\leqslant\frac{1}{2!\,s^{3/2}}|\sin^3 s|\,e^{|(\sin s)/\sqrt{s}|}. \tag{6.41}$$

Since $g(s)$ is $O(1/s^{3/2})$, the third integral on the right-hand side of (6.40) converges when $x\to\infty$, as does the first. But the second integral does *not* converge, hence neither does $h(x)$. Thus $\prod_a^\infty e^{A(s)\,ds}$ does not exist.

We shall now prove some "delicate" convergence theorems for product integrals—theorems that assert existence and nonsingularity of $\prod_a^\infty e^{A(s)\,ds}$ without assuming $A\in L^1(a,\infty)$. All these theorems have the same flavor: they assume existence of the improper integral (6.36), and in addition

ISBN 0-201-13509-4

something about the rate at which this integral converges, i.e., something about the function

$$H(x)=\operatorname{imp}\int_x^{\infty} A(s)\,ds=\operatorname{imp}\int_a^{\infty} A(s)\,ds-\int_a^{x} A(s)\,ds. \tag{6.42}$$

We note that the improper integral of (6.36) is defined if and only if $H(x)$ is defined for all $x \geqslant a$.

Our next theorem (Theorem 6.3) is actually a special case of Theorem 6.4. We state Theorem 6.3 for clarity and because of its great usefulness.

THEOREM 6.3. *Suppose that* $A:[a,\infty)\to\mathbb{C}_{n\times n}$ *is continuous. Suppose that the function* $H:[a,\infty)\to\mathbb{C}_{n\times n}$ *is well defined by* (6.42). *Then if* $HA\in L^1(a,\infty)$, *the product integral* $\prod_a^{\infty} e^{A(s)\,ds}$ *exists and is nonsingular.*

Remark. The hypothesis $HA\in L^1(A,\infty)$ states that the improper integral of (6.36) converges fast enough so that multiplication of the "improperly" integrable function A by the "remainder" $H(x)$ results in a function in $L^1(a,\infty)$.

Proof of Theorem 6.3. For any $b\in[a,\infty)$ we have

$$\prod_a^x e^{A(s)\,ds}=\prod_b^x e^{A(s)\,ds}\prod_a^b e^{A(s)\,ds}. \tag{6.43}$$

Since $\prod_a^b e^{A(s)\,ds}$ is nonsingular, to prove Theorem 6.2 it suffices to show that $\prod_b^{\infty} e^{A(s)\,ds}$ exists and is nonsingular. We note that $H(x)\to 0$ as $x\to\infty$, so we may choose b so large that

$$\|H(x)\|\leqslant \tfrac{1}{2} \qquad \text{for } x\in[b,\infty). \tag{6.44}$$

Then $(I+H(x))^{-1}$ exists for $x\in[b,\infty)$ and

$$\|(I+H(x))^{-1}\|\leqslant 2 \qquad \text{for } x\in[b,\infty). \tag{6.45}$$

Define

$$P(x)=\prod_b^x e^{A(s)\,ds} \tag{6.46}$$

and

$$Q(x)=(I+H(x))P(x). \tag{6.47}$$

Since $I+H(x)\to I$ as $x\to\infty$, existence and nonsingularity of a limit for P are equivalent to the same properties for Q. Now because $H'=-A$, we

ISBN 0-201-13509-4

have

$$Q'(x) = -A(x)P(x) + (I+H(x))A(x)P(x) = H(x)A(x)P(x) \equiv C(x)Q(x) \tag{6.48}$$

with

$$C(x) = H(x)A(x)(I+H(x))^{-1}. \tag{6.49}$$

Now because of (6.45) and the hypothesis $HA \in L^1(A,\infty)$, we have clearly $C \in L^1(A,\infty)$. Also by (6.48) we have

$$Q(x) = \prod_b^x e^{C(s)\,ds} Q(b). \tag{6.50}$$

Note that $Q(b) = (I+H(b))P(b)$ is nonsingular. Hence by Theorem 6.1, $Q(x)$ tends to a nonsingular limit Q_+ as $x \to \infty$. This proves the theorem. ■

Note that by (6.47) the limit of $P(x)$ is also Q_+.

It is simple to estimate the difference between $P(x)$ and its limit Q_+ in Theorem 6.2. We have using (6.11)

$$\begin{aligned}\|P(x)-Q_+\| &\leqslant \|P(x)-Q(x)\| + \|Q(x)-Q_+\| \\ &= \|H(x)P(x)\| + \|Q(x)-Q_+\| \\ &\leqslant \|H(x)P(x)\| + N\int_x^\infty \|C(s)\|\,ds.\end{aligned} \tag{6.51}$$

Since $P(x)$ has a limit as $x \to \infty$, $\|P(x)\|$ is bounded on $[a,\infty)$ so (6.51) yields [using also (6.49) and (6.45)]

$$\begin{aligned}\|P(x)-Q_+\| &\leqslant N'\|H(x)\| + N\int_x^\infty \|C(s)\|\,ds \\ &\leqslant N'\left\|\operatorname{imp}\int_x^\infty A(s)\,ds\right\| + 2N\int_x^\infty \|H(s)A(s)\|\,ds.\end{aligned} \tag{6.52}$$

COROLLARY to Theorem 6.3. *Suppose that* $A, B : [a,\infty) \to \mathbb{C}_{n\times n}$ *are continuous,* A *satisfies the conditions of Theorem* 6.3, *and* $B \in L^1(A,\infty)$. *Then* $\prod_a^\infty e^{\{A(s)+B(s)\}\,ds}$ *exists and is nonsingular.*

Proof. Immediate from the Corollary to Theorem 6.1. ■

As an example of the application of Theorem 6.3, consider the equation

$$Y'(x) = \begin{bmatrix} 0 & \dfrac{\cos x}{x} \\ \dfrac{\sin x}{x} & 0 \end{bmatrix} Y(x) \equiv A(x)Y(x). \tag{6.53}$$

ISBN 0-201-13509-4

In this case $A \notin L^1(a,\infty)$ but we can still prove that Y has a limit as $x \to \infty$: we have

$$H(x) = \operatorname{imp}\int_x^\infty \begin{bmatrix} 0 & \dfrac{\cos s}{s} \\ \dfrac{\sin s}{s} & 0 \end{bmatrix} ds$$

$$= \operatorname{imp}\int_x^\infty \left\{ \frac{d}{ds}\begin{bmatrix} 0 & \dfrac{\sin s}{s} \\ \dfrac{-\cos s}{s} & 0 \end{bmatrix} + \begin{bmatrix} 0 & \dfrac{\sin s}{s^2} \\ \dfrac{-\cos s}{s^2} & 0 \end{bmatrix} \right\} ds$$

$$= \begin{bmatrix} 0 & \dfrac{-\sin x}{x} \\ \dfrac{\cos x}{x} & 0 \end{bmatrix} + \int_x^\infty \begin{bmatrix} 0 & \dfrac{\sin s}{s^2} \\ \dfrac{-\cos s}{s^2} & 0 \end{bmatrix} ds. \tag{6.54}$$

It is easy to see by estimating the integral on the right-hand side of (6.54) that

$$\|H(x)\| \leqslant N/x. \tag{6.55}$$

Since also

$$\|A(x)\| \leqslant N'/x, \tag{6.56}$$

we have $HA \in L^1(a,\infty)$ so that Theorem 6.3 applies and the function

$$Y(x) = \prod_a^x e^{A(s)\,ds}\, Y(a) \tag{6.57}$$

has a limit as $x \to \infty$. The estimate (6.52) shows that the difference between Y and its limit is of the order $1/x$. One can contrast this behavior with that of the equation

$$Y(x) = \begin{bmatrix} 0 & \dfrac{1}{x} \\ \dfrac{1}{x} & 0 \end{bmatrix} Y(x) \tag{6.58}$$

whose solution is

$$Y(x) = e^{\begin{pmatrix} 0 & 1 \\ 1 & 0 \end{pmatrix}\int_a^x \frac{1}{s}ds}\, Y(x)$$

$$= e^{\begin{pmatrix} 0 & 1 \\ 1 & 0 \end{pmatrix}\log\left(\frac{x}{a}\right)}\, Y(a) = \frac{1}{2}\begin{pmatrix} 1 & 1 \\ 1 & -1 \end{pmatrix}\begin{pmatrix} x/a & 0 \\ 0 & a/x \end{pmatrix}\begin{pmatrix} 1 & 1 \\ 1 & -1 \end{pmatrix} Y(a). \tag{6.59}$$

ISBN 0-201-13509-4

Equation (6.59) shows that, depending on the choice of $Y(a)$, the function $Y(x)$ will either converge to zero or diverge to ∞ as $x \to \infty$. The importance of Theorem 6.2 and other "delicate" convergence theorems is that they consider more than just the size of $\|A(s)\|$ and thereby allow one to establish the very real difference in asymptotic behavior between an equation like (6.53) and one like (6.58).

The next theorem generalizes Theorem 6.3. In Theorem 6.3, we abandoned the hypothesis $A \in L^1(a,\infty)$, assuming only that A was improperly integrable and $HA \in L^1(a,\infty)$. The next theorem shows that it is possible to abandon the hypothesis $HA \in L^1(a,\infty)$, assuming only that HA is improperly integrable and that its indefinite improper integral times A belongs to $L^1(a,\infty)$. One can iterate this argument any number of times. A further possibility, as the theorem shows, is to assume only that HA can be split into two parts, one of which belongs to $L^1(a,\infty)$ and the other of which is improperly integrable, with its indefinite improper integral times A belonging to $L^1(a,\infty)$. This possibility, or its appropriate analogue, is considered in each iteration in the theorem below.

THEOREM 6.4. *Suppose that $A:[a,\infty)\to \mathbb{C}_{nxn}$ is continuous. Suppose that the functions defined below have the indicated properties (all improper integrals are assumed to exist):*

$$
\begin{aligned}
A(x) &= A_1(x) + R_1(x); \qquad R_1 \in L^1(a,\infty)\\
H_1(x) &\equiv \operatorname{imp}\int_x^\infty A_1(s)\,ds = A_2(x) + R_2(x); \qquad R_2A \in L^1(a,\infty)\\
H_2(x) &\equiv \operatorname{imp}\int_x^\infty A_2(s)A(s)\,ds = A_3(x) + R_3(x); \qquad R_3A \in L^1(a,\infty)\\
&\ \ \vdots\\
H_{n-1}(x) &\equiv \operatorname{imp}\int_x^\infty A_{n-1}(s)A(s)\,ds = A_n(x) + R_n(x); \qquad R_nA \in L^1(a,\infty)\\
H_n(x) &\equiv \operatorname{imp}\int_x^\infty A_n(s)A(s)\,ds; \qquad H_nA \in L^1(a,\infty).
\end{aligned}
\tag{6.60}
$$

Then $\prod_a^\infty e^{A(s)\,ds}$ exists and is nonsingular.

Proof. The proof is entirely analogous to that of Theorem 6.3. We write $P(x)$ for $\prod_a^x e^{A(s)\,ds}$ and consider the function

$$Q(x) = (I + H_1(x) + \cdots + H_n(x))P(x). \tag{6.61}$$

Since the factor $(I + H_1(x) + \cdots + H_n(x))$ approaches I as $x\to\infty$, $P(x)$ will have a nonsingular limit if and only if Q does. Choosing b large enough so

ISBN 0-201-13509-4

that $\|(I+H_1(x)+\cdots+H_n(x))^{-1}\| \leqslant 2$ for $x \geqslant b$, we find

$$\begin{aligned} Q'(x) &= (I+H_1(x)+\cdots+H_n(x))A(x)P(x) \\ &\quad -(A_1(x)+A_2(x)A(x)+\cdots+A_n(x)A(x))P(x) \\ &= (R_1(x)+R_2(x)A(x)+\cdots+R_n(x)A(x)+H_n(x)A(x))P(x) \\ &\equiv S(x)P(x) \\ &= S(x)(I+H_1(x)+\cdots+H_n(x))^{-1}Q(x). \end{aligned} \tag{6.62}$$

The proof now follows from the fact that $S \in L^1(b,\infty)$. (See the proof of Theorem 6.3.)

Our final "delicate convergence" theorem is similar to Theorem 6.3 except that the condition $HA \in L^1(a,\infty)$ is replaced by the condition $[H,A] \in L^1(a,\infty)$, where $[H,A]$ denotes the commutator

$$[H,A] = HA - AH. \tag{6.63}$$

THEOREM 6.5. *Suppose that* $A:[a,\infty)\to\mathbb{C}_{n\times n}$ *is continuous and that* $H:[a,\infty)\to\mathbb{C}_{n\times n}$ *is well defined by* (6.42). *Then if* $[H,A] \in L^1(a,\infty)$, *the product integral* $\prod_a^\infty e^{A(s)ds}$ *exists and is nonsingular.*

Proof. Let

$$P(x) = \prod_a^x e^{A(s)\,ds} \tag{6.64}$$

and

$$Q(x) = e^{H(x)}. \tag{6.65}$$

We will prove that the limit

$$\lim_{x\to\infty} Q(x)P(x) = \Pi_+ \tag{6.66}$$

exists and is nonsingular. Since $Q(x)\to I$ as $x\to\infty$, this will prove convergence of $P(x)$ to Π_+ and hence prove the theorem. By the similarity rule we have

$$\begin{aligned} Q(x)P(x) &= Q(x)\prod_a^x e^{A(s)\,ds} \\ &= \prod_a^x e^{(LQ)(s)+Q(s)A(s)Q^{-1}(s)\,ds}Q(a). \end{aligned} \tag{6.67}$$

ISBN 0-201-13509-4

Now using the fact that $H' = -A$ and Theorem 5.1 of the Appendix, we have

$$\begin{aligned} LQ(s) &\equiv Q'(s)Q^{-1}(s) \\ &= Q(s)(-A(s)+\psi(s))Q^{-1}(s), \end{aligned} \tag{6.68}$$

where

$$\psi(s) = -\sum_{k=1}^{\infty}(-1)^k \frac{(\operatorname{ad}H(s))^k}{(k+1)!}A(s). \tag{6.69}$$

Thus

$$Q(x)P(x) = \prod_a^x e^{Q(s)\psi(s)Q^{-1}(s)}Q(a). \tag{6.70}$$

If we can prove that $Q\psi Q^{-1} \in L^1(a,\infty)$, then by Theorem 6.1 it will follow that QP has a nonsingular limit, thus proving the theorem. Since $Q(x) \to I$ as $x \to \infty$, we need only prove that $\psi \in L^1[a,\infty)$. Now

$$\psi(s) = -\sum_{k=1}^{\infty}(-1)^k \frac{(\operatorname{ad}H(s))^{k-1}}{(k+1)!}[H(s),A(s)]. \tag{6.71}$$

Since $\|\operatorname{ad}H\| \leqslant 2\|H\|$ (see (A.5.3)), we have

$$\begin{aligned} \|\psi(s)\| &\leqslant \sum_{k=1}^{\infty} \frac{(2\|H(s)\|)^{k-1}}{(k+1)!}\|[H(s),A(s)]\| \\ &\leqslant e^{2\|H(s)\|}\|[H(s),A(s)]\|. \end{aligned} \tag{6.72}$$

The inequality (6.72) proves that $\psi \in L^1(a,\infty)$ and we are done.

1.7 Alternative Definitions of the Product Integral

Suppose that $A:[a,b] \to \mathbb{C}_{n\times n}$ is continuous and let $x \in [a,b]$. Let $P = \{s_0, s_1, \ldots, s_n\}$ be a partition of $[a,x]$ and write as usual $\Delta s_k = s_k - s_{k-1}$. We have

$$e^{A(s_k)\Delta s_k} = I + A(s_k)\Delta s_k + 0((\Delta s_k)^2). \tag{7.1}$$

If the mesh $\mu(P)$ is very small, then in the ordered product $\prod_{k=1}^n e^{A(s_k)\Delta s_k}$ it is plausible that the terms of order $(\Delta s_k)^2$ do not contribute much. Thus we might expect that (the order of the products is as in (0.7))

$$\prod_a^x e^{A(s)\,ds} \equiv \lim_{\mu(P)\to 0} \prod_{k=1}^n e^{A(s_k)\Delta s_k} = \lim_{\mu(P)\to 0} \prod_{k=1}^n (I + A(s_k)\Delta s_k). \tag{7.2}$$

ISBN 0-201-13509-4

This is indeed the case. In fact, if $f:\mathbb{C}_{n\times n}\to\mathbb{C}_{n\times n}$ is any function such that for small Δs_k we have

$$f(A(s_k)\Delta s_k)=I+A(s_k)\Delta s_k+0\big((\Delta s_k)^2\big), \tag{7.3}$$

then it is plausible that we will have

$$\prod_a^x e^{A(s)\,ds}=\lim_{\mu(P)\to 0}\prod_{k=1}^n f(A(s_k)\Delta s_k). \tag{7.4}$$

In this section we establish (7.4) for a class of functions f large enough to include all cases of practical interest.

DEFINITION 7.1. *Let f be a complex-valued function defined on an open connected subset of $\mathbb{C}$ containing the origin. We will say that f is a* P-function *if and only if*

(i) *f is analytic in some disk $\{z\,|\,|z|<\rho\}$ and*
(ii) *f satisfies the conditions*

$$f(0)=f'(0)=1. \tag{7.5}$$

Some examples of P-functions are e^z, $1+z$, and $(1-z)^{-1}$.

If f is a P-function analytic for $|z|<\rho$, then f has for $|z|<\rho$ an expansion of the form

$$f(z)=1+z+\sum_{n=2}^{\infty} c_n z^n. \tag{7.6}$$

If B is an $n\times n$ matrix with $\|B\|<\rho$, we can define $f(B)$ by

$$f(B)=I+B+\sum_{n=2}^{\infty} c_n B^n. \tag{7.7}$$

For $\|B\|\leqslant r_0<\rho$ we have immediately from (7.7) the estimates

$$\|f(B)-I\|\leqslant N\|B\|, \qquad \|B\|\leqslant r_0<\rho \tag{7.8}$$

and

$$\|f(B)-I-B\|\leqslant N'\|B\|^2, \qquad \|B\|\leqslant r_0<\rho \tag{7.9}$$

where N and N' are constants. From (7.8) we obtain

$$\|f(B)\|\leqslant 1+N\|B\|, \qquad \|B\|\leqslant r_0<\rho. \tag{7.10}$$

If $B_1,B_2,\ldots,B_n$ are $n\times n$ matrices all with norm less than or equal to r_0,

ISBN 0-201-13509-4

then (7.10) yields

$$\left\| \prod_{k=1}^{n} f(B_k) \right\| \leq \prod_{k=1}^{n} \|f(B_k)\| \leq \prod_{k=1}^{n} (1+N\|B_k\|)$$

$$\leq \prod_{k=1}^{n} e^{N\|B_k\|} = e^{N\sum_{k=1}^{n}\|B_k\|}. \tag{7.11}$$

Still under the same conditions on $B_1, B_2, \ldots, B_n$, we can use a telescoping sum argument to estimate the norm of the difference Δ defined by

$$\Delta = \prod_{k=1}^{n} e^{B_k} - \prod_{k=1}^{n} f(B_k). \tag{7.12}$$

Namely we have

$$\Delta = \sum_{j=1}^{n} \prod_{k=j+1}^{n} e^{B_k}\left(e^{B_j} - f(B_j)\right) \prod_{k=1}^{j-1} f(B_k). \tag{7.13}$$

In (7.13) we agree that $\prod_{n+1}^{n} f(B_k) = \prod_{1}^{0} f(B_k) = I$. The estimate (7.11) holds for $f(z) = e^z$ with N replaced by unity. Letting M denote the larger of N and unity, we easily obtain from (7.13) the estimate

$$\|\Delta\| \leq e^{M\sum_{k=1}^{n}\|B_k\|} \sum_{j=1}^{n} \|e^{B_j} - f(B_j)\|$$

$$\leq e^{M\sum_{k=1}^{n}\|B_k\|} M' \sum_{i=1}^{n} \|B_i\|^2 \tag{7.14}$$

where the last inequality is obtained by noting that the first two terms of the expansions of e^{B_j} and $f(B_j)$ cancel each other, and using (7.9) for $f(B)$ and the similar estimate for e^B. We can now prove

THEOREM 7.1. *Let f be a P-function. Let $A:[a,b] \to \mathbb{C}_{n\times n}$ be continuous. Then for $x,y \in [a,b], y < x$, the product integral $\prod_y^x e^{A(s)ds}$ can be evaluated as follows:*

$$\prod_{y}^{x} e^{A(s)ds} = \lim_{\mu(P)\to 0} \prod_{k=1}^{n} f(A(s_k)\Delta s_k), \tag{7.15}$$

where $P = \{s_0, s_1, \ldots, s_n\}$ denotes a partition of $[y,x]$.

Proof. We assume that f is analytic for $|z| < \rho$, and that $r_0 < \rho$ has been fixed, so that (7.8) and (7.9) hold for $\|B\| \leq r_0$. Let $\|A\|_\infty$ denote the

ISBN 0-201-13509-4

maximum value of $\|A(x)\|$ on $[a,b]$. We shall consider only partitions P such that $\|A\|_\infty \mu(P) \leqslant r_0$. For any $k = 1, 2, \ldots, n$ we then have

$$\|A(s_k)\Delta s_k\| \leqslant \|A\|_\infty \mu(P) \leqslant r_0 \tag{7.16}$$

so that $f(A(s_k)\Delta s_k)$ is defined and we can apply (7.14) with $B_k = A(s_k)\Delta s_k$. The result is

$$\left\| \prod_{k=1}^{n} e^{A(s_k)\Delta s_k} - \prod_{k=1}^{n} f(A(s_k)\Delta s_k) \right\| \leqslant e^{M \sum_{k=1}^{n} \|A(s_k)\|\Delta s_k} M' \sum_{j=i}^{n} \|A(s_i)\Delta s_i\|^2. \tag{7.17}$$

Replacing $\|A(s_k)\|$ by $\|A\|_\infty$ wherever it occurs and noting that $\Sigma_{k=1}^{n} \Delta s_k = x - y \leqslant b - a$, we have

$$\left\| \prod_{k=1}^{n} e^{A(s_k)\Delta s_k} - \prod_{k=1}^{n} f(A(s_k)\Delta s_k) \right\| \leqslant M' e^{M\|A\|_\infty (b-a)} \|A\|_\infty^2 \sum_{j=1}^{n} (\Delta s_i)^2$$
$$\equiv C \sum_{j=1}^{n} (\Delta s_i)^2 \leqslant C \sum_{j=1}^{n} \mu(P)\Delta s_i \leqslant C\mu(P)(b-a). \tag{7.18}$$

This estimate shows that as $\mu(P) \to 0$ the difference between $\Pi_{k=1}^{n} e^{A(s_k)\Delta s_k}$ and $\Pi_{k=1}^{n} f(A(s_k)\Delta s_k)$ approaches zero, proving the theorem. When a P-function $f(z)$ other than e^z is used to define the product integral, the latter is often denoted $\Pi_y^x f(A(s)\,ds)$ instead of $\Pi_y^x e^{A(s)\,ds}$. Thus one sees the notations $\Pi_y^x(I + A(s)\,ds)$, $\Pi_y^x(I - A(s)\,ds)^{-1}$, etc. ■

Theorem 7.1 offers a wide variety of methods for defining product integrals. The authors believe that the choice $f(z) = e^z$, as in the first section of this chapter, gives the easiest treatment for the case of a function $A:[a,b] \to \mathbb{C}_{n \times n}$. However, if one wishes to discuss product integrals of more general objects than matrices, other choices of f may be appropriate. (Of course, if objects more general than matrices are considered, the analogue of Theorem 7.1 may be false, so the choice of f may make a real difference.) The choice $f(x) = (1 - z)^{-1}$ is used, for instance, in the papers of Webb [GW1, GW2], and is a natural choice for the nonlinear, unbounded operators A he considers. This same choice is exploited in Section 4 of Chapter 4.

We now turn to a simple extension of the theory of this chapter.

ISBN 0-201-13509-4

1.8 Lebesgue-Integrable Functions

In this section we briefly discuss product integration for functions $A:[a,b]\to\mathbb{C}_{n\times n}$ all of whose entries are Lebesgue integrable over $[a,b]$. The main point of our discussion will be this: all the theorems we have proved for continuous A have natural analogues for Lebesgue-integrable A. Mainly the analogues are obtained by replacing the notion of "differential equation" by that of "integral equation" and the notion of "continuously differentiable" by that of "absolutely continuous." There is a subtlety concerning Theorem 7.1, as we shall see.

We write $A\in L^1(a,b)$ to indicate that $A:[a,b]\to\mathbb{C}_{n\times n}$ is (entrywise) Lebesgue integrable over $[a,b]$. If $A\in L^1(a,b)$, we write as before

$$\|A\|_1=\int_a^b\|A(s)\|\,ds. \tag{8.1}$$

Naturally any step-function belongs to $L^1(a,b)$. The basis of our treatment of product integrals in this section will be the following elementary fact: the step-functions are dense in $L^1(a,b)$. That is, if $A\in L^1(a,b)$, then there is a sequence $\{A_n\}$ of step-functions converging to A in the L^1 sense:

$$\lim_{n\to\infty}\|A_n-A\|_1=0. \tag{8.2}$$

We make the following definition:

DEFINITION 8.1. *Let* $A\in L^1(a,b)$. *For* $x\in[a,b]$ *the product integral* $\prod_a^x e^{A(s)\,ds}$ *is defined by*

$$\prod_a^x e^{A(s)\,ds}=\lim_{n\to\infty}E_{A_n}(x), \tag{8.3}$$

where $\{A_n\}$ *is any sequence of step-functions convergent to* A *in the* L^1 *sense and* E_{A_n} *is defined for the step-function* A_n *as in Definition* 1.3.

As remarked after Theorem 1.1, continuity of A was not used there in the proof of existence of the limit corresponding to Eq. (8.3). In fact, all that was needed was the fact that the sequence $\{A_n\}$ was Cauchy in the L^1 sense. Thus the proof of Theorem 1.1 establishes the existence of the limit in (8.3) for $A\in L^1(a,b)$.

In particular, (8.3) defines the product integral of A if A is a step-function. In this case, choosing the sequence $\{A_n\}$ such that $A_n=A$ for all n, one finds

$$\prod_a^x e^{A(s)\,ds}=E_A(x)\qquad(A\text{ a step-function}). \tag{8.4}$$

ISBN 0-201-13509-4

As seen in Theorem 1.1, the convergence in (8.3) is uniform for $x \in [a,b]$. Writing $F(x,a) = \prod_a^x e^{A(s)ds}$, it follows that $F(x,a)$ is continuous, since it is the uniform limit of continuous functions. The integral equation

$$F(x,a) = I + \int_a^x A(s)F(s,a)\,ds \tag{8.5}$$

follows immediately as in Theorem 1.2. From (8.5) one concludes that $F(s,a)$ is absolutely continuous (instead of continuously differentiable as before), and that

$$\frac{dF(s,a)}{dx} = A(x)F(x,a), \qquad \text{a.e. } x \in [a,b]. \tag{8.6}$$

Theorems 1.3 and 1.4 hold for $A \in L^1(a,b)$ with very minor modifications in the proofs: one replaces point-value approximants by a general sequence $\{A_n\}$ tending to A in the L^1 sense. Theorem 1.5 (the multiplicative property) also holds for $A \in L^1(a,b)$. Theorem 1.6 holds if the differential Eqs. (1.35) and (1.36) are replaced by the corresponding integral equations [or if (1.35) and (1.36) are interpreted as holding almost everywhere]. Theorem 2.1 holds if (2.1) is replaced by

$$Y(x) = Y_0 + \int_a^x A(s)Y(s)\,ds. \tag{8.7}$$

Theorem 2.2 holds if (2.7) is replaced by

$$Y(x) = Y(a) + \int_a^x A(s)Y(s)\,ds. \tag{8.8}$$

The fundamental theorem (Theorem 3.1) holds if P is assumed to be absolutely continuous on $[a,b]$ and the definition

$$(LP)(x) = P'(x)P^{-1}(x) \tag{8.9}$$

is interpreted as holding almost everywhere. The sum rule (Theorem 3.2) holds for $A, B \in L^1(a,b)$. The similarity rule (Theorem 3.3) holds for $B \in L^1(a,b)$ and P nonsingular and absolutely continuous. Theorem 3.4 on product integration of special matrices holds without change. Theorem 3.5 on change of variable generalizes to the case in which $A \in L^1(\text{ran}(\phi))$ and ϕ is absolutely continuous and monotone on $[a,b]$. (See [FR, p. 55] for the corresponding theorem on ordinary integrals.) Theorems 4.1 through 4.3 on bounds and equality with the time-ordered exponential can be taken over without change for $A \in L^1(a,b)$. Theorem 5.1 (Duhamel's formula) holds without change for Lebesgue integrable functions. Theorems 5.2, 5.3, and 5.4 continue to hold under the hypothesis that $A(\cdot,\lambda) \in L^1(a,b)$ for

ISBN 0-201-13509-4

each $\lambda \in [c,d]$. (And, of course, the other hypotheses of these theorems.) All of the theorems on existence of improper integrals hold without change for functions which are integrable rather than continuous. Theorem 7.1 continues to hold for any $A \in L^1(a,b)$ with certain modifications. Namely, if B is a step-function on $[y,x]$ taking the value B_k on the interval (s_{k-1}, s_k) one defines

$$\Pi(f,B) = \prod_{k=1}^{n} f(B_k \Delta s_k). \tag{8.10}$$

It is then $\Pi(f,A_n)$ that converges to $\prod_y^x e^{A(s)\,ds}$ when A_n tends to A in the L^1 sense, *but* one must consider only approximating sequences with the following property: if P_n is the partition used for the function A_n, then the mesh of P_n tends to zero as n tends to ∞. (Of course, this is easily arranged: if A_n is a step-function the intervals of whose partition are "too big," one can subdivide these intervals while leaving the values of A unchanged.) The reason for this requirement can be seen by taking $A(s) = I$ and $[x,y] = [a,b]$. Then

$$\prod_a^b e^{A(s)\,ds} = e^{I(b-a)}. \tag{8.11}$$

If we choose the sequence of step-function approximants $A_n(x) = I$ with corresponding partitions $P_n = \{a,b\}$ (i.e., P_n consists of exactly two points), then A_n clearly converges to A in the L^1 sense, but if $f(z) = 1 + z$ then

$$\Pi(f,A_n) = I + I(b-a) \qquad \text{for all } n, \tag{8.12}$$

so that $\Pi(f,A_n)$ does not converge to $\prod_a^b e^{A(s)\,ds}$. If we had taken the same values for A_n but had used for P_n the partition of $[a,b]$ into n equal parts, we would have had

$$\Pi(f,A_n) = \left(I + I\frac{b-a}{n}\right)^n \tag{8.13}$$

which *does* tend to $\prod_a^b e^{A(s)\,ds}$ as $n \to \infty$.

We omit the rather technical proof that Theorem 7.1 (as modified above) holds for $A \in L^1(a,b)$.

Notes to Chapter 1

The product integral concept first appeared in the papers by Volterra [VV1–VV3]; the results of these papers and some additional material can be found in the monograph [VV4] by Volterra and Hostinsky (abbreviated as V-H) [VV4]. Most of the material of our Chapter 1 appears in one form or another in V-H, with the exception of our Sections 6 and 8. The presentation in V-H is difficult to follow

ISBN 0-201-13509-4

largely due to their notation. For example, their matrices and matrix equations are always written in terms of components, when the use of single symbols for matrices would have greatly simplified many formulas and results. Furthermore, V-H use for product integrals a notation identical with the notation for ordinary integrals; they denote by $\int_a^b \{a_{ij}\}\,ds$ what we have denoted by $\prod_a^b e^{A(s)\,ds}$. In a similar vein, V-H denote what we write as $A'(x)A^{-1}(x)$ by $(d/dx)\{a_{ij}\}$. As might be expected, this leads rapidly to some formulas that look wrong and some that are wrong. Our Section 6 contains some new results and ideas. Theorem 6.1 and its applications have been known for some time, although the systematic use of product integration appears to smooth the way considerably in various of these applications. Theorem 6.3 and the circle of ideas concerning product integration of conditionally integrable functions are new and due to the present authors. The example of existence of $\int_a^\infty A(x)\,ds$ with nonexistence of $\prod_a^\infty e^{A(s)\,ds}$ is due to Michael Wolfe (unpublished). Theorem 6.4, which generalizes Theorem 6.3, was proved by Barry Simon. Product integration of Lebesgue-integrable functions has been considered by several authors. Birkhoff [GB] considers a certain class of Lebesgue integrable functions (not necessarily matrix valued) that are product integrable in the sense considered by him; his discussion is complicated and difficult to follow. Schlesinger [L-S3] develops product integration of matrix-valued functions whose entries are *bounded* Lebesgue-integrable functions; he claims the boundedness assumption may be removed. Schmidt [GS] develops product integration of a class of integrable Banach space operator-valued functions; his results include (as a special case) existence of the product integral of a Lebesgue-integrable matrix-valued function. The methods of these authors are much more complicated than that of our Section 8. (See the notes to Chapter 5 for further comments on Schmidt's method.) V-H only consider product integration of matrix-valued functions whose entries are *bounded* and *Riemann* integrable, which is natural in view of their definition of the product integral via what we have called point-value approximants.

We mention that V-H define double product integrals that are the analogue of ordinary iterated integrals of functions of two variables. In itself, this theory does not seem especially central, and we do not discuss it. V-H make use of double integrals in the statement of an analogue of Green's theorem for product integrals; see the notes to Chapter 2 for some discussion of this. They use this result in discussing complex contour integrals. We make no use of such theorems in our discussion of contour integrals, which is the subject of our next chapter.

ISBN 0-201-13509-4

CHAPTER 2

Contour Product Integration

2.0 Introduction

In this chapter we shall develop the notion of product integration along a contour, from a point of view that is substantially parallel to the ordinary complex variable theory of contour integrals. As we have seen, the product integral from x_0 to x_1 of a matrix valued function $A(x)$ gives the evolution of the solution of $Y(x)=A(x)Y(x)$ in passing from x_0 to x_1. In the same way, the product integral of an analytic matrix valued function $A(z)$ along a contour gives the continuation of the solution of the equation $dY/dz = A(z)Y(z)$ along the contour. As might be expected, continuations along different contours with the same initial and terminal points may differ; an analysis of this situation leads to theorems for product integrals that are analogous to the Cauchy integral theorem and the residue theorem for ordinary contour integrals. These theorems are closely related to results giving the form of, or information concerning, the solutions of equations in the neighborhood of singular points; we shall elaborate on this connection. A reference for most of the ideas of the present chapter is [EC].

2.1 The Definition of Contour Product Integrals

In the theory of functions of a complex variable ordinary contour integrals may be defined as follows: If $\gamma(t)$ is a continuously differentiable mapping from $a \leqslant t \leqslant b$ to the complex numbers $\mathbb{C}$ and $f(z)$ is a function continuous on the range Γ of $\gamma(t)$, then we may define

$$\int_\Gamma f(z)\,dz = \int_a^b f(\gamma(t))\cdot\gamma'(t)\,dt. \tag{1.1}$$

We shall follow this definition in defining contour product integrals. It will

ENCYCLOPEDIA OF MATHEMATICS and Its Applications, Gian-Carlo Rota (ed.). Vol. 10: John D. Dollard and Charles N. Friedman, Product Integration with Applications to Differential Equations

ISBN 0-201-13509-4

be convenient to include contours with "bends" or "corners" in our definition, and we do so.

DEFINITION 1.1. *A* contour *is a continuous, piecewise continuously differentiable function* $\gamma(t)$ *from an interval* $a \leqslant t \leqslant b$ *of the real numbers to the complex numbers* $\mathbb{C}$.

(This means that $\gamma(t)$ is continuous on $[a,b]$, and there is a partition $P=\{s_0,s_1,\ldots,s_n\}$ of $[a,b]$ such that $\gamma(t)$ is continuously differentiable on each subinterval $[s_i,s_{i+1}]$, the endpoint derivatives being right or left derivatives.) If Γ is the image of $\gamma(t)$, we denote the contour by $(\Gamma,\gamma(t))$ or sometimes just Γ if the particular mapping $\gamma(t)$ has been identified or is unimportant. The points $\gamma(a)$, $\gamma(b)$ are the initial and terminal points of the contour. If $(\Gamma,\gamma(t))$, $a \leqslant t \leqslant b$, is a contour, we denote by Γ^{-1} the contour defined by $\gamma(a+b-t)$, $a \leqslant t \leqslant b$; Γ^{-1} is "the contour Γ with direction reversed." If $\gamma(a)=\gamma(b)$ and $(\Gamma,\gamma(t))$ has no other self-intersections (so that Γ is a Jordan curve, i.e., a homeomorphic image of a circle), we say that Γ has *positive orientation* if for some (and hence every) point z_0 in the "interior" of Γ (i.e., the bounded component of $\mathbb{C}\backslash\Gamma$) we have

$$\frac{1}{2\pi i}\int_\Gamma \frac{dz}{z-z_0}=1.$$

For example, the positive orientation of a circle is given by the counterclockwise direction. Finally, if $\gamma(t)$ is a contour, we define $\dot{\gamma}(t)$ to be the derivative of γ at t if this exists and 0 otherwise.

DEFINITION 1.2. *Suppose that* $D \subseteq \mathbb{C}$ *is a domain* (*i.e.,* D *is nonempty, open, and connected*) *and* $A(z)$ *is a continuous function from* D *to the* $n\times n$ *complex matrices,* $\mathbb{C}_{n\times n}$. *Let* $(\Gamma,\gamma(t))$ *be a contour in* D. *We define*

$$\prod_{\Gamma} e^{A(z)\,dz}=\prod_{a}^{b} e^{A(\gamma(t))\dot{\gamma}(t)\,dt}. \tag{1.2}$$

The right-hand side of this equation is defined as in Chapter 1. According to Definition 1.2, the product integral along a contour is seemingly dependent on the particular parametrization $\gamma(t)$ of the path Γ; actually this is not the case, and we shall make a few remarks to clarify the situation. Suppose first that $\gamma:[a,b]\rightarrow\mathbb{C}$ is a continuously differentiable function. Let $P=\{t_0,t_1,\ldots,t_n\}$ be a partition of $[a,b]$ and put

$$\begin{aligned} z_k &= \gamma(t_k), \\ \Delta z_k &= z_k - z_{k-1}, \qquad k=1,2,\ldots,n, \\ \Delta t_k &= t_k - t_{k-1}. \end{aligned} \tag{1.3}$$

ISBN 0-201-13509-4

Write $\gamma(t)$ as

$$\gamma(t)=\gamma_1(t)+i\gamma_2(t), \qquad \gamma_1,\gamma_2 \text{ real.}$$

By the mean value theorem, for $k=1,2,\ldots,n$, there exist points $t_k',t_k''\in[t_{k-1},t_k]$ such that

$$\Delta z_k=\gamma(t_k)-\gamma(t_{k-1})=\{\gamma_1'(t_k')+i\gamma_2'(t_k'')\}\Delta t_k. \tag{1.4}$$

Let $A(t)$ be continuous on $\gamma([a,b])$. Corresponding to the partition P, define the step function B_P that, on $[t_{k-1},t_k]$ takes the value

$$B_{P_k}=A(\gamma(t_k))\{\gamma_1'(t_k')+i\gamma_2'(t_k'')\}. \tag{1.5}$$

Since γ is continuously differentiable and A is continuous, it is trivial that B_P converges to $A(\gamma(t))\gamma'(t)$ in the L^1 sense (even uniformly) as $\mu(P)\to0$. Thus

$$\prod_b^a e^{A(\gamma(t))\gamma'(t)\,dt}=\lim_{\mu(P)\to0}\prod_{k=1}^n e^{B_{P_k}\Delta t_k}=\lim_{\mu(P)\to0}\prod_{k=1}^n e^{A(z_k)\Delta z_k}. \tag{1.6}$$

Now suppose that γ and $\eta:[a,b]\to\mathbb{C}$ are $1-1$, continuously differentiable functions with the same image and $\gamma(a)=\eta(a)$, $\gamma(b)=\eta(b)$. Let $P_1=\{t_0,t_1,\ldots,t_n\}$ be a partition of $[a,b]$, and let $P_2=\{s_0,s_1,\ldots,s_n\}$ be the partition defined by

$$s_k=\eta^{-1}\circ\gamma(t_k). \tag{1.7}$$

Since η^{-1} and γ are continuous,

$$\mu(P_1)\to0\Leftrightarrow\mu(P_2)\to0.$$

Now we have

$$\begin{aligned}\prod_{k=1}^n e^{A(\gamma(t_k))\{\gamma_1'(t_k')+i\gamma_2'(t_k'')\}\Delta t_k}&=\prod_{k=1}^n e^{A(z_k)\Delta z_k}\\&=\prod_{k=1}^n e^{A(\eta(s_k))\{\eta_1'(s_k')+i\eta_2'(s_k'')\}\Delta s_k},\end{aligned} \tag{1.8}$$

where s_k', $s_k''\in[s_{k-1},s_k]$ and we have again used the mean value theorem as in (1.4). Taking limits as $\mu(P_1)\to0$ or $\mu(P_2)\to0$, we find

$$\prod_a^b e^{A(\gamma(t))\gamma'(t)\,dt}=\prod_a^b e^{A(\eta(t))\eta'(t)\,dt}. \tag{1.9}$$

ISBN 0-201-13509-4

Hence the product integrals of A along the contours parametrized by γ and η are equal. This result may be extended to contours which are not $1-1$ by splitting the contours into successive pieces which are $1-1$.

We mention that by Definition 1.4 and Theorem 3.5 of Chapter 1, it follows easily that

$$\prod_{\Gamma^{-1}} e^{A(z)\,dz} = \left(\prod_{\Gamma} e^{A(z)\,dz} \right)^{-1}. \tag{1.10}$$

2.2 The Product Integral of an Analytic Function and the Analogues of Cauchy's Integral Theorem

To pursue our development of contour product integrals we shall now consider the contour product integral of an *analytic* matrix valued function.

DEFINITION 2.1. *Let $D \subseteq \mathbb{C}$ be a domain. An $n \times n$ complex matrix valued function $A(z): D \to \mathbb{C}_{n\times n}$ is analytic if for each $z \in D$, $\lim_{h\to 0} h^{-1}(A(z+h)-A(z))$ exists.*

We shall now write the time-ordered exponential formula (see Theorem 1.4.3) for contour integrals. This will be needed in the proof of the next theorem. Suppose D is a domain, $A(z)$ analytic in D. Let $(\Gamma, \gamma(t))$, $a \leqslant t \leqslant b$, be a contour in D, with $\gamma(a) = z_0$. For $z \in \Gamma$ with $\gamma(t) = z$ we denote $\int_a^t A(\gamma(s))\gamma'(s)\,ds$ by $I_\Gamma^{(1)}(z)$. We may write $I_\Gamma^{(1)}(z) = \int_{z_0}^z A(\zeta)\,d\zeta$ with the understanding that the integration is performed along Γ. We define inductively $I_\Gamma^{(n)}(z) = \int_{z_0}^z A(\zeta) I^{(n-1)}(\zeta)\,d\zeta$, $n = 2, 3, \ldots$, where the integration is performed along Γ. The time-ordered exponential formula then takes the form

$$\prod_{\Gamma} e^{A(z)\,dz} = I + \sum_{n=1}^{\infty} I_\Gamma^{(n)}(\gamma(b)). \tag{2.1}$$

As in the case of ordinary contour integrals, the product integral of an analytic function along a contour is invariant under homotopic deformation of the contour (with endpoints fixed). We first prove a special case in the following theorem; the general case is considered in Theorem 2.2.

THEOREM 2.1. *Let D be a simply connected domain. Let $A(z)$ be analytic in D, and let $(\Gamma_1, \gamma_1(t))$, $(\Gamma_2, \gamma_2(t))$, $a \leqslant t \leqslant b$ be two contours in D with $\gamma_1(a) = \gamma_2(a) = z_0$, $\gamma_1(b) = \gamma_2(b) = z_1$. Then*

$$\prod_{\Gamma_1} e^{A(z)\,dz} = \prod_{\Gamma_2} e^{A(z)\,dz}. \tag{2.2}$$

ISBN 0-201-13509-4

Proof. Define $I^{(1)}(z)=\int_{z_0}^{z}A(\zeta)d\zeta$, where the integration is along any contour. Inductively, define $I^{(n)}(z)=\int_{z_0}^{z}A(\zeta)I^{(n-1)}(\zeta)d\zeta$, $n=2,3,\ldots$. Each $I^{(n)}(z)$ is well defined and analytic in D by Cauchy's integral theorem. Now for $z\in\Gamma_1$, we have $I^{(1)}(z)=I^{(1)}_{\Gamma_1}(z)$. Assume inductively that $I^{(k)}(z)=I^{(k)}_{\Gamma_1}(z)$ for $z\in\Gamma_1$. We have $I^{(k+1)}(z)=\int_{z_0}^{z}A(\zeta)I^{(k)}(\zeta)d\zeta$, where the integration is along any contour; if we require this contour to be Γ_1, this shows that $I^{(k+1)}(z)=I^{(k+1)}_{\Gamma_1}(z)$. Hence $I^{(n)}(z)=I^{(n)}_{\Gamma_1}(z)$ for $n=1,2,\ldots,z\in\Gamma_1$. The same is true with Γ_1 replaced by Γ_2 and so $I^{(n)}_{\Gamma_1}(z_1)=I^{(n)}_{\Gamma_2}(z_1)$ for $n=1,2,\ldots$. This together with (2.1) proves (2.2). ∎

COROLLARY 2.1.1. *Suppose D is simply connected, $A(z)$ analytic in D, and $z_0\in D$. For any $z\in D$, define $U(z)=\prod_{z_0}^{z}e^{A(\zeta)d\zeta}$, where the product integration is along any contour in D with initial point z_0 and terminal point z. Then $U(z)$ is well defined and analytic in D and $dU(z)/dz=A(z)U(z)$.*

Proof. $U(z)$ is well defined by the theorem. That the stated differential equation holds is an easy consequence of the representation (2.1), which gives explicitly

$$U(z)=I+\int_{z_0}^{z}A(\zeta)d\zeta+\int_{z_0}^{z}A(\zeta)\left\{\int_{z_0}^{\eta}A(\eta)d\eta\right\}d\zeta+\cdots,$$

which is a series of analytic functions uniformly convergent on compact sets.

We remark that a number of theorems of Chapter 1 were proved using the fact that the product integral is the (unique) solution of a differential equation. Many of these theorems have analogues for contour integrals which are proved in exactly the same way or are easy consequences of earlier theorems and the definition of contour product integrals. For example, if D is simply connected and $T(z)$ is analytic and invertible in D, then

$$\prod_{z_0}^{z_1}e^{T'(z)T^{-1}(z)dz}=T(z_1)T^{-1}(z_0).$$

This is the analogue of the fundamental theorem of product integration (Theorem 3.1, Chapter 1). The sum formula and similarity formula (Theorems 1.3.2 and 1.3.3) also have obvious analogues for contour integrals. The assumption of simple connectivity is of course necessary if we wish to talk about product integrals from z_0 to z_1 as well-defined quantities. We now prove a stronger version of Theorem 2.1. ∎

THEOREM 2.2. *Let $D\subseteq\mathbb{C}$ be a domain, $A(z)$ analytic in D, and $(\Gamma_1,\gamma_1(t))$, $(\Gamma_2,\gamma_2(t))$, $a\leqslant t\leqslant b$, two contours in D with $\gamma_1(a)=\gamma_2(a)=z_0$, $\gamma_1(b)=\gamma_2(b)=z_1$. Suppose Γ_1 and Γ_2 are homotopic with fixed endpoints in D. (This means*

ISBN 0-201-13509-4

that there is a continuous function $h(t,u):[a,b]\times[0,1]\to D$ *with* $h(t,0)=\gamma_1(t)$, $h(t,1)=\gamma_2(t)$, $h(a,u)=z_0$, $h(b,u)=z_1$.) *Then*

$$\prod_{\Gamma_1} e^{A(z)\,dz} = \prod_{\Gamma_2} e^{A(z)\,dz}. \tag{2.3}$$

Proof. The proof is somewhat technically complicated, but follows a standard proof for the Cauchy integral theorem for ordinary contour integrals. Let $h(t,u):[a,b]\times[0,1]\to D$ be the homotopy between γ_1 and γ_2. We claim that there is a continuous $\mathbb{C}_{n\times n}$ valued function $\mathcal{B}(t,u)$ in $S=[a,b]\times[0,1]$ such that for each point $(t_0,u_0)\in S$, there is a disk U about $h(t_0,u_0)$ and a $\mathbb{C}_{n\times n}$ valued nonsingular analytic function B defined on U with $\mathcal{B}(t,u)=B(h(t,u))$ for $h(t,u)\in U$ and $B'B^{-1}=A$ in U. Note that $\mathcal{B}$ is defined on all of S, but each B is only defined in a disk $U\subset D$. Once this claim has been established the theorem will follow, for then

$$\prod_{\Gamma_1} e^{A(z)\,dz} = \mathcal{B}(b,0)\mathcal{B}^{-1}(a,0) = \mathcal{B}(b,1)\mathcal{B}^{-1}(a,1) = \prod_{\Gamma_2} e^{A(z)\,dz}. \tag{2.4}$$

[The first and last equalities follow from the fundamental theorem of product integration applied to chains of overlapping disks in which $B'B^{-1}=A$ as above, and the middle equality is true because $h(b,0)=h(b,1)$, $h(a,0)=h(a,1)$.]

To prove the existence of the function $\mathcal{B}(t,u)$, subdivide $[a,b]$ by points t_i and $[0,1]$ by points u_j so that for each pair i,j, $h([t_i,t_{i+1}]\times[u_j,u_{j+1}])$ is contained in an open disk $U_{ij}\subset D$. By Theorem 2.1 there exists an analytic function B_{ij} in U_{ij} with $B_{ij}'B_{ij}^{-1}=A$. (Take B_{ij} to be the product integral of A from some fixed initial point z_0 to z.) Fix j. Since $U_{ij}\cap U_{i+1,j}$ is nonempty and connected, we can multiply each B_{ij} (j fixed) on the right by an element of $\mathbb{C}_{n\times n}$ so that B_{ij} and $B_{i+1,j}$ agree in $U_{ij}\cap U_{i+1,j}$. This is possible due to the fact that solutions of the equation $B'B^{-1}=A$ in a disk are determined up to right multiplication by an element of $\mathbb{C}_{n\times n}$. (This follows from Theorem 2.1 of Chapter 1.) For $u\in[u_j,u_{j+1}]$ define $\mathcal{B}_j(t,u)=B_{ij}(h(t,u))$, $t\in[t_i,t_{i+1}]$. $\mathcal{B}_j(t,u)$ is continuous in $[a,b]\times[u_j,u_{j+1}]$. We remark that it may happen that U_{ij} and $U_{i+2,j}$ intersect and that $B_{ij}, B_{i+2,j}$ do not agree in the intersection, but $\mathcal{B}_j(t,u)$ is well defined by the above prescription. Finally, we can multiply each $\mathcal{B}_j$ on the right by an element of $\mathbb{C}_{n\times n}$ so that $\mathcal{B}_j$ and $\mathcal{B}_{j+1}$ agree when $u=u_{j+1}$. Then put $\mathcal{B}(t,u)=\mathcal{B}_j(t,u)$ for $u\in[u_j,u_{j+1}]$. This finishes the proof of the theorem. ■

COROLLARY 2.2.1. *If* Γ *is null-homotopic in* D, *then* $\prod_\Gamma e^{A(z)\,dz}=I$.

Proof.

$$I=\prod_{z_0}^{z_0} e^{A(z)\,dz}.$$

ISBN 0-201-13509-4

We prove one more corollary, which describes the relation between the product integrals of an analytic function $A(z)$ along "concentric" contours. We first recall a familiar situation in ordinary complex analysis. Suppose D is a domain, and Γ_1, Γ_2 are contours in D which are Jordan curves with initial points z_1, z_2 respectively. Suppose that Γ_1, Γ_2 are positively oriented, Γ_2 is contained in the interior of Γ_1 and that the region between Γ_1 and Γ_2 is contained in D. Suppose $A(z)$ is analytic in D. Then

$$\int_{\Gamma_1} A(\zeta)\,d\zeta = \int_{\Gamma_2} A(\zeta)\,d\zeta$$

by a simple application of Cauchy's theorem. For product integrals we do not obtain simple equality. We have in fact ■

COROLLARY 2.2.2. *If Γ_1, Γ_2 are as above, then $P_1 = \prod_{\Gamma_1} e^{A(\zeta)\,d\zeta}$ and $P_2 = \prod_{\Gamma_2} e^{A(\zeta)\,d\zeta}$ are similar matrices, i.e., there exists an invertible matrix S with $P_1 = SP_2S^{-1}$.*

Proof. Consider the diagram of Figure 2.1. We adopt the following notation: If Γ_i, Γ_j are contours with terminal point of Γ_i equal to the initial point of Γ_j, then $\Gamma_j\Gamma_i$ denotes the contour consisting of Γ_i followed by Γ_j. Now from the figure and the theorem we have

$$\prod_{\Gamma_1\Gamma_3} e^{A(\zeta)\,d\zeta} = \prod_{\Gamma_3\Gamma_2} e^{A(\zeta)\,d\zeta}. \tag{2.5}$$

Using Theorem 1.5 of Chapter 1, we can write this as

$$\prod_{\Gamma_1} e^{A(\zeta)\,d\zeta} = \prod_{\Gamma_3} e^{A(\zeta)\,d\zeta} \prod_{\Gamma_2} e^{A(\zeta)\,d\zeta} \left(\prod_{\Gamma_3} e^{A(\zeta)\,d\zeta} \right)^{-1}, \tag{2.6}$$

which proves the corollary. ■

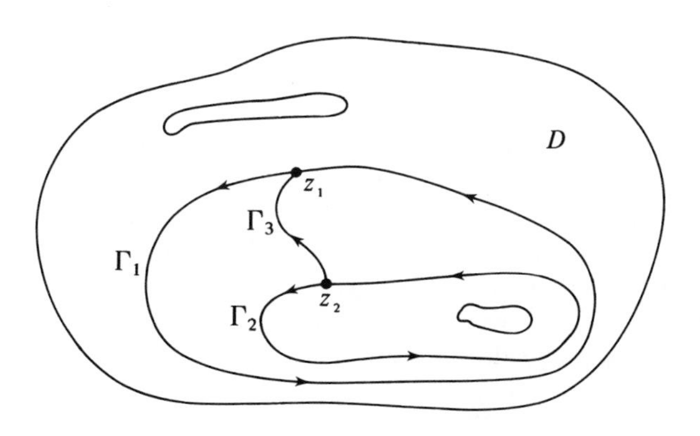

Figure 2.1

Theorems 2.1 and 2.2 show that the continuation of the solution of the differential equation

$$U'(z)=A(z)U(z), \qquad U(z_0)=U_0, \qquad z\in D \tag{2.7}$$

along a contour from z_0 to z leads to a value depending only on the homotopy class of the contour. If D is simply connected, the value $U(z)$ is independent of the contour selected.

A special case of (2.7) arises from nth-order equations of the form

$$\frac{d^n u}{dz^n}+a_{n-1}(z)\frac{d^{n-1}u}{dz^{n-1}}+\cdots+a_1(z)\frac{du}{dz}+a_0(z)u(z)=0, \tag{2.8}$$

where $u(z)$ is a complex valued function, and $a_0(z),\dots,a_{n-1}(z)$ are analytic in D. Namely, (2.8) can be put in the form (2.7) by taking

$$U(z)=\begin{bmatrix} u(z)\\ u'(z)\\ \vdots\\ u^{(n-1)}(z)\end{bmatrix}, \qquad A(z)=\begin{bmatrix} 0 & 1 & 0 & & \cdots & 0 & 0\\ 0 & 0 & 1 & 0 & \cdots & 0 & 0\\ \vdots & & & & & & \\ -a_0(z) & & -a_1(z) & & \cdots & & -a_{n-1}(z)\end{bmatrix}. \tag{2.9}$$

If the $a_i(z)$ have singularities at a point z_0, then z_0 is said to be a singular point of the equation in (2.8). In the next section we shall prove a theorem that gives some information concerning this case.

2.3 A Cauchy Integral Formula for Product Integrals

In this section we shall prove some results concerning the product integral of a function $A(z)$ having a pole of *order 1* at some point z_0. Of course, the product integral of $A(z)$ gives a solution of the equation

$$U'(z)=A(z)U(z). \tag{3.1}$$

The fact that z_0 is a pole of *order 1* of $A(z)$ is usually expressed by the statement that z_0 is a *regular singular point* of Eq. (3.1). It might be expected that this case arises from equations of the form (2.8) where the $a_i(z)$ may have poles of order at most one. However, many equations of form (2.7) in which certain $a_i(z)$ have poles of higher order give rise to matrix equations (3.1) with regular singular points. To give just one example of this we consider the equation

$$z^2\frac{d^2u}{dz^2}+za(z)\frac{du}{dz}+b(z)u=0, \tag{3.2}$$

where $a(z),b(z)$ are analytic in some domain D containing $z=0$.

ISBN 0-201-13509-4

If we define

$$U(z)=\begin{bmatrix} u(z) \\ z\dfrac{du}{dz} \end{bmatrix}, \qquad A(z)=\begin{pmatrix} 0 & 1/z \\ -b(z)/z & (-a(z)+1)/z \end{pmatrix}, \tag{3.3}$$

then

$$U'(z)=A(z)U(z) \tag{3.4}$$

and $A(z)$ has a pole of order at most one at $z=0$. The significance of the condition that $A(z)$ have a pole of order at most one at some z, say $z=0$, can be seen from the following. Consider Eq. (3.1) with

$$A(z)=A_{-1}/z+A_0+A_1z+\cdots. \tag{3.5}$$

It can be shown that there exist solutions to (3.1) and (3.5) of the form

$$U(z)=\left(\sum_{n=0}^{\infty} U_nz^n\right)\cdot z^R, \tag{3.6}$$

where R and $U_n, n=0,1,2,\ldots,$ are constant matrices. However, if $A(z)$ has a pole of order greater than one at $z=0$, then a solution of the form (3.6) is not even formally possible because $U'(z)$ and $A(z)U(z)$ would have singularities of different orders at $z=0$. The determination of series solutions of (3.1) in this situation is quite difficult; the series have in general essential singularities at $z=0$, and the coefficients are not determined by finite recursion relations. Correspondingly, it is very difficult to analyze the behavior of product integrals of functions with higher-order poles; for example, no general analogue of the residue theorem for ordinary contour integrals is available for product integrals except in the case of regular singular points. (No such result is known to the present authors in any case.)

In the present section we shall frequently consider the following situation (the notation used here will be used throughout the section and will not be defined again): $D\subseteq\mathbb{C}$ is a simply connected domain, Γ is a positively oriented contour in D, z_0 is a point of D lying inside Γ, and $A(z)$ is an analytic $\mathbb{C}_{n\times n}$ valued function in $D\backslash\{z_0\}$ (the complement of z_0 in D). See Figure 3.1.

In this situation, the residue theorem gives a prescription for evaluating $\int_\Gamma A(\zeta)d\zeta$. Since $\int_\Gamma A(\zeta)d\zeta$ is unchanged if the contour Γ is "shrunk" toward the point z_0, we may assume that Γ is contained in an annulus in D in which $A(z)$ has the convergent Laurent expansion

$$A(z)=\sum_{n=-\infty}^{\infty} A_n(z-z_0)^n. \tag{3.7}$$

ISBN 0-201-13509-4

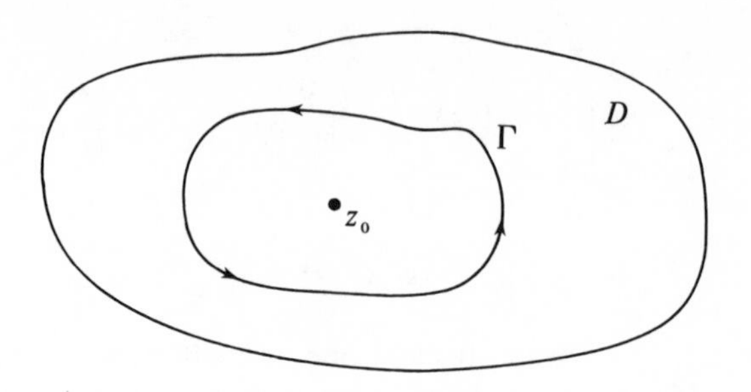

Figure 3.1

Then $\int_\Gamma A(\zeta)d\zeta = 2\pi i A_{-1}$. If $\{A(z)\}_{z\in\Gamma}$ is a commutative family, then by Theorem 1.3, Chapter 1, we have

$$\prod_\Gamma e^{A(\zeta)d\zeta} = e^{\int_\Gamma A(\zeta)d\zeta} = e^{2\pi i A_{-1}}. \tag{3.8}$$

If the commutativity assumption is dropped, it is no longer generally true that $\prod_\Gamma e^{A(\zeta)d\zeta}$ and $e^{2\pi i A_{-1}}$ are equal. However, if $A(z)$ has a pole of order at most one at z_0, and if a technical condition concerning the spectrum of A_{-1} is satisfied, then $\prod_\Gamma e^{A(\zeta)d\zeta}$ and $e^{2\pi i A_{-1}}$ are *similar* matrices, as will be shown. In the general case, it seems extremely difficult to compute $\prod_\Gamma e^{A(\zeta)d\zeta}$ (even up to similarity) or even to determine whether $\prod_\Gamma e^{A(\zeta)d\zeta} = I$, the identity matrix. The latter question is equivalent to asking when solutions of

$$\frac{du}{dz} = A(z)u(z); \qquad u(z): D\setminus\{z_0\} \to \mathbb{C}^n \tag{3.9}$$

are single valued; this need not be the case even when $A(z) = \sum_{n=-1}^{\infty} A_n z^n$ and $e^{2\pi i A_{-1}} = I$. We give an example illustrating some of these remarks.

Example. 3.1. Let

$$S(z) = \begin{pmatrix} z & 0 \\ 0 & 1 \end{pmatrix}, \qquad R = \begin{pmatrix} r_1 & r_2 \\ r_3 & r_4 \end{pmatrix},$$

and let $T(z)$ be the (multivalued in general) function $T(z) = S(z)z^R = S(z)e^{R\log z}$ defined for $z \neq 0$. Set

$$A(z) = \frac{dT}{dz}\cdot T^{-1}(z) = z^{-2}\begin{pmatrix} 0 & 0 \\ r_3 & 0 \end{pmatrix} + z^{-1}\begin{pmatrix} 1+r_1 & 0 \\ 0 & r_4 \end{pmatrix} + \begin{pmatrix} 0 & r_2 \\ 0 & 0 \end{pmatrix}.$$

$A(z)$ is analytic for $z \neq 0$. Let Γ be the contour given by $\gamma(t) = e^{it}, 0 \leqslant t \leqslant 2\pi$.

ISBN 0-201-13509-4

By picking a branch of the logarithm along $\gamma(t), 0<t<2\pi$ and applying the fundamental theorem of product integration we find

$$\prod_{\Gamma} e^{A(\zeta)d\zeta} = \prod_{\Gamma} e^{T'(\zeta)T^{-1}(\zeta)d\zeta} = T(e^{2\pi i})T^{-1}(e^{0}) = e^{2\pi i R}.$$

We have

$$A_{-1} = \begin{pmatrix} 1+r_1 & 0 \\ 0 & r_4 \end{pmatrix}$$

so

$$e^{2\pi i A_{-1}} = \begin{pmatrix} e^{2\pi i r_1} & 0 \\ 0 & e^{2\pi i r_4} \end{pmatrix}.$$

In general, $e^{2\pi i A_{-1}}$ and $e^{2\pi i R}$ are not equal or similar. For example, if $r_3 = 0, r_1 = r_4 = -\frac{1}{2}$, then

$$A_{-1} = \begin{bmatrix} \frac{1}{2} & 0 \\ 0 & -\frac{1}{2} \end{bmatrix},$$

$$e^{2\pi i A_{-1}} = -I,$$

and

$$e^{2\pi i R} = e^{i\pi\begin{pmatrix} -1 & 2r_2 \\ 0 & -1 \end{pmatrix}} = e^{i\pi\begin{pmatrix} -1 & 0 \\ 0 & -1 \end{pmatrix}} e^{\begin{pmatrix} 0 & 2\pi i r_2 \\ 0 & 0 \end{pmatrix}}$$

$$= -I\left(I + \begin{pmatrix} 0 & 2\pi i r_2 \\ 0 & 0 \end{pmatrix}\right) = \begin{pmatrix} -1 & -2\pi i r_2 \\ 0 & -1 \end{pmatrix}$$

(we have used the fact that

$$\begin{pmatrix} -1 & 0 \\ 0 & -1 \end{pmatrix} \quad \text{and} \quad \begin{pmatrix} 0 & 2\pi i r_2 \\ 0 & 0 \end{pmatrix}$$

commute) which is not similar to $-I$ unless $r_2 = 0$. Note that in this case $A(z)$ has only a pole of order one at $z=0$; we also note for later reference that $\|A_{-1}\| = \frac{1}{2}$. If we take $r_3 = 0, r_1 = r_4 = 1$, we get an example with $e^{2\pi i A_{-1}} = I$, $A(z)$ has a pole of order one at $z=0$, and

$$e^{2\pi i R} = \begin{pmatrix} 1 & 2\pi i r_2 \\ 0 & 1 \end{pmatrix},$$

which is not similar to I unless $r_2 = 0$.

The example indicates that to prove an analogue of the Cauchy integral formula for product integrals even for the case $A(z) = A_{-1}/(z - z_0) + B(z)$, $B(z)$ analytic in D, additional assumptions are necessary. One assumption

ISBN 0-201-13509-4

which is sufficient is that the difference set $\sigma(A_{-1})-\sigma(A_{-1})$ of the spectrum $\sigma(A_{-1})$ of A_{-1} does not contain positive integers. We state and prove a technical result concerning this condition.

LEMMA 3.1. *Let* $A\in\mathbb{C}_{n\times n}$ *and let* $\mathrm{ad}A:\mathbb{C}_{n\times n}\to\mathbb{C}_{n\times n}$ *be the linear operator defined by* $\mathrm{ad}A(T)=AT-TA=[A,T]$ *for* $T\in\mathbb{C}_{n\times n}$. *Let* $\sigma(A)$ *and* $\sigma(\mathrm{ad}A)$ *denote the spectra (sets of eigenvalues) of* A *and* $\mathrm{ad}A$, *respectively. Then* $\sigma(\mathrm{ad}A)=\sigma(A)-\sigma(A)=\{\lambda-\mu:\lambda,\mu\in\sigma(A)\}$.

Proof. Let $\sigma(A)=\{\lambda_1,\dots,\lambda_l\}$ and let A^t be the transpose of A. A and A^t have the same spectrum since their characteristic polynomials are the same. Choose column vectors $\{X_k\}_{k=1,\dots,l}$ and $\{Y_k\}_{k=1,\dots,l}$ with $AX_k=\lambda_k X_k, A^tY_k=\lambda_k Y_k$ (or $Y_k^tA=\lambda_k Y_k^t$). Let Z_{rs} be the nonzero $n\times n$ matrix $X_rY_s^t$. (Y_s^t is the *row* vector with the same components as Y_s.) Then $(\mathrm{ad}A)(Z_{rs})=AX_rY_s^t-X_rY_s^tA=(\lambda_r-\lambda_s)Z_{rs}$. This shows $\sigma(A)-\sigma(A)\subseteq\sigma(\mathrm{ad}A)$. Next we prove the reverse inclusion. The space $\mathbb{C}^n$ on which A acts can be decomposed as $\mathbb{C}^n=V_1\oplus\cdots\oplus V_l$ where in each V_k there is a basis $\{f_1,\dots,f_{r_k}\}$ with

$$
\begin{aligned}
Af_1&=\lambda_kf_1+f_2,\\
Af_2&=\lambda_kf_2+f_3,\\
&\ \vdots \qquad\qquad\qquad \text{(Jordan decomposition)}\\
Af_{r_k}&=\lambda_kf_{r_k}.
\end{aligned}
$$

Let μ be an eigenvalue of $\mathrm{ad}A$ with eigenmatrix Z. Assume $\mu+\lambda_k\notin\sigma(A)$. Then $A-\mu-\lambda_k$ is invertible. We have $AZ=ZA+\mu Z$, so $(A-\mu-\lambda_k)Z=Z(A-\lambda_k)$ or

$$Z=(A-\mu-\lambda_k)^{-1}Z(A-\lambda_k). \tag{3.10}$$

Applying Z to vectors in V_k and using (3.10) we find

$$
\begin{aligned}
Zf_1&=(A-\mu-\lambda_k)^{-1}Zf_2=(A-\mu-\lambda_k)^{-2}Zf_3=\cdots\\
&=(A-\mu-\lambda_k)^{-r_k+1}Zf_{r_k}=0,
\end{aligned}\tag{3.11}
$$

where the last equality holds by (3.10) and the fact that $(A-\lambda_k)f_{r_k}=0$. Thus Z annihilates V_k. Z cannot annihilate all the V_k since $Z\neq 0$, so for some k, $\mu+\lambda_k\in\sigma(A)$ or $\mu\in\sigma(A)-\sigma(A)$. This proves $\sigma(\mathrm{ad}A)\subseteq\sigma(A)-\sigma(A)$. ■

We now return to the situation described immediately preceding Figure 3.1. We want to consider the contour product integral of $A(z)$ around the contour Γ; the value of this product integral depends in general on the starting point on Γ. Accordingly, we give

ISBN 0-201-13509-4

DEFINITION 3.1. *For $z \neq z_0$, we define*

$$\Lambda(z) = \prod_{\Gamma_z} e^{A(\zeta)\,d\zeta}, \tag{3.12}$$

where Γ_z is any positively oriented contour in D starting at z, circling z_0 once, and returning to z. See Figure 3.2.

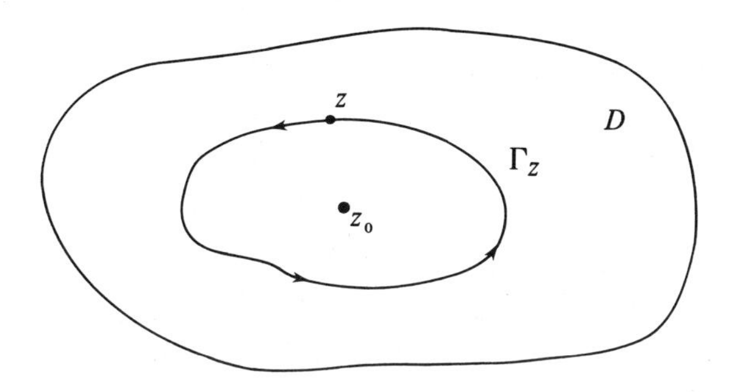

Figure 3.2

By theorem 2.2 $\Lambda(z)$ is independent of which such contour Γ_z is chosen, i.e., $\Lambda(z)$ is well defined. The next theorem describes the behavior of $\Lambda(z)$ as z approaches z_0, a singular point for $A(z)$, in a special case.

THEOREM 3.1. *$\Lambda(z)$ is analytic in $D \setminus \{z_0\}$. If $A(z) = A_{-1}/(z-z_0) + B(z)$, $B(z)$ analytic in D, then $\Lambda(z)$ has a removable singularity at z_0 and $\lim_{z \to z_0} \Lambda(z) = e^{2\pi i A_{-1}}$. [The term "removable singularity" means that $\Lambda(z)$ can be made analytic in D by properly defining $\Lambda(z_0)$, i.e., in this case $\Lambda(z_0) = e^{2\pi i A_{-1}}$.]*

Proof. Consider the following diagram:

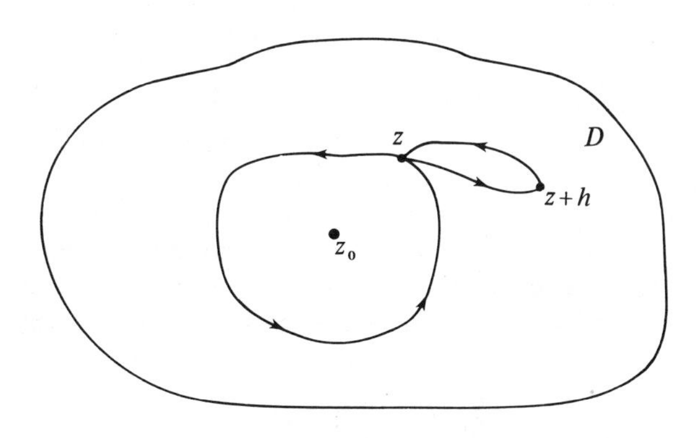

Figure 3.3

ISBN 0-201-13509-4

From Figure 3.3 and Theorem 1.5 of Chapter 1 we have

$$\Lambda(z+h)=\prod_{z}^{z+h} e^{A(\zeta)d\zeta}\Lambda(z)\prod_{z+h}^{z} e^{A(\zeta)d\zeta} \tag{3.13}$$

(along the indicated contours), so

$$\Lambda(z+h)-\Lambda(z)=\left(\prod_{z}^{z+h} e^{A(\zeta)d\zeta}-I\right)\Lambda(z)\prod_{z+h}^{z} e^{A(\zeta)d\zeta}$$
$$+\Lambda(z)\left(\prod_{z+h}^{z} e^{A(\zeta)d\zeta}-I\right). \tag{3.14}$$

Dividing by h and letting $h\to 0$ yields $d\Lambda(z)/dz=A(z)\Lambda(z)-\Lambda(z)A(z)=[A(z),\Lambda(z)]$ for $z\neq z_0$, which proves $\Lambda(z)$ analytic in $D\setminus\{z_0\}$. Finally, if we let Γ_z be a circle $z_0+re^{i\theta},\theta_0\leqslant\theta\leqslant 2\pi$ (where $z=z_0+re^{i\theta_0}$), then

$$\Lambda(z)=\prod_{\Gamma_z} e^{A(\zeta)d\zeta}=\prod_{\theta_0}^{\theta_0+2\pi} e^{\{iA_{-1}+O(r)\}d\theta} \tag{3.15}$$

where the $O(r)$ denotes terms bounded by r. We also have

$$\prod_{\Gamma_z} e^{iA_{-1}d\theta}=e^{2\pi iA_{-1}}. \tag{3.16}$$

From (3.15) and (3.16) it follows easily using Theorems 3.2 and 4.1 of Chapter 1 that $\lim_{z\to z_0}\Lambda(z)=e^{2\pi iA_{-1}}$. This finishes the proof of the theorem. ■

We remark that for $z\neq z_0$, $z'\neq z_0$, $\Lambda(z)$ and $\Lambda(z')$ are similar; this follows from Corollary 2.2.2 of Theorem 2.2. In particular, if $\Lambda(z)=I$ for some $z\neq z_0$, then $\Lambda(z)\equiv I$ for all $z\neq z_0$, and if in addition $A(z)=A_{-1}/(z-z_0)+B(z)$, then $e^{2\pi iA_{-1}}=I$. The condition $\Lambda(z)=I$ means that all solutions of the equation $du/dz=A(z)u(z)$ are single valued in $D\setminus\{z_0\}$, i.e., have at worst poles or essential singularities at z_0 and not branch points. Note that if $A(z)=A_{-1}/(z-z_0)+B(z)$, then $e^{2\pi iA_{-1}}=I$ is a necessary but not sufficient condition for single valuedness as we have seen in Example 3.1. In this example, the $\Lambda(z)$ are all similar to each other but not to their limit $e^{2\pi iA_{-1}}$. To aid the reader's intuition, we give a simple example of a family of matrices all similar to a fixed matrix, but whose limit is not similar to that matrix: for $a\neq 0$, the matrices $\begin{pmatrix}1 & a\\ 0 & 1\end{pmatrix}$ are all similar to $\begin{pmatrix}1 & 1\\ 0 & 1\end{pmatrix}$, but

$$\lim_{a\to 0}\begin{pmatrix}1 & a\\ 0 & 1\end{pmatrix}=\begin{pmatrix}1 & 0\\ 0 & 1\end{pmatrix}$$

is not similar to $\begin{pmatrix}1 & 1\\ 0 & 1\end{pmatrix}$. Thus the conditions

ISBN 0-201-13509-4

(1) $\Lambda(z)$ all similar for $z \neq z_0$,
(2) $\lim_{z \to z_0} \Lambda(z) = e^{2\pi i A_{-1}}$,
do not guarantee that $\Lambda(z)$ is similar to $e^{2\pi i A_{-1}}$.

We now prove an analogue of the Cauchy integral formula for product integrals.

THEOREM 3.2 (Cauchy Integral Formula). *Consider again the situation described immediately preceding Figure* 3.1. *Suppose* $A(z) = A_{-1}/(z - z_0) + B(z)$, *with* $B(z)$ *analytic in* D. *Suppose further that* $\sigma(A_{-1}) - \sigma(A_{-1})$ *does not contain positive integers. Then* $\prod_\Gamma e^{A(\zeta)d\zeta}$ *and* $e^{2\pi i A_{-1}}$ *are similar matrices.* (*The initial point of* Γ *is arbitrary.*)

Remarks. (1) In Volterra and Hostinsky [VV4], the result of this theorem is apparently claimed without the hypothesis on $\sigma(A_{-1})$. Some hypothesis is necessary as Example 3.1 shows.

(2) The result of the theorem in a different guise can be found in [EC]; the proof given there differs from the present one. Our proof generalizes immediately to the case that the $A(z)$ are bounded operators on a Banach space, as we shall remark later.

Proof of Theorem 3.2. We assume without loss of generality that $z_0 = 0$ to simplify the notation. Now $\prod_\Gamma e^{A(\zeta)d\zeta} = \Lambda(z')$ for some $z' \in \Gamma$. Since $\Lambda(z)$, $\Lambda(z')$, $z, z' \neq 0$ are similar, it suffices to show that $\Lambda(z)$ is similar to $e^{2\pi i A_{-1}}$ for $|z|$ sufficiently small. Now Theorem 3.3 of Chapter 1 (in the form given in (1.3.22)) implies that if $T(z)$ is analytic and invertible along a contour C in D with initial and terminal points z_1, z_2 and $S(z)$ is continuous along C, then

$$\prod_C e^{S(z)\,dz} = T(z_2) \prod_C e^{\{T^{-1}(z)S(z)T(z) - T^{-1}(z)T'(z)\}\,dz} \cdot T^{-1}(z_1). \tag{3.17}$$

Suppose that a single-valued analytic $T(z)$ in a neighborhood of $z = 0$ can be found with

$$T^{-1}(z)\frac{A_{-1}}{z}T(z) - T^{-1}(z)T'(z) = \frac{A_{-1}}{z} + B(z). \tag{3.18}$$

Then by (3.17) with $S(z) = A_{-1}/z$, $C = \Gamma$, $z_1 = z_2 = z$, we have

$$e^{2\pi i A_{-1}} = T(z)\Lambda(z)T^{-1}(z), \tag{3.19}$$

which is the result claimed in the theorem. We thus attempt to solve (3.18), which is equivalent to the system

$$\left\{ \begin{array}{l} T'(z) = [A_{-1}/z, T(z)] - T(z)B(z) \\ T(z) \quad \text{invertible near} \quad 0 \end{array} \right\}. \tag{3.20}$$

ISBN 0-201-13509-4

We look for $T(z)$ in the form

$$T(z) = I + \sum_{n=1}^{\infty} T_n z^n. \tag{3.21}$$

The invertibility near $z=0$ is automatic for $T(z)$ of this form, so we substitute (3.21) in Eq. (3.20), which yields the formal relation

$$\sum_{n=1}^{\infty} nT_n z^{n-1} = \sum_{n=1}^{\infty} [A_{-1}, T_n] z^{n-1} - \left(I + \sum_{n=1}^{\infty} T_n z^n\right)\left(\sum_{n=0}^{\infty} B_n z^n\right). \tag{3.22}$$

Identifying equal powers of z yields

$$(n - \mathrm{ad}A_{-1}) T_n z^{n-1} = -(B_0 T_{n-1} + \cdots + B_{n-1}) z^{n-1} \qquad \text{for} \quad n \geqslant 1, \tag{3.23}$$

and under the hypothesis on $\sigma(A_{-1})$, this is recursively solved by

$$T_n = -(n - \mathrm{ad}A_{-1})^{-1}(B_0 T_{n-1} + \cdots + B_{n-1}). \tag{3.24}$$

Now $(n-\mathrm{ad}A_{-1})^{-1} = O(n^{-1})$ as $n \to \infty$, and for $|z|$ sufficiently small, $\|B_n z^n\| = O(1)$ as $n \to \infty$ since $\Sigma_{n=0}^{\infty} B_n z^n$ is convergent. Hence (3.24) yields

$$\|T_n z^n\| = O(n^{-1}) \cdot \sum_{j=0}^{n-1} \|T_j z^j\| \tag{3.25}$$

for $|z|$ sufficiently small and where $T_0 = I$. From (3.25) we obtain inductively

$$\|T_n z^n\| = O(1) \qquad \text{as } n \to \infty \text{ (for } |z| \text{ small).} \tag{3.26}$$

Hence for some $\tilde{z}$, $\|T_n \tilde{z}^n\|$ is bounded uniformly in n, and so for $|z| < |\tilde{z}|$ we have

$$\|T_n z^n\| = \|T_n (z/\tilde{z})^n \tilde{z}^n\| = \|T_n \tilde{z}^n\| \, |z/\tilde{z}|^n \leqslant \text{const.} |z/\tilde{z}|^n. \tag{3.27}$$

This estimate shows that (3.21) converges and finishes the proof of the theorem. ■

Remarks. (1) By a slight refinement of the above argument it can be proved that the result of the theorem is still true in the case that $B(z)$ has a zero of order $\geqslant s$ at $z = z_0$ and $\sigma(A_{-1}) - \sigma(A_{-1})$ contains no integer $> s$. [In this case the T_n may still be determined from (3.23).] This condition on

ISBN 0-201-13509-4

$\sigma(A_{-1})-\sigma(A_{-1})$ will hold, for example, if $\|A_{-1}\|<(s+1)/2$, since then $\|\mathrm{ad}A_{-1}\|<s+1$ and $(n-\mathrm{ad}A_{-1})^{-1}$ exists for $n\geqslant s+1$. Hence the result of the theorem is always true if $\|A_{-1}\|<\frac{1}{2}$; recall, however, that Example 3.1 showed that the result of the theorem fails to hold in a case where $\|A_{-1}\|=\frac{1}{2}$.

(2) Since $\Lambda(z)$ and $\Lambda(z')$ are similar for $z,z'\neq z$, the characteristic polynomial $[P(\lambda)=\det(\Lambda(z)-\lambda I)]$ of $\Lambda(z)$, $z\neq z_0$, is independent of z and thus equals the characteristic polynomial of $e^{2\pi i A_{-1}}$. [Because $\lim_{z\to z_0}\Lambda(z)=e^{2\pi i A_{-1}}$.] Hence the eigenvalues of $\Lambda(z)$ and $e^{2\pi i A_{-1}}$ are the same, and if these are all distinct, then the result of the theorem holds. (Because two $n\times n$ matrices with the same n distinct eigenvalues are similar.)

(3) As remarked earlier, if the family $\{A(z)\}_{z\neq z_0}$ is commutative, the result of the theorem is always true with "similar" replaced by "equal."

(4) If the condition on $\sigma(A_{-1})$ does not hold, then it is possible to show that there is an invertible analytic function $T(z)$ in $\mathbb{C}\backslash\{z_0\}$ such that if $\tilde{A}(z)=T^{-1}(z)A(z)T(z)-T^{-1}(z)T'(z)$, then $\tilde{A}(z)$ has a pole of order 1 at z_0 and $\tilde{A}_{-1}$ *does* satisfy the spectrum condition of the theorem. (See Theorem 4.2 and the preceding lemma in Chapter 4 of [EC].) Hence by the similarity formula (1.3.22), $\prod_\Gamma e^{A(\zeta)d\zeta}$ is similar to $\prod_\Gamma e^{\tilde{A}(\zeta)d\zeta}$, which is similar to $e^{2\pi i\tilde{A}_{-1}}$ by Theorem 3.2. However, it is difficult to give an explicit formula for $\tilde{A}_{-1}$ since the construction of $T(z)$ involves the transformations which put various successively determined matrices in Jordan normal form. It is, however, not difficult to show that $e^{2\pi i\tilde{A}_{-1}}$ and $e^{2\pi i A_{-1}}$ may differ only in the respect that the Jordan canonical form for $e^{2\pi i\tilde{A}_{-1}}$ may contain more 1's above the diagonal than the Jordan form for $e^{2\pi i A_{-1}}$. The following example may serve to illustrate the theorem and the exceptional case in which the spectrum condition is not satisfied.

Example. 3.2. We take $D=\{z\in\mathbb{C}:|z|<2\}, z_0=0$, and suppose that $A(z)=A_{-1}/z+B(z)$ is a 2×2 matrix valued function with $B(z)$ analytic in D. Let Γ be the contour given by $\gamma(t)=e^{it}, 0\leqslant t\leqslant 2\pi$. Then $\prod_\Gamma e^{A(\zeta)d\zeta}$ is some invertible matrix and can thus be written in the form $\prod_\Gamma e^{A(\zeta)d\zeta}=e^{2\pi iR}$, where R is a 2×2 matrix (see Appendix). It follows easily that the function defined by analytic continuation of $\prod_1^z e^{A(\zeta)d\zeta}\cdot z^{-R}$ is single valued in $\{0<|z|<2\}$, and so

$$\prod_1^z e^{A(\zeta)d\zeta}=S(z)\cdot z^R, \tag{3.28}$$

where $S(1)=I$ and $S(z)$ is analytic and single valued in $\{0<|z|<2\}$, i.e., $S(z)$ is given by a convergent Laurent series. [So far the fact that $A(z)$ has only a pole of order 1 at $z=0$ has not been used.]

The fact that $A(z)$ has a pole of order 1 at $z=0$, together with an estimate of the norm of $\prod_1^z e^{A(\zeta)d\zeta}$ as $z\to0$, shows easily that in fact $S(z)$ can have only a pole of finite order at $z=0$ (i.e., not an essential singularity).

ISBN 0-201-13509-4

Suppose now that $\sigma(A_{-1})-\sigma(A_{-1})$ does not contain positive integers. Then by Theorem 3.2, R may be taken to be similar to A_{-1}. Now the Jordan form of A_{-1} is one of the three possible types

$$\begin{pmatrix} \lambda_1 & 0 \\ 0 & \lambda_2 \end{pmatrix} \quad \begin{matrix} \lambda_1 \neq \lambda_2 \\ \lambda_1-\lambda_2 \neq \text{integer} \end{matrix}, \quad \begin{pmatrix} \lambda_1 & 0 \\ 0 & \lambda_1 \end{pmatrix}, \quad \text{or} \quad \begin{pmatrix} \lambda_1 & 1 \\ 0 & \lambda_1 \end{pmatrix}, \tag{3.29}$$

and corresponding to these three possibilities the matrix z^R is similar to

$$\begin{pmatrix} z^{\lambda_1} & 0 \\ 0 & z^{\lambda_2} \end{pmatrix}, \quad \begin{pmatrix} z^{\lambda_1} & 0 \\ 0 & z^{\lambda_1} \end{pmatrix}, \quad \text{or} \quad \begin{pmatrix} z^{\lambda_1} & z^{\lambda_1}\log z \\ 0 & z^{\lambda_1} \end{pmatrix}. \tag{3.30}$$

Next suppose that the spectrum conditions fail and A_{-1} has the Jordan form

$$\begin{pmatrix} \lambda_1 & 0 \\ 0 & \lambda_2 \end{pmatrix}, \quad \text{where } \lambda_2-\lambda_1=n=\text{positive integer}, \tag{3.31}$$

then

$$e^{2\pi i A_{-1}}=\begin{pmatrix} e^{2\pi i\lambda_1} & 0 \\ 0 & e^{2\pi i\lambda_1} \end{pmatrix}. \tag{3.32}$$

Now $e^{2\pi i R}$ must have the same eigenvalues as $e^{2\pi i A_{-1}}$. However, the Jordan form of $e^{2\pi i R}$ may be

$$\begin{pmatrix} e^{2\pi i\lambda_1} & 0 \\ 0 & e^{2\pi i\lambda_1} \end{pmatrix} \quad \text{or} \quad \begin{pmatrix} e^{2\pi i\lambda_1} & 1 \\ 0 & e^{2\pi i\lambda_1} \end{pmatrix}. \tag{3.33}$$

Corresponding to these two possibilities the Jordan forms for R would be, respectively,

$$\begin{pmatrix} \lambda_1 & 0 \\ 0 & \lambda_1 \end{pmatrix} \quad \text{or} \quad \begin{pmatrix} \lambda_1 & 1 \\ 0 & \lambda_1 \end{pmatrix}, \tag{3.34}$$

and z^R would be similar, respectively, to

$$\begin{pmatrix} z^{\lambda_1} & 0 \\ 0 & z^{\lambda_1} \end{pmatrix} \quad \text{or} \quad \begin{pmatrix} z^{\lambda_1} & z^{\lambda_1}\log z \\ 0 & z^{\lambda_1} \end{pmatrix}. \tag{3.35}$$

In any of the above cases the matrix elements of $\prod_1^z e^{A(\zeta)d\zeta}$ are of the form

$$\alpha(z)z^{\lambda_1}+\beta(z)z^{\lambda_2}$$

ISBN 0-201-13509-4

or

$$\alpha(z)z^{\lambda_1}+\beta(z)z^{\lambda_1}\log z, \tag{3.36}$$

where $\alpha(z),\beta(z)$ are analytic in $0<|z|<2$. In the case that the product integral arises in the solution of an equation

$$z^2y''(z)+za(z)y'(z)+b(z)y(z)=0 \tag{3.37}$$

as in (3.2)–(3.4), our discussion shows that the solutions of (3.37) are of the form (3.36), as is shown in books on ordinary differential equations (e.g., [EC]).

Finally we remark that in certain cases a product integral formula for the similarity $T(z)$, which occurs in the proof of Theorem 3.2, can be given. Suppose that the hypotheses of the theorem hold and $z_0=0$. Then by (3.19) we have

$$\Lambda(z)=T^{-1}(z)e^{2\pi i A_{-1}}T(z). \tag{3.38}$$

We consider $z=1$ for definiteness. It can be shown [JD4] that provided either

(a) A_{-1} is skew-adjoint, i.e., $A^*_{-1}=-A_{-1}$ or

(b) $B(z)$ has a zero of order $\geqslant k$ at $z=0$ and $\|A_{-1}\|<(k+1)/2$, then

$$T(1)=\prod_1^0 e^{\{s^{-A_{-1}}B(s)s^{A_{-1}}\}ds}, \tag{3.39}$$

where the path of integration is the line segment from 1 to 0 and the product integral is an improper product integral in the sense of Chapter 1. We note that the condition (b) always holds if $\|A_{-1}\|<\frac{1}{2}$.

2.4 Generalizations

Virtually all of the results of this chapter continue to hold if we replace $\mathbb{C}^n$ by an arbitrary complex Banach space $\mathfrak{X}$ and $\mathbb{C}_{n\times n}$ by the space $\mathfrak{L}(\mathfrak{X})$ of bounded linear operators on $\mathfrak{X}$ with the norm topology. (See [JD4] for a development of the theory in this setting.) The major difference is that the spectrum $\sigma(A)$ of an operator in $\mathfrak{L}(\mathfrak{X})$ does not necessarily consist solely of eigenvalues. However, the proofs of all the theorems are unchanged. Lemma 3.1 is still true, but the proof is more difficult.

Notes to Chapter 2

Contour product integrals are considered in V-H [VV4]. They prove our Theorem 2.1, by methods differing from ours, but do not state the more general result of our Theorem 2.2 (although Corollary 2.2.2 can be found in V-H). Our Theorem 3.2 is stated in V-H, without the spectrum condition on A_{-1}; as mentioned in the

ISBN 0-201-13509-4

chapter this omission makes their theorem incorrect. V-H derive their results on complex contour integrals from earlier material on curvilinear integrals and a "Green's theorem" relating curvilinear integrals with double product integrals. We sketch some of the ideas involved, using our own notation but without changing the essential ideas. Consider a differential form

$$w = A(x,y)\,dx + B(x,y)\,dy, \tag{i}$$

where A and B are $\mathbb{C}_{n\times n}$ valued continuously differentiable functions in a domain D of the xy plane. If Γ is a path in D given by $\gamma(t) = (x(t),y(t))$ for $a \leqslant t \leqslant b$, then we may define the product integral of w along Γ to be

$$\prod_{\Gamma} e^{w} = \prod_{a}^{b} e^{\left\{A(x(t),y(t))\frac{dx}{dt} + B(x(t),y(x))\frac{dy}{dt}\right\}dt}. \tag{ii}$$

If $C(x,y)$ is a twice continuously differentiable and invertible $\mathbb{C}_{n\times n}$ valued function in D we define $\mathfrak{D}C$ to be

$$\mathfrak{D}C = \frac{\partial C}{\partial x}\cdot C^{-1}dx + \frac{\partial C}{\partial y}\cdot C^{-1}dy. \tag{iii}$$

A form w as in (i) is called *exact* if $w = \mathfrak{D}C$. It is an easy consequence of our fundamental theorem of product integration that if w is exact and Γ is a closed path, then $\prod_{\Gamma} e^{w} = I$. A form w as in (i) is called *closed* if

$$\Delta(w) = \Delta(A,B) = \frac{\partial B}{\partial x} - \frac{\partial A}{\partial y} - [A,B] = 0, \tag{iv}$$

where $[A,B] = AB - BA$.

An elementary calculation shows that an exact form w is closed. V-H prove that in a *simply connected* domain, the converse is true, i.e., *a closed form is exact*. This is proved by showing that a curvilinear integral of a form along a closed path can be identified with a certain complicated double product integral containing the expression $\Delta(w)$; if w is closed, then $\Delta(w) = 0$ and the curvilinear integral is equal to the identity, I. The cited result follows from these facts. We make no use of these results and the proof of our Theorem 2.1 is quite simple and direct; for these reasons we have not discussed curvilinear integrals or double product integrals except in these notes. A development of the above ideas also appears in Schlesinger's paper [L-S4].

Results equivalent to our Theorems 2.1 and 3.2 (but in a different context) are stated and proved in Coddington and Levinson [EC]. Many of our results including our Theorem 3.2 are proved by Rasch [GR]; however, his methods and calculations seem considerably more complicated than ours.

We add finally that it would be of great interest to have an analogue of the residue theorem more general than our Theorem 3.2, i.e., that would allow calculation or partial calculation of the product integral of a Laurent series $\sum_{n=-\infty}^{\infty} A_n z^n$ along a path circling $z = 0$. We have not been able to find a satisfactory analogue. This problem is discussed in Rasch's paper [GR] where formulas are given for the individual terms arising in the time-ordered exponential series for the product integral of a Laurent series. Rasch's formulas are cumbersome and do not seem to give much information about the values of such product integrals or their similarity classes.

ISBN 0-201-13509-4

CHAPTER 3

Strong Product Integration

0. Introduction

This chapter stands at a higher level of abstraction than those that have preceded it. Some familiarity with the theory of Banach spaces is presupposed. In particular, the reader is assumed to be familiar with the various standard topologies on the set of bounded linear operators on a Banach space (norm topology, strong topology, and weak topology). For background material, we refer the reader to any standard text on functional analysis, for example, [KY3].

The situation to be considered in this chapter is as follows: X is a Banach space over the complex numbers and A is a function defined on the real interval $[a,b]$ and taking values in the set $\mathfrak{B}(X)$ of all bounded linear operators from X to itself. The problem will be to define the product integral of A and establish properties analogous to those found in Chapter 1 for product integrals of matrix-valued functions. As we shall see, for certain types of functions A the work of Chapter 1 generalizes directly. For other types the generalization is not so immediate and will require a considerable development.

In the last section of this chapter, we shall consider a problem involving unbounded operators.

Notation and Terminology

If φ is an element of the Banach space X, then $\|\varphi\|$ denotes the norm of φ. X^* denotes the dual of X (all bounded linear maps from X to $\mathbb{C}$). If

ISBN 0-201-13509-4

ENCYCLOPEDIA OF MATHEMATICS and Its Applications, Gian-Carlo Rota (ed.). Vol.10: John D. Dollard and Charles N. Friedman, Product Integration with Applications to Differential Equations

$B \in \mathfrak{B}(X)$, the norm $\|B\|$ of B is defined by

$$\|B\| = \sup_{\substack{\varphi \in X \\ \|\varphi\|=1}} \|B\varphi\|. \tag{0.1}$$

With this norm, $\mathfrak{B}(X)$ is a Banach space. The identity operator in $\mathfrak{B}(X)$ is denoted by I. An operator $B \in \mathfrak{B}(X)$ will be called *nonsingular* if and only if B has a two-sided inverse in $\mathfrak{B}(X)$.

A sequence $\{B_n\}$ in $\mathfrak{B}(X)$ is said to converge to $B \in \mathfrak{B}(X)$ *in norm* if $\lim_{n\to\infty}\|B_n - B\| = 0$; *strongly* if $\lim_{n\to\infty}\|B_n\varphi - B\varphi\| = 0$ for each $\varphi \in X$; and *weakly* if $\lim_{n\to\infty}\lambda(B_n\varphi) = \lambda(B\varphi)$ for each $\varphi \in X$ and $\lambda \in X^*$. Strong and weak limits are denoted "s-lim" and "w-lim," respectively. If $[a,b]$ is an interval in $\mathbb{R}$ and $A:[a,b]\to\mathfrak{B}(X)$ is a function, the above concepts of sequential convergence define corresponding concepts of continuity for A. Thus, for example, we speak of norm-continuous functions from $[a,b]$ to $\mathfrak{B}(X)$. Similar remarks apply to the notion of differentiability. In any topological statement concerning $\mathfrak{B}(X)$ we shall always specify which topology is meant. In the Banach space X itself, we shall use *only* the norm topology. Thus the statement "$f:[a,b]\to X$ is continuous" means that $\lim_{x\to y}\|f(x)-f(y)\| = 0$ for $x,y \in [a,b]$. For emphasis, we shall occasionally write statements such as "$f:[a,b]\to X$ is norm-continuous."

If $C \in \mathfrak{B}(X)$, we define the exponential $e^C \in \mathfrak{B}(X)$ by the usual power series

$$e^C = \sum_{n=0}^{\infty} \frac{C^n}{n!}. \tag{0.2}$$

This series converges in the norm topology on $\mathfrak{B}(X)$. For $t \in \mathbb{R}$, the function $t \to e^{Ct}$ is differentiable in the norm topology and

$$\frac{d}{dt} e^{Ct} = Ce^{Ct} = e^{Ct}C. \tag{0.3}$$

3.1 Direct Extensions of the Results of Chapter 1

If $A:[a,b]\to\mathfrak{B}(X)$ is a norm-continuous function, then a verbatim repetition of the arguments of Chapter 1 shows that the product integral of A can be defined as in that chapter: for $x,y \in [a,b]$ with $y < x$, we have

$$\prod_y^x e^{A(s)\,ds} = \text{norm} - \lim_{\mu(P)\to 0} \prod_{k=1}^{n} e^{A(s_k)\Delta s_k}, \tag{1.1}$$

where the notation is as in Chapter 1. Instead of (1.1), one can define the product integral of A using any sequence of step-functions converging to A

ISBN 0-201-13509-4

"in the L^1 sense" (understanding the phrase in quotation marks requires a knowledge of integration of operator-valued functions, which will be discussed presently). Virtually all results obtained for continuous matrix-valued functions in Chapter 1 continue to hold in the new setting. The only exceptions are results the *statement* of which makes explicit reference to matrix concepts like "trace," "determinant," "columns," etc. (It sometimes happens that the *statement* makes no reference to matrix concepts, but such concepts have been used in the *proof*. In these cases, the results can be established in the new setting, using different proofs.) The generalization from continuous functions to Lebesgue integrable functions in Chapter 1 also has an analogue in the present case: namely, one can easily generalize from norm-continuous functions to Bochner-integrable functions. The reader interested in a comprehensive account of Bochner integration should consult one of the standard references on the subject [KB]. Because we shall make extensive use of Bochner integration later, we give here a brief outline of the theory formulated in a manner appropriate to our purposes. This formulation is equivalent to the usual ones, although somewhat different in appearance.

Bochner Integration: Let Y denote a fixed Banach space equipped with its natural (norm) topology. By a *step-function* $F:[a,b]\to Y$ we shall mean just what we meant in Chapter 1:

DEFINITION 1.1. *A function* $F:[a,b]\to Y$ *is called a* step-function *if and only if there is a partition* $P=\{s_0,s_1,\dots,s_n\}$ *of* $[a,b]$ *such that* F *is constant on each open subinterval* (s_{k-1},s_k). *We then say that* F *is* associated *with* P *and write* F_k *for the value of* F *on* (s_{k-1},s_k). *As usual, we write* $\Delta s_k=s_k-s_{k-1}$.

If F is a step-function on $[a,b]$, then with notation as above we define the *Bochner integral* of F over $[a,b]$ by

$$\int_a^b F(s)\,ds=\sum_{k=1}^{n} F_k\Delta s_k. \tag{1.2}$$

We also define the Bochner norm $\|F\|_B$ of F by

$$\|F\|_B=\int_a^b \|F(s)\|\,ds=\sum_{k=1}^{n} \|F_k\|\Delta s_n. \tag{1.3}$$

We define two step-functions on $[a,b]$ to be *equivalent* if and only if they are equal almost everywhere. (Of course, this entails equality except at a finite number of points.) Then the collection of equivalence classes of step-functions $F:[a,b]\to Y$, with the norm defined by (1.3), is a normed vector space. (We follow tradition in speaking of the elements of this space

ISBN 0-201-13509-4

as if they were actually functions.) The set of *Bochner-integrable* functions on $[a,b]$ can be viewed as the completion of this space. In more detail: just as in the Lebesgue theory of integration, it turns out [KB] that if $\{F_n\}$ is a sequence of step-functions which is Cauchy in the norm (1.3), then there is a subsequence converging almost everywhere on $[a,b]$ to a function $F:[a,b]\to Y$. Furthermore, the nonnegative real-valued function $\|F_n(x)-F(x)\|$ is Lebesgue integrable over $[a,b]$ for each n and

$$\lim_{n\to\infty}\int_a^b \|F_n(s)-F(s)\|\,ds=0. \tag{1.4}$$

For any n,m we then have

$$\begin{aligned}\left\|\int_a^b F_n(s)\,ds-\int_a^b F_m(s)\,ds\right\| &\leqslant \int_a^b \|F_n(s)-F_m(s)\|\,ds\\ &\leqslant \int_a^b \|F_n(s)-F(s)\|\,ds+\int_a^b \|F(s)-F_m(s)\|\,ds \underset{n,m\to\infty}{\to} 0.\end{aligned} \tag{1.5}$$

Thus the sequence $\{\int_a^b F_n(s)\,ds\}$ has a limit, which is by definition the Bochner integral of F:

$$\int_a^b F(s)\,ds=\lim_{n\to\infty}\int_a^b F_n(s)\,ds. \tag{1.6}$$

The collection of all (equivalence classes of) functions F obtained in the above manner is the set of *Bochner-integrable* functions on $[a,b]$ to Y, and will be denoted $L^1(a,b;Y)$. By construction, $L^1(a,b;Y)$ is a Banach space with the norm

$$\|F\|_B=\int_a^b \|F(s)\|\,ds. \tag{1.7}$$

We wish to stress the fact that the Bochner theory of integration closely parallels the usual Lebesgue theory. The concept of measurability appropriate to the Bochner theory is the following:

DEFINITION 1.2. *A function $F:[a,b]\to Y$ is* strongly measurable *if and only if there is a sequence $\{F_n\}$ of step functions which converges to F almost everywhere on $[a,b]$.*

A standard result in the Bochner theory illustrates the parallelism mentioned above:

Proposition 1.1. A function $F:[a,b]\to Y$ belongs to $L^1(a,b;Y)$ if and only if the following conditions hold:

ISBN 0-201-13509-4

(*i*) *F is strongly measurable on* $[a,b]$. (*It then follows that* $\|F\|$ *is Lebesgue measurable on* $[a,b]$.)

(*ii*) *We have*

$$\int_a^b \|F(s)\|\,ds < \infty. \tag{1.8}$$

We shall not give the proof of this fact, which can be found in [KB]. The reader unfamiliar with the Bochner theory can in most cases rely on the idea that it is like the Lebesgue theory with a different concept of measurability. In particular, if $Y=\mathbb{C}$ the theories are identical, and if $Y=\mathbb{C}_{n\times n}$, the Bochner theory is just the theory of $n\times n$ matrix-valued functions all of whose entries are Lebesgue integrable. For general Y, most results of the Lebesgue theory still hold. For example: (i) any continuous function is Bochner integrable; (ii) an indefinite Bochner integral is absolutely continuous and differentiable almost everywhere; (iii) the obvious generalization of Lebesgue's dominated convergence theorem is valid. Essentially the only surprise in the theory is that an absolutely continuous function need not be an indefinite integral [KY2]. Henceforth, we shall use properties of Bochner integrals without further explanation.

If X is a Banach space we may take in particular $Y=\mathfrak{B}(X)$ and obtain the class $L^1(a,b;\mathfrak{B}(X))$ of $\mathfrak{B}(X)$-valued Bochner-integrable functions on $[a,b]$. Note that by Proposition 1.1 each such function F is almost everywhere the limit in the *norm* topology on $\mathfrak{B}(X)$ of a sequence of $\mathfrak{B}(X)$-valued step-functions, and that the integral of F is the limit in the norm of step-function integrals. A little thought will convince the reader that the proof of existence of the product integral for such functions is immediate. Recall that in Chapter 1 what was needed to prove existence was nothing more than a sequence of step-functions converging in the L^1 sense to the given function. For Bochner-integrable functions such a sequence exists by definition. Thus no difficulties are encountered, and the product integral is obtained as a norm limit of the functions E_{A_n} used in Theorem 1.1.1. The properties of the product integral parallel those cited for Lebesgue integrable functions in Chapter 1. We shall not dwell on them because they are special cases of the more general results to be developed.

3.2 Generalization

It is fair to ask why one should generalize the theory of product integration described in the last section. To understand this, we recall the situation for $\mathbb{C}_{n\times n}$-valued functions: for continuous $A:[a,b]\to\mathbb{C}_{n\times n}$, the product integral gives the solution of the initial-value problem

$$P'(x)=A(x)P(x) \qquad P(a)=I. \tag{2.1}$$

ISBN 0-201-13509-4

For integrable $A:[a,b]\to\mathbb{C}_{n\times n}$, the product integral gives the solution of the integral equation

$$P(x)=I+\int_a^x A(s)P(s)\,ds. \tag{2.2}$$

Now (and for the rest of this chapter) let X denote a fixed Banach space. Suppose that $A:[a,b]\to\mathfrak{B}(X)$ is a function. We can generalize the initial-value problem (2.1) in various ways: if we assume A is *norm*-continuous and the derivative in (2.1) is in the sense of the *norm* topology, then the product integral as defined above for norm-continuous functions solves (2.1). If we assume A is *strongly* continuous and the derivative in (2.1) is a *strong* derivative, then the theory of the last section is no longer adequate. Even the results on Bochner-integrable functions are of no use, because a strongly continuous function is not in general Bochner integrable. The situation is even worse if we pass to the weak topology. A similar situation pertains with respect to the integral equation (2.2). If A is Bochner integrable and the integral in (2.2) is interpreted as the Bochner integral of $AP\in L^1(a,b;\mathfrak{B}(X))$, then the theory of the last section is adequate. If one passes to weaker hypotheses on A, it is not. Now it is quite commonplace to study equations like (2.1) under the assumption that A is strongly continuous, and (as we shall see) Bochner integrability of A is not necessary in order for (2.2) to have a solution. This indicates the desirability of a more general theory of the product integral.

In this chapter we shall focus attention on the solution of the integral equation (2.2). This will allow us to make the most general assumptions on the function A, and results on the initial value problem (2.1) can be obtained by specializing the results on (2.2). We now discuss the assumptions to be made on the function A.

As we have stated, a strongly continuous function $A:[a,b]\to\mathfrak{B}(X)$ need not be Bochner integrable. The problem in this case is not concerned with unboundedness: If A is strongly continuous on $[a,b]$, then for each $\varphi\in X$ the function $\|A(x)\varphi\|$ is continuous and hence bounded on $[a,b]$. By the uniform boundedness principle the norm $\|A(x)\|$ is thus bounded on $[a,b]$. The difficulty concerning Bochner integrability is that A may not be strongly measurable. However, there is a weaker condition that *is* satisfied by strongly continuous functions A. It is the following:

Condition I. For each $\varphi\in X$, the X-valued function $A(x)\varphi$ is strongly measurable; that is, there is a sequence $\{\varphi_n\}$ of X-valued step-functions such that $\lim_{n\to\infty}\|\varphi_n(x)-A(x)\varphi\|=0$ for almost every $x\in[a,b]$. (The exceptional set can depend on φ.)

Condition I will be one of the two fundamental conditions imposed on the functions A which we shall product integrate. The second condition is

ISBN 0-201-13509-4

intuitively intended to express the notion that the total impact of the "driving term" A in Eqs. (2.1) and (2.2) should be finite. We have seen in Chapter 1 that the size of a product integral can be estimated in terms of $\int_a^b \|A(s)\|\,ds$, and it is natural to think that our second condition should require finiteness of this integral. However, a small technical point arises: under Condition I, it is not necessarily true (unless X is separable) that the norm $\|A(s)\|$ is a measurable function on $[a,b]$. Thus the integral $\int_a^b \|A(s)\|\,ds$ need not have a sense. This difficulty can be overcome by passing from the integral to the *upper* integral, whose definition we recall:

DEFINITION 2.1. *Let f be a nonnegative function on the interval $[a,b]$. The* upper integral *of f is defined by the prescription*

$$\overline{\int_a^b} f(s)\,ds = \inf_g \int_a^b g(s)\,ds, \tag{2.3}$$

where the infimum is taken over all Lebesgue-integrable functions g such that $g(s) \geqslant f(s)$ on $[a,b]$. (If there are no such Lebesgue-integrable functions g, the infimum is by definition ∞.)

Of course the upper integral agrees with the usual integral when both are defined. The virtue of the upper integral is that it does not require measurability of f. The following property of upper integrals [KB] is fundamental: if f has a finite upper integral on $[a,b]$, then there is a Lebesgue-integrable function h such that $h(s) \geqslant f(s)$ on $[a,b]$ and for $a \leqslant y \leqslant x \leqslant b$, we have

$$\overline{\int_y^x} f(s)\,ds = \int_y^x h(s)\,ds. \tag{2.4}$$

From this property one can deduce many useful facts about upper integrals. For instance, the upper integral is an absolutely continuous set function (with respect to Lebesgue measure). It is of course not linear in f but satisfies

$$\overline{\int_a^x} \{f_1(s) + f_2(s)\}\,ds \leqslant \overline{\int_a^x} f_1(s)\,ds + \overline{\int_a^x} f_2(s)\,ds. \tag{2.5}$$

This follows directly from the definition. In the rest of this chapter, all properties of upper integrals to be used are either immediate consequences of Definition 2.1 or else follow directly from the fundamental property (2.4). We can now formulate

Condition II. The function $A : [a,b] \to \mathfrak{B}(X)$ satisfies

$$\overline{\int_a^b} \|A(s)\|\,ds < \infty. \tag{2.6}$$

ISBN 0-201-13509-4

In the next section we develop the theory of functions satisfying Conditions I and II.

3.3 The Space $L_s^1(a,b;\mathfrak{B}(X))$

DEFINITION 3.1. *Let* $A:[a,b]\to\mathfrak{B}(X)$ *be a function. We say that* A *is* strongly integrable *on* $[a,b]$ *and write* $A\in L_s^1(a,b;\mathfrak{B}(X))$ *if and only if* A *satisfies conditions I and II above, that is:*

(*I*) *For each* $\varphi\in X$, *the function* $A(x)\varphi$ *is strongly measurable and*

(*II*) $\|A(x)\|$ *has finite upper integral on* $[a,b]$.

In practice, we shall frequently write $L_s^1(a,b)$ instead of $L_s^1(a,b;\mathfrak{B}(X))$. We shall primarily be interested in integrals of functions $A(x)\varphi$, where $A\in L_s^1(a,b)$. We therefore introduce the natural notion of equivalence: two functions $A,B\in L_s^1(a,b)$ are *equivalent* if and only if $A(x)\varphi$ agrees with $B(x)\varphi$ almost everywhere for each $\varphi\in X$. (The exceptional set can depend on φ.) In the theory of product integration to be developed, equivalent functions in $L_s^1(a,b)$ will play virtually identical roles. Nevertheless, we do not identify equivalent functions, since if A and B are such functions the upper integrals of their norms may not be the same, and these upper integrals will figure in our theory. Thus the elements of $L_s^1(a,b)$ are *functions*, not equivalence classes. It is easy to see that $L_s^1(a,b)$ is a linear space, because the sum of strongly measurable functions is strongly measurable and

$$\bar{\int_a^b}\|A(s)+B(s)\|\,ds\leqslant\bar{\int_a^b}\{\|A(s)\|+\|B(s)\|\}\,ds$$
$$\leqslant\bar{\int_a^b}\|A(s)\|\,ds+\bar{\int_a^b}\|B(s)\|\,ds. \tag{3.1}$$

If $A\in L_s^1(a,b)$, we define the number $\|A\|_1$ by

$$\|A\|_1=\bar{\int_a^b}\|A(s)\|\,ds. \tag{3.2}$$

We now remark that if $A\in L_s^1(a,b)$ and $\varphi\in X$, then $A(s)\varphi$ is Bochner integrable over $[a,b]:A(s)\varphi$ is strongly measurable by hypothesis, so we need only verify the second condition of Proposition 1.1. But this is trivial, since

$$\int_a^b\|A(s)\varphi\|\,ds\leqslant\bar{\int_a^b}\|A(s)\|\,\|\varphi\|\,ds=\|A\|_1\|\varphi\|. \tag{3.3}$$

We note that by (3.3) the condition $\|A\|_1=0$ implies that $\|A(s)\varphi\|=0$ for a.e. $s\in[a,b]$ and any $\varphi\in X$, so that A is equivalent to the zero function. However, the reverse is not true in general: A can be equivalent to zero

ISBN 0-201-13509-4

and yet not satisfy $\|A\|_1=0$.

DEFINITION 3.2. *If $A\in L_s^1(a,b)$ then $\int_a^b A(s)\,ds$ is defined to be the strong integral of A over $[a,b]$. That is, $\int_a^b A(s)\,ds$ is that unique linear operator J on X such that*

$$J\varphi=\int_a^b A(s)\varphi\,ds \qquad (\varphi\in X), \tag{3.4}$$

where the integral in (3.4) is a Bochner integral. We note that if A and B are equivalent functions, then $\int_a^b A(s)\,ds=\int_a^b B(s)\,ds$.

It follows immediately from Definition 3.2 and the inequality (3.3) that

$$\left\|\int_a^b A(s)\,ds\right\|\leqslant \bar{\int_a^b}\|A(s)\|\,ds=\|A\|_1. \tag{3.5}$$

Our first theorem lists some types of functions belonging to $L_s^1(a,b)$.

THEOREM 3.1. *Let $A:[a,b]\to\mathfrak{B}(X)$ be a function. Then*
(i) If A is Bochner integrable on $[a,b]$, then $A\in L_s^1(a,b)$. In particular, any norm-continuous function and any step-function belong to $L_s^1(a,b)$.
(ii) If A is strongly continuous, then $A\in L_s^1(a,b)$.
(iii) If A is weakly continuous and X is separable, then $A\in L_s^1(a,b)$.

Proof. In part (i), if $\{A_n\}$ is a sequence of step-functions such that $\|A_n(s)-A(s)\|$ approaches 0 for a.e. s as $n\to\infty$, then for any $\varphi\in X$ the norm $\|A_n(s)\varphi-A(s)\varphi\|$ satisfies the same condition. Since $A_n(s)\varphi$ is a step-function, Condition I of Definition 3.1 holds. Condition II is automatic by Proposition 1.1 on Bochner integrals. Thus (i) holds. Note that in this case the strong and Bochner integrals of A agree, and $\|A\|_1$ is the Bochner norm $\|A\|_B$ of A. To prove part (ii) note that in this case, for each $\varphi\in X$ the X-valued function $A(s)\varphi$ is (norm-) continuous and hence strongly measurable. In fact, if $P=\{s_0,s_1,\ldots,s_n\}$ is a partition of $[a,b]$ and A_P is the step-function taking the value $A(s_k)$ on the kth subinterval of P, then $A_P(s)\varphi$ converges *uniformly* to $A(s)\varphi$ when $\mu(P)\to 0$. Also, as remarked earlier, $\|A(s)\|$ is in this case bounded on $[a,b]$ and thus has a finite upper integral. Thus (ii) holds. The proof of (iii) is a bit more technical: a theorem of Pettis [KY3] guarantees that in the separable case, weak continuity implies strong measurability. Also, a double application of the uniform boundedness principle shows that $\|A(s)\|$ is bounded on $[a,b]$. Thus (iii) holds. ■

ISBN 0-201-13509-4

We shall need a convenient concept of convergence in $L_s^1(a,b)$, which is provided by

DEFINITION 3.3. *Let* $\{A_n\}$ *be a sequence in* $L_s^1(a,b)$ *and let* $A \in L_s^1(a,b)$. *We shall say that* $\{A_n\}$ converges to A in the L_s^1 sense *and write* $A_n \xrightarrow{L_s^1} A$ *if and only if the following two conditions hold*:

(*i*) *For each* $\varphi \in X$ *we have*

$$\lim_{n\to\infty} \int_a^b \|A_n(s)\varphi - A(s)\varphi\| \, ds = 0. \tag{3.6}$$

(*That is,* $A_n\varphi$ *converges to* $A\varphi$ *in* $L^1(a,b;X)$.)

(*ii*) *There is an* $M \geqslant 0$ *such that*

$$\|A_n\|_1 = \int_a^b \|A_n(s)\| \, ds \leqslant M, \qquad n = 1, 2, \dots. \tag{3.7}$$

We note that if $A_n \xrightarrow{L_s^1} A$ and $\varphi \in X$, then by condition (i) we have

$$\lim_{n\to\infty} \int_a^b A_n(s)\varphi \, ds = \int_a^b A(s)\varphi \, ds. \tag{3.8}$$

Equation (3.8) states that $\int_a^b A_n(s)\,ds$ converges to $\int_a^b A(s)\,ds$ in the strong operator topology.

As in Chapter 1, the theory of product integration to be presented here will begin with step-functions. We shall need to know that functions in $L_s^1(a,b)$ can be obtained as L_s^1-limits of sequences of step-functions. For this purpose we first remark that if A is strongly integrable on $[a,b]$, then directly from Definition 3.1 it follows that A is strongly integrable on any subinterval $[c,d]$ of $[a,b]$ so that $\int_c^d A(s)\,ds$ is defined. We now make

DEFINITION 3.4. *Let* $A \in L_s^1(a,b)$ *and let* $P = \{s_0, s_1, \dots, s_n\}$ *be a partition of* $[a,b]$. *Write* $\Delta s_k = s_k - s_{k-1}$. *Then*

(*i*) The mean-value approximant $\bar{A}_P$ *associated to* A *and* P *is the step-function whose value* $\bar{A}_{P_k}$ *on* $(s_{k-1}, s_k]$ *is the mean value of* A *on that interval. That is*

$$\bar{A}_{P_k} = \frac{1}{\Delta s_k} \int_{s_{k-1}}^{s_k} A(s) \, ds. \tag{3.9}$$

(*ii*) *The* point-value approximant A_P *associated to* A *and* P *is the step-function whose value on* $(s_{k-1}, s_k]$ *is* $A(s_k)$.

We shall mainly be concerned with mean-value approximants. It should be remarked that *equivalent functions in* $L_s^1(a,b)$ *have equal mean-value*

ISBN 0-201-13509-4

approximants corresponding to any given partition P. We also note that

$$\|\bar{A}_P\|_1 = \int_a^b \|\bar{A}_P(s)\|\,ds = \sum_{k=1}^{n} \|\bar{A}_{Pk}\|\Delta s_k$$

$$= \sum_{k=1}^{n} \left\| \int_{s_{k-1}}^{s_k} A(s)\,ds \right\| \leqslant \sum_{k=1}^{n} \int_{s_{k-1}}^{s_k} \|A(s)\|\,ds = \|A\|_1. \tag{3.10}$$

The same reasoning shows that if $\varphi \in X$, then

$$\int_a^b \|\bar{A}_P(s)\varphi\|\,ds \leqslant \int_a^b \|A(s)\varphi\|\,ds. \tag{3.11}$$

Clearly the maps $A \to \bar{A}_P$ and $A \to A_P$ are linear on $L_s^1(a,b)$. Our basic approximation theorem is

THEOREM 3.2. *Let $A \in L_s^1(a,b)$ and let $\{P_n\}$ be a sequence of partitions of $[a,b]$ with mesh $\mu(P_n)$ tending to zero. Then*

(*i*) $\bar{A}_{P_n} \xrightarrow{L_s^1} A$ *and*

(*ii*) *if A is strongly continuous, then* $A_{P_n} \xrightarrow{L_s^1} A$.

Proof. Part (i). Because of (3.10) the functions $\bar{A}_{P_n}$ satisfy the second condition for L_s^1 convergence. It remains to verify that

$$\lim_{n\to\infty} \int_a^b \|\bar{A}_{P_n}(s)\varphi - A(s)\varphi\|\,ds = 0 \qquad (\varphi \in X). \tag{3.12}$$

To prove (3.12) it is enough to show that if $f \in L^1(a,b;X)$ and $\bar{f}_P$ is the mean-value approximant associated with f and P (defined as for $\bar{A}_P$), then

$$\lim_{n\to\infty} \int_a^b \|\bar{f}_{P_n}(s) - f(s)\|\,ds \equiv \lim_{n\to\infty} \|\bar{f}_{P_n} - f\|_B = 0. \tag{3.13}$$

Now if f is continuous it is quite simple to verify (3.13) using elementary estimates. Furthermore, ([KB]), the continuous functions are dense in $L^1(a,b;X)$. If f is a general element of $L^1(a,b;X)$ and g is continuous, we have

$$\|\bar{f}_{P_n} - f\|_B \leqslant \|\bar{f}_{P_n} - \bar{g}_{P_n}\|_B + \|\bar{g}_{P_n} - g\|_B + \|g - f\|_B. \tag{3.14}$$

Now $\|\bar{f}_{P_n} - \bar{g}_{P_n}\|_B \leqslant \|f-g\|_B$ [the reasoning is as in (3.11)]. From (3.14) and the result for continuous functions, (3.13) follows for any $f \in L^1(a,b;X)$.

Part (ii). The proof of this part is entirely elementary, and we omit it.

Let $A \in L_s^1(a,b)$. Then as mentioned before, A is strongly integrable on any subinterval of $[a,b]$. Let $f:[a,b] \to X$ be a step-function taking the value

ISBN 0-201-13509-4

f_k on the kth subinterval (s_{k-1}, s_k) of its partition. Since A is strongly integrable on (s_{k-1}, s_k), the function $A(s)f_k$ is Bochner integrable on $[s_{k-1}, s_k]$, and since this holds for each k the function $A(s)f(s)$ is Bochner integrable on $[a,b]$. It is important to have this result also for continuous f. This and some properties of products of operator functions are proved in the next theorem. ■

THEOREM 3.3.

(i) If $A \in L_s^1(a,b)$ and $f:[a,b] \to X$ is continuous, then Af is Bochner integrable on $[a,b]$.

(ii) If $A \in L_s^1(a,b)$ and $B:[a,b] \to \mathfrak{B}(X)$ is strongly continuous, then AB and BA belong to $L_s^1(a,b)$.

Proof. Part (i). Since f is continuous, we can in the usual manner find a sequence $\{f_n\}$ of step-functions which converges to f uniformly on $[a,b]$. The functions Af_n are all Bochner integrable, and

$$\int_a^b \|A(s)f_m(s) - A(s)f_n(s)\| ds \leqslant \overline{\int_a^b} \|A(s)\| \, \|f_n(s) - f_m(s)\| ds$$

$$\leqslant \|A\|_1 \max_{s \in [a,b]} \|f_n(s) - f_m(s)\| \underset{m,n \to \infty}{\to} 0. \tag{3.15}$$

This shows that the sequence $\{Af_n\}$ is Cauchy in the space $L^1(a,b;X)$ of Bochner-integrable functions from $[a,b]$ to X. Since $L^1(a,b;X)$ is a Banach space, the sequence $\{Af_n\}$ has a limit $F \in L^1(a,b;X)$. As in the corresponding situation in the Lebesgue theory, there is a subsequence of $\{Af_n\}$ converging to F almost everywhere. But by construction $\{Af_n\}$ converges to Af everywhere on $[a,b]$. Thus

$$A(x)f(x) = F(x) \quad \text{a.e.}, \tag{3.16}$$

so Af is Bochner integrable. This proves part (i).

Part (ii): To show $AB \in L_s^1(a,b)$, note that if $\varphi \in X$, then $B(x)\varphi$ is continuous. By part (i), $A(x)B(x)\varphi$ is thus Bochner integrable and hence strongly measurable. Also $\|B(x)\|$ is bounded by some number M for $x \in [a,b]$, so

$$\overline{\int_a^b} \|A(s)B(s)\| ds \leqslant M \overline{\int_a^b} \|A(s)\| ds < \infty. \tag{3.17}$$

Thus $AB \in L_s^1(a,b)$. To show $BA \in L_s^1(a,b)$, note that the upper integral of $\|BA\|$ is finite by the argument just given. We need only show that $BA\varphi$ is strongly measurable. By hypothesis there are step-functions A_n such that $A_n\varphi$ tends to $A\varphi$ almost everywhere on $[a,b]$. Let $\{B_n\}$ be a sequence of

ISBN 0-201-13509-4

step-functions tending strongly to B everywhere on $[a,b]$. (For instance, a sequence of point-value approximants will do.) Then $B_nA_n\varphi$ is a sequence of step-functions tending to $BA\varphi$ almost everywhere on $[a,b]$, and we are done. ■

If $f:[a,b]\to X$ is a step-function and $A_n \xrightarrow{L_s^1} A$, then it follows simply from the definition of L_s^1 convergence that

$$\lim_{n\to\infty}\int_a^b \|A_n(s)f(s)-A(s)f(s)\|\,ds=0, \tag{3.18}$$

that is: A_nf converges to Af in $L^1(a,b;X)$. We will need also

THEOREM 3.4. *Suppose that* $f:[a,b]\to X$ *is continuous, and* $A_n \xrightarrow{L_s^1} A$. *Then* (3.18) *holds, so that*

$$\lim_{n\to\infty}\int_a^b A_n(s)f(s)\,ds=\int_a^b A(s)f(s)\,ds. \tag{3.19}$$

Proof. Let f_m be a sequence of step-functions converging uniformly to f on $[a,b]$. (For example, a sequence of point-value approximants.) Writing

$$\Delta_n(s)=A_n(s)-A(s), \tag{3.20}$$

we have

$$\begin{aligned}\int_a^b \|\Delta_n(s)f(s)\|\,ds &\leqslant \bar{\int_a^b} \|\Delta_n(s)\|\,\|f(s)-f_m(s)\|\,ds+\int_a^b\|\Delta_n(s)f_m(s)\|\,ds\\ &\leqslant \sup_{s\in[a,b]}\|f(s)-f_m(s)\|\,\sup_n \bar{\int_a^b}\|\Delta_n(s)\|\,ds+\int_a^b\|\Delta_n(s)f_m(s)\|\,ds.\end{aligned} \tag{3.21}$$

Because of (3.7) the supremum over n on the right-hand side of (3.21) is finite. Letting $\varepsilon>0$, we can thus choose a fixed m for which the first term on the right-hand side is less than $\varepsilon/2$, and for this m the second term can be made less than $\varepsilon/2$ by choosing n large enough. This completes the proof. ■

We now define a notion of differentiability appropriate to our purposes. The idea is to say that a function is differentiable if it is an indefinite integral. In the ordinary Lebesgue theory, the concept we are looking for would just be absolute continuity. In the context of Bochner integration, as

ISBN 0-201-13509-4

mentioned before, an absolutely continuous function need not be an indefinite integral. We accordingly make

DEFINITION 3.5.

(*i*) *Let* $A:[a,b]\to\mathfrak{B}(X)$ *be a function. We say that* A *is* L_s^1-differentiable *if and only if there is a function* $B\in L_s^1(a,b)$ *such that*

$$A(x)=A(c)+\int_c^x B(s)\,ds \qquad (x,c\in[a,b]). \tag{3.22}$$

In this case, B *is called an* L_s^1 derivative *of* A *and written* A'.

(*ii*) *Let* $f:[a,b]\to X$ *be a function. We say that* f *is* L^1-differentiable *if and only if there is a function* $g\in L^1(a,b;X)$ *such that*

$$f(x)=f(c)+\int_c^x g(s)\,ds \qquad (x,c\in[a,b]). \tag{3.23}$$

In this case, g *is called the* L^1-derivative *of* f *and written* f'.

We remark that if (3.22) or (3.23) holds for one $c\in[a,b]$, then it holds for all $c\in[a,b]$. We also note the following: if f is L^1-differentiable, then by (3.23) and the theory of Bochner integration, it follows that f is differentiable a.e. with derivative g. Thus the L^1-derivative g is uniquely determined a.e. Furthermore, if A is L_s^1-differentiable, then applying (3.22) to an element $\varphi\in X$ we find by the same argument that $B\varphi$ is uniquely determined a.e., and thus any two L_s^1-derivatives of A are equivalent in the sense explained earlier. The L_s^1-derivative is unique only to this extent. In the L_s^1 sense, the equation $A'=B$ means: B is one of the (equivalent) L_s^1-derivatives of A. For either sort of differentiability defined, a function is constant if and only if its derivative is equivalent to zero. (In the case of L^1-differentiability, the derivative would then be identified with zero.)

If A is L_s^1-differentiable, then for $x,y\in[a,b]$ with $y\leqslant x$ we have

$$\|A(x)-A(y)\|=\left\|\int_y^x A'(s)\,ds\right\|\leqslant\bar{\int_y^x}\|A'(s)\|\,ds. \tag{3.24}$$

This estimate together with the absolute continuity of upper integrals shows that A is norm-continuous (even norm-absolutely continuous). Similarly, if f is L^1-differentiable, then f is continuous.

To relate L_s^1-differentiability to a more familiar concept, suppose that $A:[a,b]\to\mathfrak{B}(X)$ is strongly differentiable with strongly continuous derivative A'. For each $\varphi\in X$ we then have, by the analogue for Bochner integration of the fundamental theorem of calculus,

$$A(x)\varphi=A(c)\varphi+\int_c^x A'(s)\varphi\,ds \qquad (x,c\in[a,b]). \tag{3.25}$$

ISBN 0-201-13509-4

As we know, $A' \in L_s^1(a,b)$, so that (3.25) proves that A is L_s^1-differentiable with L_s^1-derivative A'. If X is separable, the same result holds if A is weakly differentiable with weakly continuous derivative A'.

We shall have need of the product rule for differentiation:

THEOREM 3.5. *Let* $A,B:[a,b]\to\mathfrak{B}(X)$ *and* $f:[a,b]\to X$ *be functions. Then*:

(*i*) *If both A and B are* L_s^1*-differentiable, it follows that AB is* L_s^1*-differentiable and*

$$(AB)' = A'B + AB'. \tag{3.26}$$

(*ii*) *If A is* L_s^1*-differentiable and f is* L^1*-differentiable, it follows that Af is* L^1*-differentiable and*

$$(Af)' = A'f + Af'. \tag{3.27}$$

Proof. Since the two parts of the theorem are so similar, we prove only the first: Since A and B are L_s^1-differentiable, they are norm-continuous and A' and B' belong to $L_s^1(a,b)$. Theorem 3.2 then shows that $A'B + AB' \in L_s^1(a,b)$. Thus the strong integral

$$G(x) = \int_a^x (A'(s)B(s) + A(s)B'(s))\,ds \tag{3.28}$$

exists. The problem is to prove that

$$G(x) = A(x)B(x) - A(a)B(a). \tag{3.29}$$

To see this, let $\{A_n'\}$ and $\{B_n'\}$ be sequences of step-functions tending to A' and B' in the L_s^1 sense. Let

$$A_n(x) = A(a) + \int_a^x A_n'(s)\,ds,$$

$$B_n(x) = B(a) + \int_a^x B_n'(s)\,ds. \tag{3.30}$$

The functions A_n and B_n are piecewise linear, and elementary calculus shows that

$$A_n(x)B_n(x) - A(a)B(a) = \int_a^x (A_n'(s)B_n(s) + A_n(s)B_n'(s))\,ds. \tag{3.31}$$

It is also easy to see from (3.30) that as $n\to\infty$, A_n and B_n, respectively, tend strongly to A and B, uniformly on $[a,b]$. Equation (3.29) is obtained by letting n tend to infinity in (3.31) and making a few manipulations and estimates to show that the right-hand side of (3.31) tends strongly to that of (3.28). ■

ISBN 0-201-13509-4

3.4 Solution of Integral Equations

In this section we consider the equations

$$U(x,a)=I+\int_a^x A(s)U(s,a)\,ds \tag{4.1}$$

and

$$V(x,a)=I-\int_a^x V(s,a)A(s)\,ds, \tag{4.2}$$

where $A\in L_s^1(a,b)$. If A were a Lebesgue integrable $n\times n$ matrix-valued function, then by the work of Chapter 1, the solutions would be

$$U(x,a)=\prod_a^x e^{A(s)\,ds}, \qquad V(x,a)=U(a,x)=(U(x,a))^{-1}. \tag{4.3}$$

We shall prove the same formulas for $A\in L_s^1(a,b)$ after we have defined the product integral for such functions. In this section we shall establish that (4.1) and (4.2) have solutions when $A\in L_s^1(a,b)$, using the method of successive iteration to obtain a time-ordered exponential series. Before doing so, we discuss the precise meaning of equations of the above type. To avoid clutter, we shall discuss only (4.1). The discussion applies to (4.2) after obvious modifications.

The integral on the right-hand side of (4.1) is to be understood in the strong sense. That is, for any $\varphi\in X$ the function $A(s)U(s,a)\varphi$ is to be Bochner integrable and (4.1) is to hold when applied to φ. The equation

$$U(x,a)\varphi=\varphi+\int_a^x A(s)U(s,a)\varphi\,ds \tag{4.4}$$

then shows that $U(x,a)\varphi$ is the indefinite integral of a Bochner-integrable function and is hence continuous. Thus $U(x,a)$ is a strongly continuous operator-valued function. Since $A\in L_s^1(a,b)$ it follows by Theorem 3.3 that $A(s)U(s,a)$ is in $L_s^1(a,b)$. Definition 3.5 then shows that $U(x,a)$ is L_s^1-differentiable with L_s^1 derivative $A(s)U(s,a)$. [Note in particular that $U(x,a)$ must therefore be norm continuous.] Thus if U satisfies (4.1), then it satisfies the initial value problem

$$\begin{aligned} &U'(x,a)=A(x)U(x,a) \qquad (L_s^1 \text{ sense}),\\ &U(a,a)=I. \end{aligned} \tag{4.5}$$

On the other hand, according to Definition 3.5, the conditions (4.5) imply (4.1). Thus for $A\in L_s^1(a,b)$, the integral equation (4.1) is equivalent to the initial value problem (4.5) *with the derivative understood in the L_s^1 sense.*

ISBN 0-201-13509-4

When we discuss product integration of functions in $L_s^1(a,b)$, we will often make use of this equivalence of integral equations with L_s^1-differential equations.

THEOREM 4.1. *Let $A \in L_s^1(a,b)$. Then eqs.* (4.1) *and* (4.2) *are solved by the time-ordered exponential series obtained by successive iteration*:

$$U(x,a) = I + \int_a^x A(s_1)\,ds_1 + \int_a^x A(s_1)\left\{\int_a^{s_1} A(s_2)\,ds_2\right\}ds_1 + \cdots \tag{4.6}$$

and

$$V(x,a) = I - \int_a^x A(s_1)\,ds_1 + \int_a^x \left\{\int_a^{s_1} A(s_2)\,ds_2\right\}A(s_1)\,ds_1 - \cdots. \tag{4.7}$$

Remark. Uniqueness of the solutions is not being asserted here. This will be proved in the next theorem.

Proof. We prove only the result for $U(x,a)$, since the proof for $V(x,a)$ is entirely analogous. Let

$$U_0(x,a) = I, \tag{4.8}$$

and for $n \geqslant 1$ let

$$U_n(x,a) = \int_a^x A(s)\,U_{n-1}(s,a)\,ds \qquad \text{(strong integral)}. \tag{4.9}$$

To see that $U_n(x,a)$ is well defined, we note that $U_0(x,a)$ is well defined and norm-continuous. Assuming for induction that $U_{n-1}(x,a)$ is well defined and norm-continuous, Theorem 3.3 implies that $A(x)U_{n-1}(x,a)$ belongs to $L_s^1(a,b)$. Thus the integral on the right-hand side of (4.9) makes sense and $U_n(x,a)$ is well defined and L_s^1-differentiable, hence norm-continuous. Thus by induction, $U_n(x,a)$ is defined and norm-continuous for $n = 0, 1, 2, \ldots$. Let

$$M(x) = \bar{\int_a^x} \|A(s)\|\,ds. \tag{4.10}$$

Then

$$\|U_1(x,a)\| = \left\|\int_a^x A(s)\,ds\right\| \leqslant \frac{M(x)}{1!}. \tag{4.11}$$

Assume for induction that

$$\|U_n(x,a)\| \leqslant \frac{M(x)^n}{n!} \tag{4.12}$$

ISBN 0-201-13509-4

for some $n \geqslant 1$. We then have

$$\|U_{n+1}(x,a)\| = \left\| \int_a^x A(s)U_n(s,a)\,ds \right\| \leqslant \bar{\int}_a^x \|A(s)\| \frac{M(s)^n}{n!}\,ds. \quad (4.13)$$

We now prove that the right-hand side of (4.13) is bounded by $M(x)^{n+1}/(n+1)!$: let g be an integrable function such that

$$g(x) \geqslant \|A(x)\| \qquad (x \in [a,b]). \quad (4.14)$$

Let

$$M_g(x) = \int_a^x g(s)\,ds. \quad (4.15)$$

Then

$$M_g(x) \geqslant M(x) \qquad (x \in [a,b]). \quad (4.16)$$

So

$$\bar{\int}_a^x \|A(s)\| \frac{M(x)^n}{n!}\,ds \leqslant \int_a^x g(s) \frac{M_g(s)^n}{n!}\,ds = \int_a^x d\left(\frac{M_g(s)^{n+1}}{(n+1)!} \right)$$
$$= \frac{M_g(x)^{n+1}}{(n+1)!}. \quad (4.17)$$

Now $M(x)$ is by definition the infimum of $M_g(x)$ over all integrable g satisfying (4.14). Taking the infimum of (4.17) over all such g, we obtain (4.17) with M_g replaced by M on the right-hand side. Thus by (4.13) we have

$$\|U_{n+1}(x,a)\| \leqslant \frac{M(x)^{n+1}}{(n+1)!}. \quad (4.18)$$

Hence by induction (4.12) holds for all $n \geqslant 1$, and clearly it holds also for $n=0$. Now by the definition (4.10) of M, we have

$$M(x) \leqslant \bar{\int}_a^b \|A(s)\|\,ds \equiv \|A\|_1. \quad (4.19)$$

Thus

$$\|U_n(x,a)\| \leqslant \frac{\|A\|_1^n}{n!}, \qquad x \in [a,b], n = 0,1,2,\ldots. \quad (4.20)$$

ISBN 0-201-13509-4

This estimate shows that the time-ordered exponential series

$$U(x,a) \equiv \sum_{n=0}^{\infty} U_n(x,a) \tag{4.21}$$

converges in the norm topology, uniformly on $[a,b]$. Since each term on the right in (4.21) is norm continuous and the limit is uniform, $U(x,a)$ is norm continuous. Thus by Theorem 3.3, $A(x)U(x,a)$ belongs to $L_s^1(a,b)$ and the integral on the right of (4.1) is well defined. Now let

$$S_n(x,a) = \sum_{k=0}^{n} U_k(x,a). \tag{4.22}$$

Then by the definition of $U_n(x,a)$ we have for $n \geqslant 0$

$$S_{n+1}(x,a) = I + \int_a^x A(s) S_n(s,a)\, ds. \tag{4.23}$$

Equation (4.1) is now obtained by applying (4.23) to an element φ of X, letting n tend to infinity and using uniform convergence of S_n to U. ■

We now establish an important property of the solutions of (4.1) and (4.2):

THEOREM 4.2. *Let $A \in L_s^1(a,b)$. Then the solutions of* (4.1) *and* (4.2) *are unique and are inverses of one another.*

Proof. Let $U(x,a)$ and $V(x,a)$ denote any solutions of (4.1) and (4.2), respectively. We will show that

$$V(x,a)U(x,a) = U(x,a)V(x,a) = I. \tag{4.24}$$

This will establish all assertions of the theorem. To prove (4.24) note that first, since both V and U are L_s^1-differentiable, the function VU is L_s^1-differentiable and the product rule gives

$$(VU)'(x) = V(x,a)(A(x) - A(x))U(x,a) = 0. \tag{4.25}$$

Thus VU is a constant, and since $V(a,a)U(a,a) = I$ the constant is I. Now let W be defined by

$$W(x) = U(x,a)V(x,a) - I. \tag{4.26}$$

The product rule gives

$$\begin{aligned} W'(x) &= A(x)U(x,a)V(x,a) - U(x,a)V(x,a)A(x) \\ &= A(x)W(x) - W(x)A(x). \end{aligned} \tag{4.27}$$

ISBN 0-201-13509-4

Also $W(a)=0$, so

$$W(x)=\int_a^x \{A(s)W(s)-W(s)A(s)\}\,ds. \tag{4.28}$$

From (4.28) we obtain

$$\|W(x)\| \leqslant 2\bar{\int_a^x} \|A(s)\|\,\|W(s)\|\,ds \leqslant \left(2\bar{\int_a^x}\|A(s)\|\,ds\right)\sup_{s\in[a,x]}\|W(s)\|. \tag{4.29}$$

Now let $d>a$. For $x\leqslant d$ we obtain from (4.29)

$$\|W(x)\| \leqslant \left(2\bar{\int_a^d}\|A(s)\|\,ds\right)\sup_{s\in[a,d]}\|W(s)\| \qquad (a\leqslant x\leqslant d). \tag{4.30}$$

Let $M_d=\sup_{s\in[a,d]}\|W(s)\|$. Taking the sup of (4.30) over $x\in[a,d]$ we find

$$M_d \leqslant \left(2\bar{\int_a^d}\|A(s)\|\,ds\right)M_d. \tag{4.31}$$

By the absolute continuity of upper integrals, there is an $\varepsilon>0$ such that if $a\leqslant y\leqslant z\leqslant b$ and $z-y\leqslant\varepsilon$, then $\bar{\int_y^z}\|A(s)\|\,ds<1/2$. Choosing $d=a+\varepsilon$ in (4.31) we find $M_d=0$ so that $W(x)=0$ in $[a,a+\varepsilon]$. Thus a can be replaced by $a+\varepsilon$ in (4.28). Continuing the argument in the obvious way, we obtain $W(x)=0$ on $[a,b]$. Thus (4.24) is proved.

3.5 Product Integration of Functions in $L_s^1(a,b)$

Our discussion of product integration for functions in $L_s^1(a,b)$ follows closely the discussion of Chapter 1. At points where definitions or proofs are identical with those of Chapter 1 we usually refer the reader to this chapter.

If $B:[a,b]\to\mathfrak{B}(X)$ is a step-function, then $E_B:[a,b]\to\mathfrak{B}(X)$ is defined precisely as in Definition 1.1.3. We repeat this fundamental definition for easy reference:

DEFINITION 5.1. *Let $B:[a,b]\to\mathfrak{B}(X)$ be a step function, associated with the partition $P=\{s_0,s_1,\ldots,s_n\}$ of $[a,b]$. Writing B_k for the value of B on (s_{k-1},s_k), and setting $\Delta s_k=s_k-s_{k-1}$, we define $E_B:[a,b]\to\mathfrak{B}(X)$ as follows:*

$$\begin{aligned} E_B(x) &= e^{B_1(x-s_0)} && (x\in[s_0,s_1])\\ &= e^{B_2(x-s_1)}e^{B_1\Delta s_1} && (x\in[s_1,s_2])\\ &\ \ \vdots \\ &= e^{B_n(x-s_{n-1})}\cdots e^{B_2\Delta s_2}e^{B_1\Delta s_1} && (x\in[s_{n-1},s_n]). \end{aligned}$$

ISBN 0-201-13509-4

As in Chapter 1 (Sec. 8), $E_B(x)$ will turn out to be the product integral of the step-function B over the interval $[a,x]$.

LEMMA 5.1. *Let* $B:[a,b]\rightarrow\mathfrak{B}(X)$ *be a step-function. Then*
(*i*) $E_B(a)=I$.
(*ii*) $E_B(x)$ *is nonsingular for each* $x\in[a,b]$.
(*iii*) E_B *and* E_B^{-1} *are* L_s^1-*differentiable, and*

$$E_B'(x)=B(x)E_B(x), \qquad (E_B^{-1})'(x)=-E_B^{-1}(x)B(x). \tag{5.1}$$

(*iv*) E_B *and* E_B^{-1} *satisfy the bounds*

$$\|E_B(x)\|\leqslant e^{\int_a^x\|B(s)\|ds} \qquad \|E_B^{-1}(x)\|\leqslant e^{\int_a^x\|B(s)\|ds}. \tag{5.2}$$

Proof. We discuss only part (iii), since the other parts are proved precisely as in Lemma 1.1.1. To prove (iii), according to Definition 3.5 one must verify that the right-hand sides of (5.1) are in $L_s^1(a,b)$ and that

$$E_B(x)=I+\int_a^x B(s)E_B(s)\,ds \tag{5.3}$$

and

$$E_B^{-1}(x)=I-\int_a^x E_B^{-1}(s)B(s)\,ds. \tag{5.4}$$

We discuss only (5.3). $E_B(x)$ is norm-continuous in x by inspection and B is a step-function and hence belongs to $L_s^1(a,b)$. By Theorem 3.3, $BE_B\in L_s^1(a,b)$. The discussion is now completed as in Lemma 1.1.1, using the fact that, except at the division points of the partition associated with B, the function E_B is norm-differentiable with norm-continuous derivative BE_B. This completes our discussion. ∎

We are now in a position to prove existence of the product integral for functions $A\in L_s^1(a,b)$. Our proof will use the fact that we already know (Theorem 4.1) that the integral equation (4.1) has a solution $U(x,a)$ and that this solution is norm-continuous.

We will see that the solution of (4.1) coincides with the product integral of A. It might be asked why we should bother to construct a solution of (4.1) by the product-integral method when we have already solved (4.1) using the time-ordered exponential. The reason is this: there is nontrivial information in the statement that the solution of (4.1) can be constructed via product integration. This becomes most obvious when one passes from the study of bounded operators ($A(s)\in\mathfrak{B}(X)$ for $s\in[a,b]$) to the study of unbounded operators in (4.1). In this case the time-ordered exponential series (4.6) becomes most intractable—its study is plagued by questions

ISBN 0-201-13509-4

about domains. The product integral approach, however, can yield fruitful results in this case, as many authors have found [TK1, TK3, GW1] and as we shall see in the next section.

We recall that if $A \in L_s^1(a,b)$, then there is at least one sequence of step-functions converging to A in the L_s^1 sense on $[a,b]$ (Theorem 3.2). We now prove

THEOREM 5.1. *Let* $A \in L_s^1(a,b)$. *Let* $\{A_n\}$ *be any sequence of step-functions such that* $A_n \xrightarrow{L_s^1} A$. *Then the functions* $E_{A_n}(x)$ *converge strongly, uniformly on* $[a,b]$, *to a function which will be called* the (strong) product integral of A over $[a,x]$, *denoted* $\prod_a^x e^{A(s)ds}$. *This function coincides with the unique solution of Eq.* (4.1).

Proof. Let $U(x,a)$ be the unique solution of (4.1). As we have seen, $U(x,a)$ is L_s^1-differentiable. By Lemma 4.1, $E_{A_n}^{-1}(x)$ is also L_s^1-differentiable. By the product rule for L_s^1-differentiation, $E_{A_n}^{-1}(x)U(x,a)$ is also L_s^1-differentiable, and taking account of (4.5) and (5.1) we find

$$\left(E_{A_n}^{-1}U\right)'(x) = E_{A_n}^{-1}(x)(A(x)-A_n(x))U(x,a) \qquad \left(L_s^1 \text{ sense}\right). \quad (5.5)$$

By the meaning of L_s^1 differentiation, (5.5) is equivalent to (note $(E_{A_n}^{-1}U)(a)=I$)

$$E_{A_n}^{-1}(x)U(x,a) = I + \int_a^x E_{A_n}^{-1}(s)(A(s)-A_n(s))U(s,a)\,ds. \quad (5.6)$$

Multiplying (5.6) on the left by $E_{A_n}(x)$, we find

$$U(x,a) - E_{A_n}(x) = E_{A_n}(x)\int_a^x E_{A_n}^{-1}(s)(A(s)-A_n(s))U(s,a)\,ds. \quad (5.7)$$

We now apply (5.7) to an element $\varphi \in X$ and estimate the norm of the integral by the integral of the norm. This yields

$$\|(U(x,a)-E_{A_n}(x))\varphi\| \leqslant \|E_{A_n}(x)\| \int_a^x \|E_{A_n}^{-1}(s)\|\,\|(A(s)-A_n(s))U(s,a)\varphi\|\,ds. \quad (5.8)$$

The estimate (5.2) on $\|E_B(x)\|$ and $\|E_B^{-1}(x)\|$ show that each of these norms is bounded by $e^{\|B\|_1}$, where $\|B\|_1$ is defined as in (3.2) [since B is a step-function, the upper integral of (3.2) is of course an ordinary integral]. Inserting these estimates in (5.8) we have

$$\begin{aligned}\|(U(x,a)-E_{A_n}(x))\varphi\| &\leqslant e^{2\|A_n\|_1}\int_a^x \|(A(s)-A_n(s))U(s,a)\varphi\|\,ds \\ &\leqslant e^{2\|A_n\|_1}\int_a^b \|(A(s)-A_n(s))U(s,a)\varphi\|\,ds. \quad (5.9)\end{aligned}$$

ISBN 0-201-13509-4

Note that the estimate on the right-hand side of (5.9) is independent of x. Also: (i) the sequence $\{\|A_n\|_1\}$ is bounded in n by the definition of L_s^1 convergence and (ii) the function $U(s,a)\varphi$ is continuous, so by Theorem 3.4 the integral of the right-hand side of (5.9) approaches zero as $n\to\infty$. Thus the theorem is proved. ■

Remark. If A and B are *equivalent* L_s^1 functions, then as previously remarked they have identical mean-value approximants, and it follows that they have identical product integrals.

COROLLARY 1. *If $A:[a,b]\to\mathfrak{B}(x)$ is a step-function, then*

$$\prod_a^x e^{A(s)\,ds}=E_A(x). \tag{5.10}$$

Proof. The sequence $A,A,A,\ldots$ converges to A in the L_s^1 sense, and this establishes the result. ■

COROLLARY 2. *If $A:[a,b]\to\mathfrak{B}(x)$ is strongly continuous, then*

$$\prod_a^x e^{A(s)\,ds}=s-\lim_{\mu(P)\to 0}\prod_{k=1}^n e^{A(s_k)\Delta s_k}, \tag{5.11}$$

where P is a partition of $[a,x]$ and notation is as in Chapter 1 [*see* (1.0.8)].

Proof. By Theorem 3.2, the point-value approximant A_P converges to A in the L_s^1 sense on $[a,x]$. Theorem 1 (applied to the interval $[a,x]$) shows that $E_{A_P}(x)$ converges strongly to $\prod_a^x e^{A(s)\,ds}$ when $\mu(P)\to 0$. But $E_{A_P}(x)$ has precisely the form of the product on the right-hand side of (5.11). This completes the proof. ■

COROLLARY 3. *Suppose that $A\in L_s^1(a,b)$ and suppose that A is* commutative. *That is, suppose there is a set of measure zero $T\subset[a,b]$ such that*

$$A(s)A(s')=A(s')A(s) \qquad \textit{for } s,s'\in[a,b]\setminus T. \tag{5.12}$$

Then

$$\prod_a^x e^{A(s)\,ds}=e^{\int_a^x A(s)\,ds}. \tag{5.13}$$

Proof. We choose a sequence $\{A_n\}$ of mean-value approximants which approaches A in the L_s^1 sense on $[a,x]$. Using (5.12) and Definition 3.4, it is easy to see that the finitely many values taken by A_n commute with each other. Using the definition of E_{A_n} and the special form of A_n, one then

ISBN 0-201-13509-4

finds that in fact

$$E_{A_n}(x) = e^{\int_a^x A(s)\,ds}, \qquad n = 1,2,3,\ldots. \tag{5.14}$$

Since $\{E_{A_n}(x)\}$ converges to $\prod_a^x e^{A(s)ds}$, the corollary is proved. ■

The analogues of several theorems of Chapter 1 are contained in Theorem 5.1. In addition to establishing existence of the product integral, Theorem 1 shows that if $P(x,a) = \prod_a^x e^{A(s)ds}$, then

(i) $P(x,a)$ is a solution of the integral equation (4.1) [or equivalently a solution of the initial value problem (4.5)]

(ii) $P(x,a)$ is equal to the time-ordered exponential series of Theorem 4.1 [Eq. (4.21)].

In Chapter 1 we were able to prove nonsingularity of product integrals using the concepts of determinant and trace. These concepts are not available in the present context. However, we know that $\prod_a^x e^{A(s)ds}$ is the solution $U(x,a)$ of (4.1), and Theorem 4.2 shows that $\prod_a^x e^{A(s)ds}$ is nonsingular, with inverse $V(x,a)$ satisfying (4.2). The development now proceeds as in Chapter 1: we set

$$\prod_a^a e^{A(s)ds} = I \tag{5.15}$$

and

$$\prod_x^a e^{A(s)ds} = \left(\prod_a^x e^{A(s)ds}\right)^{-1} = V(x,a) \qquad (x \in [a,b]). \tag{5.16}$$

Since A is strongly integrable over $[a,b]$, it is strongly integrable over any subinterval $[y,x]$ of $[a,b]$. Thus $\prod_y^x e^{A(s)ds}$ is defined. Extending the definition as in (5.16) to the case $y > x$, we can say that $\prod_y^x e^{A(s)ds}$ is defined for any $x,y \in [a,b]$.

THEOREM 5.2. *Let $A \in L_s^1(a,b)$ and let $x,y,z \in [a,b]$. Then*

$$\prod_y^z e^{A(s)ds} \prod_x^y e^{A(s)ds} = \prod_x^z e^{A(s)ds}. \tag{5.17}$$

Proof. If $x \leqslant y \leqslant z$, take a sequence $\{A_n\}$ of mean-value approximants converging to A in the L_s^1 sense on $[x,z]$, such that each function A_n has y as one of the division points of its partition. Then (5.17) is immediate from the fact that the limit of the product of two strongly convergent sequences is the product of their limits. For other relations between x, y, and z, Eq. (5.17) follows from the definition of product integration from a larger to a smaller value of the argument. ■

ISBN 0-201-13509-4

The next theorem is a minor but useful generalization of our information on the connection between integral equations and product integrals.

THEOREM 5.3. *Let $A \in L_s^1(a,b)$. For $x,y \in [a,b]$ define*

$$F(x,y) = \prod_y^x e^{A(s)\,ds}. \tag{5.18}$$

Then $F(x,y)$ is the unique solution of either of the integral equations

$$F(x,y) = I + \int_y^x A(s)F(s,y)\,ds \tag{5.19}$$

or

$$F(x,y) = I + \int_y^x F(x,s)A(s)\,ds. \tag{5.20}$$

[*Remark: In* (5.19) *we consider x as variable and y as fixed. In* (5.20) *we do the reverse.*]

Proof. We first discuss (5.19). We note that if $y = a$ in (5.19), then the desired conclusion is known from Theorem I. For general $y \in [a,b]$, we write

$$F(x,y) = F(x,a)F(a,y). \tag{5.21}$$

We know that $F(x,a)$ satisfies (5.19) with $y = a$. Inserting this equation in (5.21) we find

$$\begin{aligned} F(x,y) &= \left\{ I + \int_a^x A(s)F(s,a)\,ds \right\} F(a,y) \\ &= \left\{ I + \int_a^y A(s)F(s,a)\,ds + \int_y^x A(s)F(s,a)\,ds \right\} F(a,y) \\ &= \left\{ F(y,a) + \int_y^x A(s)F(s,a)\,ds \right\} F(a,y) \\ &= I + \int_y^x A(s)F(s,y)\,ds, \end{aligned} \tag{5.22}$$

where we have used $F(y,a)F(a,y) = I$ and $F(s,a)F(a,y) = F(s,y)$. Thus $F(x,y)$ satisfies (5.19). Similarly, if $x = a$ in (5.18), then $F(a,y) = F(y,a)^{-1} = V(y,a)$. Equation (5.20) is then immediate from (4.2). The result for general x follows as above. To prove uniqueness of the solution of (5.19), note that if $G(x,y)$ is any solution of (5.19), then G is L_s^1-differentiable with respect to x. Using (5.20) we find that the function $F(y,x)G(x,y)$ is L_s^1-differentiable with respect to x and has derivative zero. Since $F(y,x)$

ISBN 0-201-13509-4

and $G(x,y)$ reduce to the identity for $x=y$, we thus find $F(y,x)G(x,y)=I$. Multiplying on the left by $F(x,y)$ we obtain $G(x,y)=F(x,y)$. Uniqueness for (5.20) is proved similarly. This proves the theorem. ■

COROLLARY 1. *Under the hypotheses of Theorem* 5.3 *the function* $F(x,y)$ *satisfies*

$$\frac{\partial}{\partial x}F(x,y)=A(x)F(x,y) \qquad (L_s^1 \text{ sense}) \tag{5.23}$$

and

$$\frac{\partial}{\partial y}F(x,y)=-F(x,y)A(y) \qquad (L_s^1 \text{ sense}). \tag{5.24}$$

It follows that $F(x,y)$ *is norm-continuous in either variable.*

Proof. This is an immediate consequence of (5.19), (5.20), and our earlier remarks on the equivalence of integral equations and L_s^1-differential equations. ■

COROLLARY 2. *If* $A:[a,b]\to\mathfrak{B}(X)$ *is strongly continuous, then Eqs.* (5.23) *and* (5.24) *hold everywhere on* $[a,b]$, *the derivatives being interpreted as strong derivatives. (At a and b the derivatives are one-sided.)*

Proof. Since A is strongly continuous we have $A\in L_s^1(a,b)$. By Corollary 1, $F(x,y)$ is norm-continuous in either variable. Thus the integrands in (5.19) and (5.20) are strongly continuous and the assertion of Corollary 2 follows immediately from the "fundamental theorem of calculus" for Bochner integrals. ■

COROLLARY 3. *Let* $A\in L_s^1(a,b)$ *and let* $U_0, V_0\in\mathfrak{B}(X)$. *Let* $c\in[a,b]$. *Then the equations*

$$H(x)=U_0+\int_c^x A(s)H(s)\,ds \tag{5.25}$$

and

$$K(x)=V_0+\int_c^x K(s)A(s)\,ds \tag{5.26}$$

have the unique solutions

$$H(x)=\prod_c^x e^{A(s)\,ds}U_0 \tag{5.27}$$

ISBN 0-201-13509-4

and

$$K(x) = V_0 \prod_x^c e^{-A(s)\,ds}. \tag{5.28}$$

Proof. It is immediate from Theorem 5.3 that H and K of (5.27) and (5.28) satisfy the desired equations. Uniqueness is proved as at the end of Theorem 5.3, using the invertibility of product integrals. ■

We now prove the fundamental theorem of product integration (compare Theorem 1.3.1).

THEOREM 5.4. *Suppose that $B:[a,b] \to \mathfrak{B}(X)$ has the following properties:*
(i) *B is L_s^1-differentiable and*
(ii) *B is nonsingular [that is, $B(x)$ is nonsingular for each $x \in [a,b]$].*
Let $LB:[a,b] \to \mathfrak{B}(X)$ be defined by

$$(LB)(x) = B'(x)B^{-1}(x). \tag{5.29}$$

Then for $x,y \in [a,b]$ we have

$$\prod_y^x e^{(LB)(s)\,ds} = B(x)B^{-1}(y). \tag{5.30}$$

Proof. Since B is L_s^1 – differentiable it is norm-continuous. Thus B^{-1} is norm-continuous, and by Theorem 3.3 it follows that $LB \in L_s^1(a,b)$, so that the product integral in (5.30) is defined. Now denote the right-hand side of (5.30) by $H(x,y)$. This function is L_s^1-differentiable in x, and $H(y,y) = I$, so that

$$\begin{aligned} H(x,y) &= I + \int_y^x \frac{\partial H}{\partial s}(s,y)\,ds \\ &= I + \int_y^x B'(s)B^{-1}(y)\,ds = I + \int_y^x (LB)(s)H(s,y)\,ds. \end{aligned} \tag{5.31}$$

Theorem 5.3 now shows that $H(x,y)$ is the product integral of LB over $[x,y]$. This completes the proof. ■

COROLLARY. *Under the hypotheses of Theorem 5.4, $B^{-1}(x)$ is L_s^1-differentiable with derivative $-B^{-1}(x)B'(x)B^{-1}(x)$.*

Proof. To show $B^{-1}(y)$ is L_s^1-differentiable, multiply (5.30) on the left by $B^{-1}(x)$ and use Corollary 1 of Theorem 5.3 to differentiate with respect to y in the L_s^1 sense. The result claimed follows immediately. ■

ISBN 0-201-13509-4

In Chapter 1, several theorems were proved by appealing to Theorem 1.2.1 on the uniqueness of the solution of an initial value problem. In the present chapter several results follow in an analogous way from Theorem 5.3 and its corollaries. This is illustrated by the proof of the last theorem, which should be compared with the proof of the corresponding Theorem 1.3.1. The next two theorems are proved using Theorem 5.3, and we only sketch their proofs.

THEOREM 5.5 (The sum rule). *Let $A, B \in L_s^1(a,b)$ and let $x,y \in [a,b]$. Define*

$$F(x,y) = \prod_y^x e^{A(s)\,ds}. \tag{5.32}$$

Then

$$\prod_y^x e^{\{A(s)+B(s)\}\,ds} = F(x,y) \prod_y^x e^{F(y,s)B(s)F(s,y)\,ds}. \tag{5.33}$$

Proof. Because $F(x,y)$ is norm-continuous in both variables, Theorem 3.3 shows that $F(y,s)B(s)F(s,y)$ belongs to $L_s^1(a,b)$. We now remark that the two sides of (5.33) equal I when $x=y$ and satisfy the same L_s^1-differential equation in the variable x and hence the same integral equation. By Theorem 5.3 the proof is complete. ■

THEOREM 5.6 (The similarity rule). *Let $A \in L_s^1(a,b)$ and let $T:[a,b] \to \mathcal{B}(X)$ be a function which is L_s^1-differentiable and such that $T(x)$ is nonsingular for each $x \in [a,b]$. Let $x,y \in [a,b]$. Then*

$$T(x) \prod_y^x e^{A(s)\,ds} T^{-1}(y) = \prod_y^x e^{\{(LT)(s)+T(s)A(s)T^{-1}(s)\}\,ds}. \tag{5.34}$$

Proof. The idea of the proof is the same as for the last theorem, so the proof will be omitted. We note the alternative version of (5.34) (compare 1.3.22):

$$T^{-1}(x) \prod_y^x e^{A(s)\,ds} T(y) = \prod_y^x e^{\{T^{-1}(s)A(s)T(s) - T^{-1}(s)T'(s)\}\,ds}. \tag{5.35}$$

We can obtain the change of variable formula (1.3.54) for strong product integrals under the additional assumption that the function ϕ is monotone. [The assumption of monotonicity resolves a technical point which arises in the proof that $A(\phi(s))\phi'(s)$ is in L_s^1.] We shall not give the details of the proof but we remark that taking $\phi(s) = -s$ yields

$$\prod_y^x e^{A(s)\,ds} = \prod_{-y}^{-x} e^{-A(-s)\,ds}. \tag{5.36}$$

■

ISBN 0-201-13509-4

We now prove the validity of bounds similar to those of Theorem 1.4.1:

THEOREM 5.7. *Let $A\in L_s^1(a,b)$ and let $x,y\in[a,b]$. Then*

$$\left\| \prod_y^x e^{A(s)\,ds} \right\| \leqslant e^{\bar{\int}_{\min(x,y)}^{\max(x,y)} \|A(s)\|\,ds} \tag{5.37}$$

and

$$\left\| \prod_y^x e^{A(s)\,ds} - I \right\| \leqslant e^{\bar{\int}_{\min(x,y)}^{\max(x,y)} \|A(s)\|\,ds} - 1. \tag{5.38}$$

Proof. For convenience of notation, we will prove the result for the case $y=a, x=b$. [The results for any x,y with $y\leqslant x$ follows immediately. The results for $y>x$ can then be deduced from the case $y<x$ via formula (5.36).] Let $\{A_n\}$ be a sequence of mean-value approximants converging to A in the L_s^1 sense on $[a,b]$. Using Eq. (3.10) we have

$$\int_a^b \|A_n(s)\|\,ds \leqslant \bar{\int}_a^b \|A(s)\|\,ds. \tag{5.39}$$

The estimate (5.2) yields

$$\|E_{A_n}(b)\| \leqslant e^{\int_a^b \|A_n(s)\|\,ds}. \tag{5.40}$$

Since $\prod_a^b e^{A(s)\,ds}$ is the strong limit of $E_{A_n}(b)$, the last two inequalities imply (5.37) with $y=a$ and $x=b$. To prove (5.38) with $y=a$ and $x=b$, we insert the bound (5.2) for $\|E_{A_n}(x)\|$ in the integral equation (5.3) for $E_{A_n}(x)$ to find

$$\begin{aligned} \|E_{A_n}(b) - I\| &\leqslant \int_a^b \|A_n(s)\| e^{\int_a^s \|A_n(u)\|\,du}\,ds \\ &= e^{\int_a^b \|A_n(s)\|\,ds} - 1 \leqslant e^{\bar{\int}_a^b \|A(s)\|\,ds} - 1. \end{aligned} \tag{5.41}$$

The inequality (5.38) with $y=a$ and $x=b$ now follows immediately. ■

We state the following theorem without proof because the idea of the proof is precisely the same as for Theorem 1.5.1.

THEOREM 5.8 (Duhamel's formula). *Suppose that $A,B\in L_s^1(a,b)$. For $x,y\in[a,b]$ let*

$$F(x,y) = \prod_y^x e^{A(s)\,ds} \tag{5.42}$$

ISBN 0-201-13509-4

and

$$G(x,y)=\prod_{y}^{x} e^{B(s)\,ds}. \tag{5.43}$$

Then

$$F(x,a)-G(x,a)=\int_{a}^{x} G(x,s)(A(s)-B(s))F(s,a)\,ds. \tag{5.44}$$

The results above establish the framework of the theory of product integration for functions $A \in L_s^1(a,b)$. These are the results that will be used in the next section. It is possible to prove in the present context analogues of most of the other theorems of Chapter 1, but we shall not do so. We refer the interested reader to [JD3] for results on improper strong product integrals.

3.6 Product Integrals Involving Unbounded Operators

Heretofore the functions A under consideration have taken values in the set $\mathfrak{B}(X)$ of *bounded* operators on the Banach space X. In this section we consider functions A whose values are *unbounded* operators. The generalization to unbounded operators offers rather formidable difficulties, but this generalization is essential because in applications the operators that arise are often unbounded. The premier example of this is the theory of partial differential equations, where unbounded operators arise throughout in a natural way. To cite just one example, the unbounded Laplacean operator occurs in many celebrated partial differential equations of mathematical physics (Laplace's equation, the heat equation, the wave equation, the nonrelativistic Schrödinger equation, etc.). An application of our results to a problem in partial differential equations is given in Section 5 of Chapter 4.

Material involving unbounded operators is of course technical in nature, and to define all terms occurring in the present section would lead us too far afield. For definitions of the terms to be used, we refer the reader to any standard text on functional analysis, for example [KY3].

We wish to exploit the product integral to solve the differential equation

$$\frac{dU(x)}{dx}=A(x)U(x). \tag{6.1}$$

In order to do this, we shall choose operators $A(x)$ such that the operator $e^{A(s_k)\Delta s_k}$ is defined and bounded for $\Delta s_k > 0$ [although $A(s_k)$ itself may be unbounded]. This will allow us to define Riemann products involving A, and these products will be tractable because they involve only bounded

ISBN 0-201-13509-4

operators. In order that $e^{A(s_k)\Delta s_k}$ should be defined and bounded for $\Delta s_k > 0$, it suffices that $A(s_k)$ should be the infinitesimal generator of a contraction semigroup on the Banach space X. It is possible to proceed under this assumption [JD3], but for ease of presentation we shall pass to a simpler situation here. Namely we assume that X is a Hilbert space $\mathcal{H}$ and that for each x, the operator $A(x)$ is skew-adjoint. We may thus write

$$A(x) = iB(x), \tag{6.2}$$

where $B(x)$ is self-adjoint. In order to exploit the theory developed in Section 5, we shall need to approximate our possibly unbounded operators by bounded ones. We begin with some remarks on the type of approximation to be used.

Let B be a self-adjoint operator on $\mathcal{H}$ with domain D_B. We can write

$$B = \int_{-\infty}^{\infty} \mu \, dE_\mu, \tag{6.3}$$

where E_μ is the resolution of the identity for B. Let $\lambda > 0$ and define

$$B_\lambda = \lambda^2 B(\lambda^2 + B^2)^{-1} = \int_{-\infty}^{\infty} \frac{\mu\lambda^2}{\lambda^2 + \mu^2} \, dE_\mu. \tag{6.4}$$

Since

$$\left| \frac{\mu\lambda}{\lambda^2 + \mu^2} \right| \leqslant \frac{1}{2},$$

we have that B_λ is bounded and

$$\| B_\lambda \| \leqslant \lambda/2. \tag{6.5}$$

Since

$$\frac{\lambda^2}{\lambda^2 + \mu^2} \leqslant 1,$$

we have

$$\| B_\lambda \varphi \| \leqslant \| B\varphi \| \qquad (\varphi \in D_B). \tag{6.6}$$

Clearly B_λ is self-adjoint. We also have for $\varphi \in D_B$

$$\lim_{\lambda \to \infty} B_\lambda \varphi = B\varphi, \tag{6.7}$$

ISBN 0-201-13509-4

because

$$\|B_\lambda\varphi - B\varphi\|^2 = \int_{-\infty}^{\infty} \mu^2 \left| 1 - \frac{\lambda^2}{\lambda^2+\mu^2} \right|^2 d\|E_\mu\varphi\|^2 \underset{\lambda\to\infty}{\to} 0 \tag{6.8}$$

where convergence to zero is a consequence of Lebesgue's dominated convergence theorem. Similarly, we find that

$$s-\lim_{\lambda\to\infty} e^{iB_\lambda s} = e^{iBs}, \tag{6.9}$$

uniformly for s in any bounded subset of $[0,\infty)$. We now prove some lemmas that will be needed later:

LEMMA 6.1. *Suppose B is a step-function on $[a,b]$ associated with the partition $P=\{s_0,s_1,\ldots,s_n\}$. Suppose that for each $k=1,\ldots n$, the value B_k taken by B on the interval (s_{k-1},s_k) is a (possibly unbounded) self-adjoint operator. Let B_λ be the function obtained by replacing B_k with the operator $B_{k\lambda}$ defined as in* (6.4). *Let $E_{iB}(x)$ and $E_{iB_\lambda}(x)$ be defined as in Definition* 5.1. *Then*

$$s-\lim_{\lambda\to\infty} E_{iB_\lambda}(x) = E_{iB}(x), \tag{6.10}$$

uniformly for $x\in[a,b]$.

Proof. Uniform convergence for $x\in[s_0,s_1]$ is immediate from (6.9), which holds uniformly for s in a bounded set. For $x\in[s_1,s_2]$ and $\varphi\in\mathcal{H}$ we have

$$\begin{aligned} E_{iB_\lambda}(x)\varphi &= e^{iB_{2\lambda}(x-s_1)}e^{iB_{1\lambda}\Delta s_1}\varphi \\ &= e^{iB_{2\lambda}(x-s_1)}e^{iB_1\Delta s_1}\varphi \\ &\quad + e^{iB_{2\lambda}(x-s_1)}\left(e^{iB_{1\lambda}\Delta s_1} - e^{iB_1\Delta s_1}\right)\varphi. \end{aligned} \tag{6.11}$$

The first term on the right-hand side of (6.11) converges uniformly to $E_{iB}(x)\varphi$ because of (6.9). The second term converges uniformly to zero by (6.9) and the fact that $\|e^{iB_{2\lambda}(x-s_1)}\| \leqslant 1$. Similarly, we prove uniform convergence on every other interval $[s_{k-1},s_k]$ and hence on $[a,b]$. ■

LEMMA 6.2. *Suppose that $B\in L_s^1(a,b)$ (recall that $B(x)$ is then bounded) and that $B(x)$ is self-adjoint for a.e. $x\in[a,b]$. Then for $x,y\in[a,b]$ we have that $\prod_y^x e^{iB(s)\,ds}$ is unitary.*

Proof. It is enough to give the proof for $y\leqslant x$. The proof for $y>x$ then follows because by (5.36) we have

$$\prod_y^x e^{iB(s)\,ds} = \prod_{-y}^{-x} e^{-iB(-s)\,ds}. \tag{6.12}$$

Furthermore, the lemma for general $y \leqslant x$ follows immediately from the lemma with $y=a$. Thus let $y=a$ and let $\{B_n\}$ be a sequence of mean-value approximants converging to B in the L_s^1 sense on $[a,b]$. Clearly $\{iB_n\}$ converges to iB in the L_s^1 sense, so by Theorem 5.1, $E_{iB_n}(x)$ converges strongly to $\prod_a^x e^{iB(s)ds}$.

We will prove that $E_{iB_n}(x)$ is unitary for each n. Then as the strong limit of unitary operators, $\prod_a^x e^{iB(s)ds}$ must be an isometry. Since $\prod_a^x e^{iB(s)ds}$ is also known to be nonsingular, it must then be unitary, and the lemma will be proved. To see that $E_{iB_n}(x)$ is unitary, let V_k be one of the values assumed by B_n. Then

$$V_k = \frac{1}{\Delta s_k} \int_{s_{k-1}}^{s_k} B(s)\,ds. \tag{6.13}$$

If $\varphi,\psi \in \mathcal{H}$, then

$$\begin{aligned}(\varphi, V_k\psi) &= \frac{1}{\Delta s_k} \int_{s_{k-1}}^{s_k} (\varphi, B(s)\psi)\,ds \\ &= \frac{1}{s_k} \int_{s_{k-1}}^{s_k} (B(s)\varphi,\psi)\,ds = (V_k\varphi,\psi),\end{aligned} \tag{6.14}$$

where we have used the fact that $B(s)$ is self-adjoint for a.e. s. Thus V_k is self-adjoint and $e^{iV_k t}$ is unitary for $t \in \mathbb{R}$. Since $E_{iB_n}(x)$ is a product of operators of the form $e^{iV_k t}$, $E_{iB_n}(x)$ is unitary and the lemma is proved. ■

Lemma 6.3. *Suppose that $B, C \in L_s^1(a,b)$ and that $B(x)$ is self-adjoint for a.e. $x \in [a,b]$. Then for $x,y \in [a,b]$ we have*

$$\left\| \prod_y^x e^{\{iB(s)+C(s)\}\,ds} \right\| \leqslant e^{\|C\|_1}. \tag{6.15}$$

Proof. Fix $y \in [a,b]$ and let

$$P(x) = \prod_y^x e^{iB(s)\,ds}. \tag{6.16}$$

Then

$$\prod_y^x e^{\{iB(s)+C(s)\}\,ds} = P(x) \prod_y^x e^{P^{-1}(s)C(s)P(s)\,ds}.$$

Using Lemma 6.2 and (5.37) of Theorem 5.7, we have

$$\left\| \prod_y^x e^{\{iB(s)+C(s)\}\,ds} \right\| \leqslant \|P(x)\| e^{\int_{\min(x,y)}^{\max(x,y)} \|P^{-1}(s)C(s)P(s)\|\,ds} \leqslant e^{\|C\|_1}, \tag{6.17}$$

where the last inequality is obtained using the unitarity of $P(x)$. This proves the lemma. ■

ISBN 0-201-13509-4

We can now prove the main theorem of this section.

THEOREM 6.1. *Let B be a function defined on $[a,b]$ such that*
(i) for each $s\in[a,b]$, $B(s)$ is self-adjoint;
(ii) the function $(i+B(s))^{-1}$ is L_s^1-differentiable on $[a,b]$ with derivative C;
(iii) The function $F(s)=(i+B(s))C(s)$ belongs to $L_s^1(a,b)$.
Then the operators $B(s)$ have a common domain D, and if $c\in[a,b]$ there is a unique operator valued function $U(x,c)$ defined for $x\in[a,b]$, such that $U(c,c)=I$, $U(x,c)$ is unitary for each $x\in[a,b]$, and

$$\frac{d}{dx}U(x,c)\varphi=iB(x)U(x,c)\varphi \tag{6.18}$$

for each $\varphi\in D$. [Remark: The derivative in (6.18) is the ordinary derivative, in the norm topology, of an $\mathcal{H}$-valued function. At the points a and b, the derivatives are of course one-sided.] $U(x,c)$ will be called the product integral *of iB over $[c,x]$, written as usual $\prod_c^x e^{iB(s)ds}$.*

Proof.
Step 1: All the operators $B(s)$ have the same domain D and $\|(i+B(x))(i+B(y))^{-1}\|$ is uniformly bounded as x and y range through $[a,b]$:

The domain of $B(s)$ is the range of $(i+B(s))^{-1}$, so we must show that all these ranges are the same. Let $x,y\in[a,b]$ and let

$$G(x,y)=\prod_y^x e^{-F(s)ds}, \tag{6.19}$$

where F is defined in (iii). By Theorem 5.7 we have

$$\|G(x,y)\|\leqslant e^{\bar{\int}_{\min(x,y)}^{\max(x,y)}\|F(s)\|ds}\leqslant e^{\bar{\int}_a^b\|F(s)\|ds}=e^{\|F\|_1}. \tag{6.20}$$

Also G is L_s^1-differentiable with respect to x. Let

$$H(x,y)=(i+B(x))^{-1}G(x,y). \tag{6.21}$$

Then H is L_s^1-differentiable with respect to x, and the product rule shows that its derivative is zero. Thus

$$H(x,y)=H(y,y)=(i+B(y))^{-1} \tag{6.22}$$

or

$$(i+B(x))^{-1}G(x,y)=(i+B(y))^{-1}. \tag{6.23}$$

Since $G(x,y)$ is nonsingular, Eq. (6.23) shows that the range of $(i+B(x))^{-1}$

ISBN 0-201-13509-4

equals that of $(i+B(y))^{-1}$, so that $B(x)$ and $B(y)$ have a common domain D. Also (6.23) implies

$$(i+B(x))(i+B(y))^{-1}=G(x,y) \tag{6.24}$$

and the bound (6.20) shows that $\|(i+B(x))(i+B(y)^{-1}\|$ is uniformly bounded for $x,y\in[a,b]$.

Step 2: Construction of $U_\lambda(x,y)$ and $W_\lambda(x,y)$:

For $\lambda>0$, let $B_\lambda(s)$ be defined by

$$B_\lambda(s)=\lambda^2 B(s)(\lambda^2+B(s)^2)^{-1}. \tag{6.25}$$

We intend to define U_λ as the product integral of B_λ. We must first verify that $B_\lambda\in L_s^1(a,b)$. We will in fact prove more than this: B_λ is L_s^1-differentiable, hence norm-continuous. To see this, note that for any real $\lambda\neq 0$,

$$(i\lambda+B(s))(i+B(s))^{-1}=I+i(\lambda-1)(i+B(s))^{-1}. \tag{6.26}$$

Since $(i+B(s))^{-1}$ is L_s^1-differentiable, so is the left-hand side of (6.26), and thus (by the Corollary to Theorem 5.4) so is its inverse $(i+B(s))(i\lambda+B(s))^{-1}$. Now

$$(i+B(s))(i\lambda+B(s))^{-1}=I+i(1-\lambda)(i\lambda+B(s))^{-1}. \tag{6.27}$$

By (6.27), the function $(i\lambda+B(s))^{-1}$ is L_s^1-differentiable. Furthermore, (6.27) implies that

$$B(s)(i\lambda+B(s))^{-1}=I-i\lambda(i\lambda+B(s))^{-1}, \tag{6.28}$$

so we now see that $B(s)(i\lambda+B(s))^{-1}$ is L_s^1-differentiable. This holds for any real $\lambda\neq 0$. If $\lambda>0$, then

$$B_\lambda(s)=\lambda^2 B(s)(i\lambda+B(s))^{-1}(-i\lambda+B(s))^{-1}, \tag{6.29}$$

so that by the product rule B_λ is L_s^1-differentiable. Equations (6.26) through (6.28) allow the L_s^1-derivative of B_λ to be calculated, and we write this derivative down in order to remark on one of its properties that will be needed later:

$$\begin{aligned}B_\lambda'(s)=\lambda^2\big\{&-i\lambda(i\lambda+B(s))^{-1}F(s)(i+B(s))(\lambda^2+B(s)^2)^{-1}\\&+B(s)(\lambda^2+B(s)^2)^{-1}F(s)(i+B(s))(-i\lambda+B(s))^{-1}\big\},\end{aligned} \tag{6.30}$$

ISBN 0-201-13509-4

where $F(s)$ is defined in hypothesis (iii). The property on which we wish to remark is the following: if $s,s' \in [a,b]$, then

$$\|B_\lambda'(s)(i+B(s'))^{-1}\| \leqslant M\|F(s)\|, \tag{6.31}$$

where M is a number independent of s,s' and λ. To see this, we show how to deal with the first term on the right-hand side of (6.30). The treatment of the second is similar. We have

$$\begin{aligned}
&\|\lambda^3(i\lambda+B(s))^{-1}F(s)(i+B(s))(\lambda^2+B(s)^2)^{-1}(i+B(s'))^{-1}\| \\
&\leqslant \|\lambda(i\lambda+B(s))^{-1}\|\,\|F(s)\|\,\|\lambda^2(\lambda^2+B(s)^2)^{-1}\|\,\|(i+B(s))(i+B(s'))^{-1}\| \\
&\leqslant \|F(s)\|\,\|(i+B(s))(i+B(s'))^{-1}\| \leqslant M\|F(s)\|,
\end{aligned} \tag{6.32}$$

where to obtain the last inequality we have used the results of Step 1. This completes the argument.

For $x,y \in [a,b]$ we set

$$U_\lambda(x,y) = \prod_y^x e^{iB_\lambda(s)\,ds}. \tag{6.33}$$

Since B_λ is self-adjoint for each λ, the operator $U_\lambda(x,y)$ is unitary by Lemma 6.2. We define for $x,y \in [a,b]$

$$W_\lambda(x,y) = \prod_y^x e^{\{iB_\lambda(s) - F(s)\}\,ds}. \tag{6.34}$$

By Lemma 6.3, we have

$$\|W_\lambda(x,y)\| \leqslant e^{\|F\|_1}. \tag{6.35}$$

Consider the function

$$H_\lambda(x,y) \equiv (i+B(x))^{-1}W_\lambda(x,y). \tag{6.36}$$

We have, differentiating with respect to x,

$$\begin{aligned}
H_\lambda'(x,y) &= C(x)W_\lambda(x,y) + (i+B(x))^{-1}\{iB_\lambda(x) - F(x)\}W_\lambda(x,y) \\
&= (i+B(x))^{-1}iB_\lambda(x)W_\lambda(x,y) = iB_\lambda(x)H_\lambda(x,y),
\end{aligned} \tag{6.37}$$

where the last equality holds since $B_\lambda(x)$ and $(B(x)+i)^{-1}$ commute. Now (6.37) implies (by Corollary 3 to Theorem 5.3) that

$$\begin{aligned}
H_\lambda(x,y) &= U_\lambda(x,y)H_\lambda(y,y) \\
&= U_\lambda(x,y)(i+B(y))^{-1}.
\end{aligned} \tag{6.38}$$

ISBN 0-201-13509-4

Thus we have

$$(i+B(x))^{-1}W_\lambda(x,y)=U_\lambda(x,y)(i+B(y))^{-1}. \tag{6.39}$$

This equation shows that $U_\lambda(x,y)$ maps the domain D to itself, because if $\varphi\in D$, then $\varphi=(B(y)+i)^{-1}\psi$ with $\psi\in\mathcal{H}$, and (6.39) shows that $U_\lambda(x,y)\varphi \in D$.

Step 3: Strong convergence of $U_\lambda(x,y)$ as $\lambda\to\infty$: We first discuss convergence of $U_\lambda(x,a)$: Since $iB_\lambda(s)$ is norm-continuous, its product integral can be evaluated using ordinary "Riemann products" (Corollary 2 of Theorem 5.1). That is, let $\{P_n\}$ be a sequence of partitions of $[a,b]$ with mesh tending to zero. Let $\{B_{\lambda n}\}$ be the corresponding sequence of point-value approximants to B_λ. $\{B_{\lambda n}$ takes the value $B_\lambda(s_k)$ on a typical interval $(s_{k-1},s_k]$ of the partition P_n.$\}$ Then as $n\to\infty$, $E_{iB_{\lambda n}}(x)$ converges strongly to $U_\lambda(x,a)$, uniformly on $[a,b]$. For fixed n, Lemma 6.1 shows that $E_{iB_{\lambda n}}(x)$ also has a strong limit as $\lambda\to\infty$, uniformly on $[a,b]$. Let us write

$$U_{\lambda n}(x,y)=\prod_y^x e^{iB_{\lambda n}(s)\,ds}. \tag{6.40}$$

Then $U_{\lambda n}(x,a)$ is the same as $E_{iB_{\lambda n}}(x)$, by Corollary 1 of Theorem 5.1. By Lemma 6.2, $U_\lambda(x,y)$ and $U_{\lambda n}(x,y)$ are both unitary. Now let $\varphi\in X$. We have

$$\begin{aligned}\|U_\lambda(x,a)\varphi-U_{\lambda'}(x,a)\varphi\| &\leqslant \|U_\lambda(x,a)\varphi-U_{\lambda n}(x,a)\varphi\|\\ &\quad+\|U_{\lambda n}(x,a)\varphi-U_{\lambda' n}(x,a)\varphi\|+\|U_{\lambda' n}(x,a)\varphi-U_{\lambda'}(x,a)\varphi\|.\end{aligned} \tag{6.41}$$

For fixed n, the middle term on the right of (6.41) approaches 0 as $\lambda,\lambda'\to\infty$, uniformly for $x\in[a,b]$. Suppose we can prove that

(*) The quantity $\|U_\lambda(x,a)\varphi-U_{\lambda n}(x,a)\varphi\|$ approaches zero as $n\to\infty$, *at a rate independent of x and λ.*

Then from (6.36) it will follow that $U_\lambda(x,a)$ has the Cauchy property in λ, uniformly for $x\in[a,b]$. Thus $U_\lambda(x,a)$ will converge strongly as $\lambda\to\infty$, uniformly for $x\in[a,b]$. Now to prove (*) for all $\varphi\in X$, it suffices to prove (*) for a set of φ dense in X, because U_λ and $U_{\lambda n}$ are unitary. As our dense set we choose the common domain D of the operators $B(x)$. By Duhamel's formula (Theorem 5.8) we have

$$U_\lambda(x,a)-U_{\lambda n}(x,a)=i\int_a^x U_{\lambda n}(x,s)\{B_\lambda(s)-B_{\lambda n}(s)\}U_\lambda(s,a)\,ds. \tag{6.42}$$

ISBN 0-201-13509-4

We apply this equation to $\varphi \in D$ and use (6.39) to find

$$U_\lambda(x,a)\varphi - U_{\lambda n}(x,a)\varphi = i\int_a^x U_{\lambda n}(x,s)\{B_\lambda(s) - B_{\lambda n}(s)\}$$
$$\times (i+B(s))^{-1} W_\lambda(s,a)(i+B(a))\varphi\, ds. \tag{6.43}$$

We thus have

$$\|U_\lambda(x,a)\varphi - U_{\lambda n}(x,a)\varphi\| \leqslant \int_a^b \|U_{\lambda n}(x,s)\|\, \|\{B_\lambda(s) - B_{\lambda n}(s)\}(i+B(s))^{-1}\|$$
$$\times \|W_\lambda(s,a)\|\, \|(i+B(a))\varphi\| ds$$
$$\leqslant M' \int_a^b \|\{B_\lambda(s) - B_{\lambda n}(s)\}(i+B(s))^{-1}\| ds \tag{6.44}$$

where M' is a bound for $\|W_\lambda(s,a)\|\, \|(i+B(a))\varphi\|$ [see (6.35)]. Now suppose that the partition associated with $B_{\lambda n}$ is $P_n = \{s_0, s_1, \ldots, S_m\}$. Then the integral on the right-hand side of (6.44) is

$$\sum_{k=1}^{m} \int_{s_{k-1}}^{s_k} \|\{B_\lambda(s) - B_\lambda(s_k)\}(i+B(s))^{-1}\| ds \equiv \Delta_n. \tag{6.45}$$

Since B_λ is L_s^1-differentiable, we have

$$\{B_\lambda(s) - B_\lambda(s_k)\}(i+B(s))^{-1} = -\int_s^{s_k} B'_\lambda(t)(i+B(s))^{-1} dt. \tag{6.46}$$

Using the bound (6.31) we have

$$\|\{B_\lambda(s) - B_\lambda(s_k)\}(i+B(s))^{-1}\| \leqslant M \bar{\int_s^{s_k}} \|F(t)\| dt \leqslant M \bar{\int_{s_{k-1}}^{s_k}} \|F(t)\| dt. \tag{6.47}$$

Now because of the absolute continuity of upper integrals, given $\varepsilon > 0$ there is a $\delta > 0$ such that if $s_k - s_{k-1} < \delta$ then $\int_{s_{k-1}}^{s_k} \|F(s)\| ds < \varepsilon$. Thus if $\mu(P) < \delta$, we have

$$\Delta_n \leqslant \sum_{k=1}^{m} \int_{s_{k-1}}^{s_k} M\varepsilon\, ds = M\varepsilon(b-a). \tag{6.48}$$

This shows that assertion (*) holds, so that $U_\lambda(x,a)$ converges strongly as $\lambda \to \infty$, uniformly for $x \in [a,b]$.

The above proof, applied to the interval $[y,b]$, shows that $U_\lambda(x,y)$ converges strongly as $\lambda \to \infty$, uniformly for $x \in [y,b]$. We write

$$U(x,y) = s - \lim_{\lambda \to \infty} U_\lambda(x,y). \tag{6.49}$$

ISBN 0-201-13509-4

Now if $x<y$, then writing

$$U_\lambda(x,y)=\prod_y^x e^{iB_\lambda(s)\,ds}=\prod_{-y}^{-x} e^{-iB_\lambda(-s)\,ds}, \tag{6.50}$$

we can apply our proof to the interval $[-y,-a]$ to find that $U_\lambda(x,y)$ converges uniformly for $-x\in[-y,-a]$, that is for $x\in[a,y]$. Thus combining our results, we have that $U_\lambda(x,y)$ converges strongly as $\lambda\to\infty$, uniformly for $x\in[a,b]$, to an operator $U(x,y)$. Clearly $U(x,x)=I$. The relation

$$U(x,y)\,U(y,z)=U(x,z), \qquad x,y,z\in[a,b] \tag{6.51}$$

follows immediately from the corresponding property of U_λ by taking strong limits. In particular, taking $z=x$ in (6.51) we find that $U(x,y)$ is nonsingular. Since $U(x,y)$ is the strong limit of the unitary operators $U_\lambda(x,y)$, $U(x,y)$ is an isometry, and this together with nonsingularity implies unitarity of $U(x,y)$. We also note continuity properties of U: the function $U_\lambda(x,y)$ is norm-continuous in x and converges strongly to $U(x,y)$, uniformly for x in $[a,b]$. Thus $U(x,y)$ is strongly continuous in x on $[a,b]$. Now

$$U(x,y)=U(y,x)^* \tag{6.52}$$

and $U(y,x)$ is strongly continuous in y. Because $U(y,x)$ is unitary, $U(y,x)^*$ is also strongly continuous in y. Thus $U(x,y)$ is strongly continuous in each of its variables separately.

Step 4: Convergence of $W_\lambda(x,y)$: using the sum rule in formula (6.34), we find

$$W_\lambda(x,y)=U_\lambda(x,y)\prod_y^x e^{-U_\lambda(y,s)F(s)U_\lambda(s,y)\,ds}. \tag{6.53}$$

We claim that $W_\lambda(x,y)$ converges strongly as $\lambda\to\infty$, uniformly for $x\in[a,b]$, to the function

$$W(x,y)=U(x,y)\prod_y^x e^{-U(y,s)F(s)U(s,y)\,ds}. \tag{6.54}$$

We already know that $U_\lambda(x,y)$ converges to $U(x,y)$ in the sense desired, so we need only prove convergence of the product integral in (6.53) to that in (6.54). This is a completely elementary application of Duhamel's formula which we leave to the reader. Note that $W_\lambda(x,y)$ is norm-continuous in x, and from the uniform strong convergence we find that $W(x,y)$ is strongly continuous in x.

ISBN 0-201-13509-4

Step 5: If $\varphi \in D$ and $c \in [a,b]$, then $U(x,c)\varphi$ satisfies the differential equation (6.18): we begin from the integral equation satisfied by $U_\lambda(x,c)$:

$$U_\lambda(x,c) = I + \int_c^x iB_\lambda(s)\, U_\lambda(s,c)\, ds. \tag{6.55}$$

Applying this equation to an element $\varphi \in D$ and using (6.39), we have

$$U_\lambda(x,c)\varphi = \varphi + \int_c^x iB_\lambda(s)(i+B(s))^{-1} W_\lambda(s,c)(i+B(c))\varphi\, ds. \tag{6.56}$$

Now as $\lambda \to \infty$, $W_\lambda(s,c)$ approaches $W(s,c)$ strongly and $\lambda^2(\lambda^2 + B(s)^2)^{-1}$ converges strongly to I. Also by (6.35) and some obvious estimates, the integrand in (6.56) is bounded in norm by a fixed integrable function. Letting $\lambda \to \infty$ and applying Lebesgue's dominated convergence theorem (for Bochner integrals) to take the limit under the integral sign, we find

$$U(x,c)\varphi = \varphi + \int_y^x iB(s)(i+B(s))^{-1} W(s,c)(i+B(c))\varphi\, ds. \tag{6.57}$$

Now $W(s,c)$ is strongly continuous in s, and $B(s)(i+B(s))^{-1}$ is L_s^1-differentiable [see the discussion of (6.28)], so that $B(s)(i+B(s))^{-1}$ is norm-continuous in s. It follows that $U(x,c)\varphi$ is differentiable and

$$\frac{d}{dx} U(x,c)\varphi = iB(x)(i+B(x))^{-1} W(x,c)(i+B(c))\varphi. \tag{6.58}$$

The earlier relation (6.39) immediately implies

$$(i+B(x))^{-1} W(x,c) = U(x,c)(i+B(c))^{-1}. \tag{6.59}$$

Thus (6.58) becomes

$$\frac{d}{dx} U(x,c)\varphi = iB(x)\, U(x,c)\varphi \tag{6.60}$$

as desired.

Step 6: Uniqueness: suppose that $V(x,c)$ is a function satisfying $V(c,c) = I$ and (6.60) with U replaced by V. Let

$$T(x) = U(c,x)\, V(x,c). \tag{6.61}$$

Then $T(c) = I$. Also, if $\varphi, \psi \in D$, then

$$\frac{d}{dx}(\varphi, T(x)\psi) = \frac{d}{dx}(U(x,c)\phi, V(x,c)\psi) = 0. \tag{6.62}$$

ISBN 0-201-13509-4

Thus

$$(\varphi, T(x)\psi) = (\varphi, T(c)\psi) = (\varphi, \psi) \qquad (x \in [a,b]). \tag{6.63}$$

Since D is dense in $\mathcal{H}$, we have $T(x) = I$ for all x, so

$$V(x,c) = U(x,c), \tag{6.64}$$

as desired. This completes the proof. ■

COROLLARY 1. *In addition to* (*i*) *of Theorem* 6.1, *suppose that* $(i + B(s))^{-1}$ *is strongly differentiable with derivative C. Also suppose that C and* $F(s) = (i + B(s))C(s)$ *are strongly continuous. Then the conclusion of Theorem* 6.1 *holds.*

Proof. By the remarks after Definition 3.5, we know that a strongly differentiable function with strongly continuous derivative is L_s^1-differentiable. Also, any strongly continuous function on $[a,b]$ is in $L_s^1(a,b)$ (Theorem 3.1). Thus the hypotheses of Theorem 6.1 are implied by those of Corollary 1, and we are done. ■

COROLLARY 2. *In addition to* (*i*) *of Theorem* 6.1, *suppose that* $\mathcal{H}$ *is separable,* $(i + B(s))^{-1}$ *is weakly differentiable with derivative C, and that C and* $F(s) = (i + B(s))C(s)$ *are weakly continuous. Then the conclusion of Theorem* 6.1 *holds.*

Proof. By Theorem 3.1, C and F belong to $L_s^1(a,b)$. $(i + B(s))^{-1}$ is the weak indefinite integral of C. Since $C \in L_s^1(a,b)$ the strong integral of C exists (Definition 3.2) and must agree with the weak integral. Thus $(i + B(s))^{-1}$ is the strong indefinite integral of C, and $(i + B(s))^{-1}$ is therefore L_s^1-differentiable with derivative C. Thus the hypotheses of Theorem 6.1 are satisfied, and we are done. ■

It is sometimes easier to verify conditions on B than on $(i + B(s))^{-1}$. It is possible to show that the hypotheses of Corollary 1 are equivalent to the following: condition (i) of Theorem 6.1, together with (ii′) the operators $B(s)$ all have the same domain D; and (iii′) the function B is strongly C^1 on the domain D. Corollary 2 can be similarly reformulated: it is only necessary to replace "strongly" by "weakly" in condition (iii′).

Notes to Chapter 3

The theory of product integration of L_s^1 functions developed in Chapter 3 is to the best of our knowledge more general than other previous treatments of the product integration of linear operator-valued functions. In [GS], Schmidt develops product integration of a class of Banach space operator-valued functions $A(x)$

ISBN 0-201-13509-4

which are "strongly integrable" in a certain sense with respect to a positive Borel measure λ. His notion of "strongly integrable" is (for the case of λ=Lebesgue measure) intermediate between Bochner integrability and our L_s^1 theory; in particular, he assumes that $\|A(\cdot)\|$ is measurable and integrable over its domain of definition, and for each vector v, $A(\cdot)v$ is approximable a.e. by a sequence of measurable simple functions. This class of functions does not seem to be closed under taking of sums [the *measurability* of $\|A(\cdot)\|$ being the problem] and hence would not have been sufficient for the theory we have developed.

The present authors considered the possibility of a theory of *weak* product integration analogous to the L_s^1 theory but utilizing the weak operator topology; however, it does not seem likely that a useful theory could be developed along such lines. For example, weak convergence of A_n to A does not even implv weak convergence of the corresponding exponentials.

Results analogous to those of Section 6 in the case that the operators $A(x)$ are generators of contraction semigroups on a Banach space are developed in [JD3] where the relationship of our hypotheses to those of other authors is discussed.

ISBN 0-201-13509-4

CHAPTER 4

Applications

The purpose of this chapter is to illustrate the usefulness of product integral concepts in various situations. In some sections, we apply results already obtained in earlier chapters. In others, we return to the basic idea of the product integral and apply it in a new context.

4.1 Asymptotics for the Schrödinger Equation

We begin by introducing some useful notation:

DEFINITION 1.1. *Let f and g be two complex-valued functions defined on* $(0,\infty)$. *The notation*

$$f(x)\approx g(x) \tag{1.1}$$

means

$$\lim_{x\to\infty} |f(x)-g(x)|=0. \tag{1.2}$$

Similarly, if f and g are $n\times 1$ or $n\times n$ matrix-valued functions on $(0,\infty)$, *then the notation* (1.1) *means*

$$\lim_{x\to\infty} \|f(x)-g(x)\|=0. \tag{1.3}$$

In this section we shall deal with the asymptotic behavior (for large x) of solutions of the radial Schrödinger equation

$$y''=\left(V(x)+\frac{l(l+1)}{x^2}-k^2\right)y, \qquad x\in(0,\infty). \tag{1.4}$$

ISBN 0-201-13509-4

ENCYCLOPEDIA OF MATHEMATICS and Its Applications, Gian-Carlo Rota (ed.). Vol. 10: John D. Dollard and Charles N. Friedman, Product Integration with Applications to Differential Equations

In (1.4), l is a nonnegative integer, $k>0$, and V is a real function (called the *potential*) continuous on $(0,\infty)$. Equation (1.4) arises in quantum mechanics. On physical grounds it is expected (for appropriate potentials V) that asymptotically y of (1.4) will approach a linear combination of the "plane waves" e^{ikx} and e^{-ikx}. That is, we anticipate that if y is any solution of (1.4), then there should exist constants $d_\pm$ such that

$$y(x)\approx d_+e^{ikx}+d_-e^{-ikx}. \tag{1.5}$$

Furthermore, the constants $d_\pm$ should not both be zero unless y is identically zero. In Theorem 1.6.2 we proved such a result [Eq. (1.6.18)] for the case $V=0$. After Theorem 1.6.2 it was remarked that if $V\in L^1(a,\infty)$ with $a>0$, then (1.5) could again be obtained. In fact it was found that

$$|y(x)-d_+e^{ikx}-d_-e^{-ikx}|\leqslant N\int_x^\infty |W(s)|\,ds, \tag{1.6}$$

where

$$W(s)=V(s)+\frac{l(l+1)}{s^2}. \tag{1.7}$$

If $V\notin L^1(a,\infty)$ then the right-hand side of (1.6) is infinite [note that since V is continuous on $(0,\infty)$, nonexistence of $\int_a^\infty |V(s)|\,ds$ can only occur because $|V(s)|$ fails to die off fast enough for large s]. In this case, the estimate (1.6) is worthless. In this section we shall illustrate what can be done to analyze the asymptotic behavior of solutions of (1.4) when $V\notin L^1(a,\infty)$. We will find in some cases a behavior different from that of (1.5). To avoid needless complications and notational clutter, we set $l=0$. (All the results to be obtained hold equally for $l\neq 0$. The term $l(l+1)/x^2$ does not affect the asymptotic analysis because it belongs to $L^1(a,\infty)$, and it can be handled using the Corollary to Theorem 1.6.1. See [JD1] for details.)

To begin with, we rewrite (1.4) (with $l=0$) in the usual way: setting

$$Y=\begin{pmatrix} y \\ y' \end{pmatrix}, \tag{1.8}$$

(1.4) becomes

$$Y'=AY \tag{1.9}$$

with

$$A(x)=\begin{pmatrix} 0 & 1 \\ V(x)-k^2 & 0 \end{pmatrix}. \tag{1.10}$$

ISBN 0-201-13509-4

Let $a>0$. The function $Y(x)$ is then given by

$$Y(x)=\prod_a^x e^{A(s)\,ds}Y(a), \tag{1.11}$$

and to study the asymptotic behavior of $Y(x)$ we study that of $\prod_a^x e^{A(s)\,ds}$. As in Chapter 1, there is reason to believe that our analysis will be simplified if we perform a similarity with the matrix

$$M=\begin{pmatrix} 1 & 1 \\ ik & -ik \end{pmatrix} \tag{1.12}$$

which diagonalizes

$$\begin{pmatrix} 0 & 1 \\ -k^2 & 0 \end{pmatrix}.$$

We thus put

$$\prod_a^x e^{A(s)\,ds}=M\prod_a^x e^{B(s)\,ds}M^{-1} \tag{1.13}$$

where

$$B(s)=M^{-1}A(s)M=\begin{bmatrix} ik-\dfrac{iV(s)}{2k} & 0 \\ 0 & -ik+\dfrac{iV(s)}{2k} \end{bmatrix}+\frac{iV(s)}{2k}\begin{pmatrix} 0 & -1 \\ 1 & 0 \end{pmatrix}$$

$$\equiv B_1(s)+B_2(s). \tag{1.14}$$

We have made the indicated separation in (1.14) because $B_1(s)$ is a diagonal matrix whose product integral Q_1 can be evaluated explicitly: let

$$\Theta(k,x)=\int_a^x\left(k-\frac{V(s)}{2k}\right)ds=k(x-a)-\frac{1}{2k}\int_a^x V(s)\,ds. \tag{1.15}$$

Then

$$Q_1(x)\equiv\prod_a^x e^{B_1(s)\,ds}=e^{\int_a^x B_1(s)\,ds}=\begin{pmatrix} e^{i\Theta(k,x)} & 0 \\ 0 & e^{-i\Theta(k,x)} \end{pmatrix}. \tag{1.16}$$

Thus by the sum rule

$$\prod_a^x e^{B(s)\,ds}=\prod_a^x e^{\{B_1(s)+B_2(s)\}\,ds}=Q_1(x)\prod_a^x e^{\hat{B}_2(s)\,ds}, \tag{1.17}$$

ISBN 0-201-13509-4

where

$$\hat{B}_2(s)=Q_1^{-1}(s)B_2(s)Q_1(s)=\frac{iV(s)}{2k}\begin{pmatrix} 0 & -e^{-2i\Theta(k,s)} \\ e^{2i\Theta(k,s)} & 0 \end{pmatrix}. \tag{1.18}$$

We shall presently show that, for certain potentials V, the improper product integral

$$\hat{\Pi}_+ = \prod_a^\infty e^{\hat{B}_2(s)\,ds} \tag{1.19}$$

exists and is nonsingular. To see what this will imply, let us assume that it is so. Then

$$\prod_a^x e^{\hat{B}_2(s)\,ds} \approx \hat{\Pi}_+. \tag{1.20}$$

Now a relation of the form $f(x)\approx g(x)$ can clearly be multiplied on the right or left by any bounded function. Multiplying (1.20) on the left by $Q_1(x)$ we have by (1.17)

$$\prod_a^x e^{B(s)\,ds} \approx Q_1(x)\hat{\Pi}_+. \tag{1.21}$$

Equation (1.13) then yields

$$\prod_a^x e^{A(s)\,ds} \approx MQ_1(x)\hat{\Pi}_+ M^{-1}. \tag{1.22}$$

Finally, from (1.11) we find

$$Y(x)\approx MQ_1(x)\hat{\Pi}_+ M^{-1}Y(a). \tag{1.23}$$

Equation (1.23) gives the asymptotic behavior of the function $Y(x)$. Let us write

$$\Pi_+ M^{-1}Y(a)=\begin{pmatrix} d_+ \\ d_- \end{pmatrix}. \tag{1.24}$$

By the nonsingularity of $\hat{\Pi}_+$ and M, the constants $d_\pm$ are both zero if and only if $Y(a)$ is zero, i.e., if and only if $Y(x)$ is identically zero. Using (1.24) and (1.16), Eq. (1.23) reads

$$Y(x)=\begin{pmatrix} y(x) \\ y'(x) \end{pmatrix}\approx\begin{pmatrix} d_+e^{i\Theta(k,x)}+d_-e^{-i\Theta(k,x)} \\ ikd_+e^{i\Theta(k,x)}-ikd_-e^{-i\Theta(k,x)} \end{pmatrix}. \tag{1.25}$$

This shows that the function $y(x)$ asymptotically approaches a linear

ISBN 0-201-13509-4

combination of the functions $e^{i\Theta(k,x)}$ and $e^{-i\Theta(k,x)}$. From Eq. (1.15) we now see the following: if the improper integral

$$\operatorname{imp}\int_a^\infty V(s)\,ds \equiv \lim_{x\to\infty}\int_a^x V(s)\,ds \tag{1.26}$$

exists, then

$$\Theta(k,x) \approx kx + \text{const.}, \tag{1.27}$$

so that

$$e^{\pm i\Theta(k,x)} \approx C_{\pm} e^{\pm ikx}. \tag{1.28}$$

Then (1.25) shows that $y(x)$ has the "expected" asymptotic behavior. If the limit in (1.26) does *not* exist, then (1.25) shows that $y(x)$ has an "anomalous" asymptotic behavior, i.e., $y(x)$ is not asymptotically a linear combination of e^{ikx} and e^{-ikx}. A famous case of such anomalous asymptotic behavior is provided by the Coulomb potential

$$V(x) = \lambda/x, \tag{1.29}$$

where λ is a nonzero real number. In this case

$$\Theta(k,x) = k(x-a) - (\lambda/2k)\log(x/a) \tag{1.30}$$

and it is well known [LIS] that for the Coulomb potential the solutions of (1.4) have the anomalous asymptotic behavior given by (1.25) and (1.30). We now prove existence and nonsingularity of $\hat{\Pi}_+$ [and thus the asymptotic form (1.25)] for a class of potentials including the Coulomb potential, namely potentials of the form

$$V(x) = \lambda/x^r \tag{1.31}$$

where λ is a nonzero real number and $r > 1/2$. To prove existence and nonsingularity of $\hat{\Pi}_+ = \Pi_a^\infty e^{\hat{B}_2(s)\,ds}$ we use the delicate convergence result of Theorem 1.6.3: we need only establish that the improper integral

$$\hat{H}_2(x) = \operatorname{imp}\int_x^\infty \hat{B}_2(s)\,ds \tag{1.32}$$

exists, and that $\hat{H}_2\hat{B}_2 \in L^1(a,\infty)$. Now a glance at (1.18) shows that $B_2(x)$ is $O(1/x^r)$ for large x. If we can show that $\hat{H}_2(x)$ exists and is $O(1/x^r)$, then $\hat{H}_2\hat{B}_2$ will belong to $L^1(a,\infty)$ because $r > 1/2$. To establish existence of $\hat{H}_2(x)$, note that by (1.18) $\hat{B}_2(s)$ has only two nonzero entries, which are complex conjugates of one another. We examine the integral from x to R

ISBN 0-201-13509-4

of the (2, 1) entry which, apart from irrelevant factors, is

$$I(x,R)=\int_x^R \frac{e^{2i\Theta(k,s)}}{s^r}\,ds. \tag{1.33}$$

We wish to prove that $I(x,R)$ has a limit as $R\to\infty$. Clearly this will be true for all x if it is true for one x. We assume that x is so large that the function

$$\Theta'(k,s)=\frac{d}{ds}\Theta(k,s)=k-\frac{\lambda}{2ks^r} \tag{1.34}$$

does not vanish on $[x,\infty)$. Then we may integrate by parts to find

$$I(x,R)=\left.\frac{e^{2i\Theta(k,s)}}{2is^r\Theta'(k,s)}\right|_x^R-\frac{1}{2i}\int_x^R e^{2i\Theta(k,s)}\frac{d}{ds}\left(\frac{1}{s^r\Theta'(k,s)}\right)ds. \tag{1.35}$$

Now

$$s^r\Theta'(k,s)=ks^r-\frac{\lambda}{2k}. \tag{1.36}$$

Using (1.36), one immediately finds that the integrand occurring in (1.35) is $O(1/s^{r+1})$. It follows that the limit as $R\to\infty$ can be taken in (1.35), and

$$I(x)=\lim_{R\to\infty} I(x,R)=\frac{-e^{2i\Theta(k,x)}}{2i(kx^r-\lambda/2k)}+\int_x^\infty O\left(\frac{1}{s^{r+1}}\right)ds=O\left(\frac{1}{x^r}\right). \tag{1.37}$$

Thus we have showed that the (2, 1) entry of $\hat{H}_2(x)$ exists and is $O(1/x^r)$. Since the (1,2) entry differs only by complex conjugation, we have that $\hat{H}_2(x)$ exists and is $O(1/x^r)$, as desired. This completes the proof. The estimate (1.6.52) shows that in the present case the difference between $\prod_a^x e^{\hat{B}_2(s)ds}$ and $\hat{\Pi}_+$ is $O(1/x^r)+O(1/x^{2r-1})$. The same then follows for the difference between $Y(x)$ and its asymptotic form (1.25).

An analysis similar to that given above can be carried out to prove existence and nonsingularity of $\hat{\Pi}_+$ for potentials of the form

$$V(x)=\frac{\lambda\sin\mu x^\alpha}{x^\beta}, \tag{1.38}$$

where λ is a nonzero real number, $\mu>0$, and α and β are positive real numbers satisfying $\alpha+2\beta>2$. A general discussion of such potentials is given in [JD1]. (See also [FA].) We select for examination here potentials

ISBN 0-201-13509-4

of the form

$$V(x)=\lambda\frac{\sin\mu x}{x}. \qquad (\mu>0) \tag{1.39}$$

These are of interest because, as will be seen, existence of $\hat{\Pi}_+$ can be proved for such potentials *unless* k of Eq. (1.4) has the value $\mu/2$. [The same phenomenon occurs whenever $\alpha=1$ in (1.38). For $\alpha\neq 1$, there are no exceptional values of k.] For this one exceptional value of k we are thus unable to establish the asymptotic form given in (1.25). We now prove existence of $\hat{\Pi}_+$ for a potential V of the form (1.39) when $k\neq\mu/2$: proceeding as before, we must prove existence of the function $\hat{H}_2$ defined in (1.32) and show $\hat{H}_2\hat{B}_2\in L^1(a,\infty)$. The integral from x to R of the (2,1) entry of $\hat{B}_2$ is, apart from irrelevant factors,

$$I(x,R)=\int_x^R e^{2i\Theta(k,s)}\frac{\sin\mu s}{s}\,ds, \tag{1.40}$$

where

$$\Theta(k,s)=k(s-a)-\frac{\lambda}{2k}\int_a^x\frac{\sin\mu t}{t}\,dt\equiv ks+\psi(k,s). \tag{1.41}$$

Writing $\sin\mu s$ in terms of exponentials, we find

$$I(x,R)=\frac{1}{2i}(I_+(x,R)-I_-(x,R)) \tag{1.42}$$

with

$$I_\pm(x,R)=\int_x^R e^{i(2k\pm\mu)s}\left\{\frac{e^{2i\psi(k,s)}}{s}\right\}ds. \tag{1.43}$$

If $k\neq 2\mu$, then we have

$$I_\pm(x,R)=\frac{e^{i(2k\pm\mu)s}}{i(2k\pm\mu)}\frac{e^{2i\psi(K,s)}}{s}\bigg|_x^R-\frac{1}{i(2k\pm\mu)}\int_x^R e^{i(2k\pm\mu)s}\frac{d}{ds}\left(\frac{e^{2i\psi(k,s)}}{s}\right)ds, \tag{1.44}$$

and an elementary estimate shows that the limits $\lim_{R\to\infty}I_\pm(x,R)=I_\pm(x)$ exist and are $O(1/x)$. Thus arguing as before we have $\hat{H}_2(x)=O(1/x)$, and since $\hat{B}_2(x)$ is also $O(1/x)$ we have $\hat{H}_2\hat{B}_2\in L^1(a,\infty)$, completing the proof. In the present case, the estimate (1.6.52) shows that the difference between $\prod_a^x e^{\hat{B}_2(s)\,ds}$ and $\hat{\Pi}_+$ is $O(1/x)$. The same then follows for the difference between $Y(x)$ and its asymptotic form (1.25). We note that $(\sin\mu s)/s$ is

ISBN 0-201-13509-4

improperly integrable over $[a, \infty)$, so that, by (1.26)–(1.28), the solutions of (1.4) are asymptotically linear combinations of $e^{\pm ikx}$. Thus for $k \neq \mu/2$ things are as nice as one could wish: the solutions of (1.4) have the "expected" asymptotic behavior. However, if $k = \mu/2$, then $I_-(x,R)$ of (1.43) has no limit as $R \to \infty$: because $(\sin \mu t)/t$ is improperly integrable over $[a,\infty)$, $\psi(k,s)$ approaches a constant as $s \to \infty$ and the integral $I_-(x,R)$ diverges logarithmically when $R \to \infty$. Thus the proof breaks down for $k = \mu/2$. A simple argument using the Corollary to Theorem 1.6.1 shows that exactly the same results are obtained for a potential of the form

$$V(x) = \lambda \frac{\sin \mu x}{x} + U(x), \tag{1.45}$$

where U is a real continuous function in $L^1(a,\infty)$. Again one has the "expected" asymptotic behavior for $k \neq \mu/2$, and no information for $k = \mu/2$.

The above analysis suggests that if V is as in (1.45) and $k = \mu/2$, then a solution of (1.4) might exhibit a "strange" asymptotic behavior. (That is, the solution might *not* asymptotically be a linear combination of e^{ikx} and e^{-ikx}.) Such "strange" behavior can actually occur. In fact, there is a famous example in which a solution with $k = \mu/2$ is square-integrable over $[a,\infty)$. Of course, such a solution cannot asymptotically be a linear combination of e^{ikx} and e^{-ikx}. The example is due to von Neumann and Wigner [JV]. In their example, V is as in (1.45) with $\lambda = -8$, $\mu = 2$ and the square-integrable solution occurs for $k = 1$. The analysis given above shows that square-integrable solutions cannot occur for any other value of k. These results have an interpretation in quantum mechanics—von Neumann and Wigner's example exhibits a "positive energy bound state" for $k = \mu/2$, and there are no such bound states for $k \neq \mu/2$.

We now illustrate another technique for deducing asymptotic behavior. This technique is based on the similarity rule (Theorem 1.3.3). Concerning the potential V we assume that

(i) V is continuously differentiable on $[b_0, \infty)$ for some $b_0 > 0$ and

(ii) The derivative V' satisfies

$$\int_{b_0}^{\infty} |V'(s)|\, ds < \infty. \tag{1.46}$$

A type of potential satisfying these conditions and not covered by our previous results is

$$V(x) = \lambda / x^{\beta}, \tag{1.47}$$

with λ a nonzero real number and $0 < \beta \leqslant 1/2$.

ISBN 0-201-13509-4

If V satisfies the condition (ii), then $V(x)$ has a limit V_0 as $x \to \infty$. If $k^2 > V_0$, then there will be a number $b \geqslant b_0$ such that

$$k^2 - V(x) \geqslant \alpha > 0 \qquad \text{for } x \geqslant b, \tag{1.48}$$

where α is a fixed number. Thus for $x \geqslant b$ the function

$$\lambda(x) = \sqrt{k^2 - V(x)} \tag{1.49}$$

is well defined and positive. We set

$$\sigma(x) = \int_b^x \lambda(s)\, ds. \tag{1.50}$$

We will prove that for $k^2 > V_0$ the solution y of (1.4) has the following asymptotic behavior:

$$y(x) \approx c_+ e^{i\sigma(x)} + c_- e^{-i\sigma(x)}, \tag{1.51}$$

where $c_\pm$ are both zero if and only if $y(x)$ is identically zero. This is the asymptotic behavior suggested by "WKB intuition" in quantum mechanics [LIS].

To prove (1.51), we adopt the notation of Eq. (1.8) through (1.10). We write

$$Y(x) = \prod_b^x e^{A(s)\,ds} Y(b), \tag{1.52}$$

where b is chosen so that (1.48) holds. We must then analyze the asymptotic behavior of $\prod_b^x e^{A(s)\,ds}$. This time we proceed by diagonalizing the matrix $A(s)$ directly. With $\lambda(x)$ as in (1.49), we set

$$M(s) = \begin{pmatrix} 1 & 1 \\ i\lambda(s) & -i\lambda(s) \end{pmatrix}. \tag{1.53}$$

Then

$$M^{-1}(s) A(s) M(s) = \begin{pmatrix} i\lambda(s) & 0 \\ 0 & -i\lambda(s) \end{pmatrix} \equiv i\Lambda(s) \tag{1.54}$$

Furthermore, because V is continuously differentiable, the same holds for M. Using the similarity rule in the form (1.3.22) we have

$$M^{-1}(x) \prod_b^x e^{A(s)\,ds} M(b) = \prod_b^x e^{\{i\Lambda(s) - M^{-1}(s)M'(s)\}\,ds} \equiv \prod_b^x e^{B(s)\,ds}. \tag{1.55}$$

ISBN 0-201-13509-4

Thus

$$\prod_a^x e^{A(s)ds} = M(x)\prod_b^x e^{B(s)ds} M^{-1}(b). \tag{1.56}$$

Note that

$$\lim_{x\to\infty} M(x) = M_+ = \begin{pmatrix} 1 & 1 \\ i\sqrt{k^2-V_0} & -i\sqrt{k^2-V_0} \end{pmatrix}. \tag{1.57}$$

Now the matrix $\Lambda(s)$ is diagonal. Thus

$$Q(x) \equiv \prod_b^x e^{i\Lambda(S)ds} = e^{i\int_b^x \Lambda(s)ds} = \begin{pmatrix} e^{i\sigma(x)} & 0 \\ 0 & e^{-i\sigma(x)} \end{pmatrix}, \tag{1.58}$$

where $\sigma(x)$ is given by (1.50). The sum rule now gives

$$\prod_b^x e^{B(s)ds} = Q(x)\prod_b^x e^{C(s)ds}, \tag{1.59}$$

where

$$C(s) = -Q^{-1}(s)M^{-1}(s)M'(s)Q(s). \tag{1.60}$$

Suppose we can prove that the improper product integral

$$\Pi_+ = \prod_b^\infty e^{C(s)ds} \tag{1.61}$$

exists and is nonsingular. Using (1.56), (1.57), (1.59), and (1.60) we will then have

$$\prod_b^x e^{A(s)ds} \approx M_+ Q(x)\Pi_+ M^{-1}(b). \tag{1.62}$$

Because of (1.52) and the form of $Q(x)$, it will then follow that $y(x)$ has the asymptotic form given in (1.51).

We now prove existence and nonsingularity of Π_+. For this purpose it suffices to show that $C \in L^1(b,\infty)$. Since Q is unitary we have

$$\|C(s)\| = \|Q^{-1}(s)M^{-1}(s)M'(s)Q(s)\| = \|M^{-1}(s)M'(s)\|. \tag{1.63}$$

A simple calculation yields

$$M^{-1}(s)M'(s) = \frac{\lambda'(s)}{2\lambda(s)}\begin{pmatrix} 1 & -1 \\ -1 & 1 \end{pmatrix} = -\frac{1}{4}\frac{V'(s)}{k^2-V(s)}\begin{pmatrix} 1 & -1 \\ -1 & 1 \end{pmatrix} \tag{1.64}$$

ISBN 0-201-13509-4

Since the matrix $\begin{pmatrix} 1 & -1 \\ -1 & 1 \end{pmatrix}$ has norm 2 and (1.48) holds, we find for $x \geqslant b$

$$\|C(s)\| \leqslant |V'(s)|/2\alpha. \tag{1.65}$$

Since $V' \in L^1(b,\infty)$ we have $C \in L^1(b,\infty)$, completing the proof.

The formulas just derived can be used in a different way to deduce the behavior of solutions of (1.4) for large values of k^2. Namely, if y is a solution of (1.4), then by (1.52), (1.56), and (1.59) we have

$$\begin{pmatrix} y(x) \\ y'(x) \end{pmatrix} = Y(x) = M(x)\begin{pmatrix} e^{i\sigma(x)} & 0 \\ 0 & e^{-i\sigma(x)} \end{pmatrix} \prod_b^x e^{C(s)ds} M^{-1}(b)\, Y(b). \tag{1.66}$$

The derivation above shows that this formula is valid for x in any interval $[b,c]$ such that V is continuously differentiable on $[b,c]$ and $k^2 - V(x) > 0$ on $[b,c]$. We now restrict attention to such an interval. We wish to emphasize the dependence of Y on k and will for this reason use k as a label in some formulas. Thus we write $y(x,k)$ instead of $y(x)$, $\lambda(x,k)$ instead of $\lambda(x)$, etc. By the inequality (1.4.2) of Theorem 1.4.1, we can write

$$\prod_b^x e^{C(s)ds} = I + R(x,k) \tag{1.67}$$

where

$$\|R(x,k)\| \leqslant e^{\int_b^x \|C(s)\|ds} - 1. \tag{1.68}$$

Now using (1.63) and (1.64) we have

$$\int_b^x \|C(s)\|\, ds \leqslant \int_b^c \|C(s)\|\, ds = \frac{1}{2}\int_b^c \frac{|V'(s)|}{k^2 - V(s)}\, ds. \tag{1.69}$$

The integral on the right-hand side of (1.69) is easily seen to be $O(1/k^2)$ for large k, and it follows simply that $\|R(x,k)\|$ is $O(1/k^2)$ for large k, uniformly for $x \in [b,c]$. Inserting (1.67) in (1.66) and computing the (1,1) entry we find, after some matrix manipulations,

$$y(x,k) = y(b)\cos\sigma(x,k) + \frac{y'(b)}{\lambda(b,k)}\sin\sigma(x,k) + \gamma(x,k) \tag{1.70}$$

with

$$|\gamma(x,k)| \leqslant \left\{ |y(b)| + \frac{|y'(b)|}{\lambda(b,k)} \right\}\left\{ e^{1/2\int_b^c \frac{|V'(s)|}{k^2 - V(s)}ds} - 1 \right\} = O\left(\frac{1}{k^2}\right). \tag{1.71}$$

In (1.70) we choose the initial conditions $y(b)$ and $y'(b)$ independent of k, since we wish to compare solutions with the same initial conditions and

ISBN 0-201-13509-4

different values of k. For $x \in [b,c]$, V is bounded, so that

$$\lambda(x,k) = \sqrt{k^2 - V(x)} = k\sqrt{1 - \frac{V(x)}{k^2}} = k + O\left(\frac{1}{k}\right), \tag{1.72}$$

where the term of order $1/k$ has this order uniformly for $x \in [b,c]$. Integrating (1.72), we find

$$\sigma(x,k) = k(x-b) + O(1/k). \tag{1.73}$$

Thus (1.70) yields

$$y(x,k) = y(b)\cos k(x-b) + \frac{y'(b)}{k}\sin k(x-b) + O\left(\frac{1}{k}\right). \tag{1.74}$$

Equations (1.70) and (1.74) constitute what is often called the WKB approximation giving y asymptotically for large k.

4.2 Weyl's Limit-Circle Classification

In this section we use the product integral to clarify the nature of "limit-circle" behavior for a second-order linear differential equation of the type

$$(py')' + qy = \lambda y \tag{2.1}$$

on the interval $[0,\infty)$. In Eq. (2.1), p is a nonzero complex-valued function that is continuously differentiable on $[0,\infty)$, q is a complex-valued function continuous on $[0,\infty)$, and λ is a complex constant. Equation (2.1) is said to be in the *limit-circle case* if, for some value of λ, every solution of (2.1) is in $L^2(0,\infty)$. A familiar result of Weyl [EC] states that if this condition holds for one value of λ, then it holds for every value of λ. This result is not difficult to obtain by ordinary methods, but the product integral proof is instructive and will now be given. We rewrite (2.1) as

$$y'' + \frac{p'}{p}y' + \frac{q}{p}y = \frac{\lambda}{p}y. \tag{2.2}$$

Setting

$$Y = \begin{pmatrix} y \\ y' \end{pmatrix} \tag{2.3}$$

we obtain the equivalent equation

$$Y' = A_\lambda Y, \tag{2.4}$$

ISBN 0-201-13509-4

where

$$A_\lambda = \begin{bmatrix} 0 & 1 \\ \dfrac{\lambda - q}{p} & -\dfrac{p'}{p} \end{bmatrix}. \tag{2.5}$$

We put

$$T_\lambda(x) = \prod_0^x e^{A_\lambda(s)\,ds} \tag{2.6}$$

so that

$$Y(x) = T_\lambda(x)\, Y(0). \tag{2.7}$$

For reference we note that by Theorem 1.1.4 we have

$$\det T_\lambda(x) = e^{-\int_0^x \frac{p'(s)}{p(s)}\,ds} = \frac{p(0)}{p(x)}. \tag{2.8}$$

Now by Theorem 1.2.2 the columns of $T_\lambda(x)$ are solutions of (2.4), and because of the form of A_λ any such solution has the form $\binom{f}{f'}$ with f a solution of (2.1). Thus $T_\lambda(x)$ has the form

$$T_\lambda(x) = \begin{pmatrix} f_\lambda(x) & g_\lambda(x) \\ f'_\lambda(x) & g'_\lambda(x) \end{pmatrix}, \tag{2.9}$$

where f_λ and g_λ are solutions of (2.1) so chosen that $T_\lambda(0) = I$. We have indicated explicitly that these solutions depend on λ. Since (by Theorem 1.2.2) any solution of (2.1) is a linear combination of f_λ and g_λ, square-integrability of an arbitrary solution of (2.1) is equivalent to square-integrability of f_λ and g_λ.

To compare the solutions of (2.1) for various values of λ, note that by (2.5) we have

$$A_{\lambda_2}(x) = A_{\lambda_1}(x) + (\lambda_2 - \lambda_1)\begin{bmatrix} 0 & 0 \\ \dfrac{1}{p(x)} & 0 \end{bmatrix} \equiv A_{\lambda_1}(x) + (\lambda_2 - \lambda_1) D(x). \tag{2.10}$$

Thus by the sum rule we have

$$T_{\lambda_2}(x) = T_{\lambda_1}(x) \prod_0^x e^{(\lambda_2 - \lambda_1) T_{\lambda_1}^{-1}(s) D(s) T_{\lambda_1}(s)\,ds} \equiv T_{\lambda_1}(x)\, W_{12}(x). \tag{2.11}$$

ISBN 0-201-13509-4

In studying the square-integrability of entries of T_{λ_2} or T_{λ_1}, the important feature is of course the behavior of these entries for large values of x. Weyl's result thus states that in certain circumstances there is a similarity between the asymptotic behavior of the entries of T_{λ_2} and T_{λ_1}. We ask: when can we assert that T_{λ_2} and T_{λ_1} will have a similar behavior for large x? Theorem 1.6.1 shows that if $T_{\lambda_1}^{-1}DT_{\lambda_1} \in L^1(0,\infty)$, then $W_{12}(x)$ will have a nonsingular limit W_{12}^+ as $x \to \infty$. Let us assume $T_{\lambda_1}^{-1}DT_{\lambda_1} \in L^1(0,\infty)$. Then (2.11) shows that

$$T_{\lambda_1}^{-1}(x)T_{\lambda_2}(x) \approx W_{12}^+, \tag{2.12}$$

and this equation expresses a similarity between the asymptotic behavior of T_{λ_2} and T_{λ_1}. Also

$$T_{\lambda_2}^{-1}(x)D(x)T_{\lambda_2}(x) = W_{12}^{-1}(x)T_{\lambda_1}^{-1}(x)D(x)T_{\lambda_1}(x)W_{12}(x). \tag{2.13}$$

Since $W_{12}(x)$ approaches the nonsingular limit W_{12}^+, W_{12} and W_{12}^{-1} are bounded on $[0,\infty)$. Equation (2.13) then shows that $T_{\lambda_2}^{-1}DT_{\lambda_2} \in L^1(0,\infty)$. By the same argument with λ_2 and λ_1 interchanged, we obtain

$$T_{\lambda_1}^{-1}DT_{\lambda_1} \in L^1(0,\infty) \Leftrightarrow T_{\lambda_2}^{-1}DT_{\lambda_2} \in L^1(0,\infty). \tag{2.14}$$

We now compute $T_\lambda^{-1}DT_\lambda$. Using (2.8) and (2.9) we find

$$T_\lambda^{-1}DT_\lambda = \frac{1}{p(0)}\begin{pmatrix} -g_\lambda f_\lambda & -g_\lambda^2 \\ f_\lambda^2 & g_\lambda f_\lambda \end{pmatrix}. \tag{2.15}$$

By inspection, the condition $T_\lambda^{-1}DT_\lambda \in L^1(0,\infty)$ is equivalent to the statement that both f_λ and g_λ belong to $L^2(0,\infty)$. Thus Eq. (2.1) is in the limit-circle case if and only if $T_\lambda^{-1}DT_\lambda \in L^1(a,\infty)$ for some λ, and our conclusion (2.14) is just Weyl's result in a different guise. Our proof shows that in the limit-circle case Weyl's result is intuitively attributable to the fact that for any complex λ_1 and λ_2 the product integrals T_{λ_1} and T_{λ_2} have a similar asymptotic behavior, in the sense of Eq. (2.12).

If $J(x)$ is any continuous matrix-valued function, then replacing D by J in Eq. (2.13), we find that in the limit-circle case the conditions $T_{\lambda_1}^{-1}JT_{\lambda_1} \in L^1(0,\infty)$ and $T_{\lambda_2}^{-1}JT_{\lambda_2} \in L^1(0,\infty)$ are equivalent. Thus for instance we can conclude for any continuous function h that hf_{λ_1} and hg_{λ_1} both belong to $L^2(0,\infty)$ if and only if the same holds for hf_{λ_2} and hg_{λ_2}, or that hf'_{λ_1} and hg'_{λ_1} both belong to $L^2(0,\infty)$ if and only if the same holds for hf'_{λ_2} and hg'_{λ_2}.

In closing this section we remark that product-integral methods do not seem particularly well adapted to the discussion of the *limit-point* case for Eq. (2.1). [Equation (2.1) is said to be in the limit point case if and only if it is not in the limit-circle case.] A major task in this case is to prove that for

ISBN 0-201-13509-4

Im$\lambda \neq 0$, Eq. (2.1) has a square-integrable solution. Of course there cannot be two linearly independent square-integrable solutions because then the equation would be in the limit-circle case. Now the product integral is very well suited to a discussion of properties possessed by *all* solutions of a differential equation. This is because, containing as it does a fundamental set of solutions of the equation, the product integral can deal with all solutions simultaneously. But if two solutions of a differential equation have a very different character, this will not be readily discoverable using product-integral methods, which emphasize rather what the solutions have in common. Thus proving existence of exactly one square-integrable solution for (2.1) is a problem which is unnatural from the product-integral viewpoint. The usual proof [EC] can of course be rewritten in product integral language, but nothing is gained by doing this.

4.3 The Lie Product Formula

The Lie product formula [MR2] states that for $A, B \in \mathbb{C}_{n\times n}$ we have

$$e^{A+B} = \lim_{n\to\infty} (e^{A/n} e^{B/n})^n. \tag{3.1}$$

This formula has a far-reaching and useful generalization due to Trotter [HT]. In this section we try to provide insight into (3.1) by showing that it is a special case of the sum rule for product integrals. To see this, note that

$$e^{A+B} = \prod_0^1 e^{(A+B)ds}. \tag{3.2}$$

We apply the sum rule to "pull out" the term B in (3.2). Noting that $\prod_0^t e^{Bds} = e^{Bt}$, we have

$$e^{A+B} = e^B \prod_0^1 e^{\hat{A}(s)ds} \tag{3.3}$$

with

$$\hat{A}(s) = e^{-Bs} A e^{Bs}. \tag{3.4}$$

Now let $P = \{s_0, s_1, \ldots, s_n\}$ denote a partition of $[0,1]$. Then

$$\prod_0^1 e^{\hat{A}(s)ds} = \lim_{\mu(P)\to 0} \prod_{k=1}^n e^{\hat{A}(s_k)\Delta s_k}. \tag{3.5}$$

Now by (3.4) we have

$$e^{\hat{A}(s_k)\Delta s_k} = e^{-Bs_k} e^{A\Delta s_k} e^{Bs_k}. \tag{3.6}$$

ISBN 0-201-13509-4

Thus (noting that $\Delta s_1 = s_1 - s_0 = s_1$) we have

$$\prod_{k=1}^{n} e^{\hat{A}(s_k)\Delta s_k} = e^{-Bs_n} e^{A\Delta s_n} e^{Bs_n} e^{-Bs_{n-1}} e^{A\Delta s_{n-1}} e^{Bs_{n-1}} \cdots e^{-Bs_1} e^{A\Delta s_1} e^{Bs_1}$$

$$= e^{-Bs_n} \prod_{k=1}^{n} e^{A\Delta s_k} e^{B\Delta s_k} = e^{-B} \prod_{k=1}^{n} e^{A\Delta s_k} e^{B\Delta s_k}. \tag{3.7}$$

Equations (3.3), (3.5), and (3.7) yield

$$e^{A+B} = \lim_{\mu(P)\to 0} \prod_{k=1}^{n} e^{A\Delta s_k} e^{B\Delta s_k}. \tag{3.8}$$

Equation (3.1) is obtained from (3.8) by taking $\Delta s_k = 1/n$ for all k. We can obtain an estimate of the rate of convergence in (3.8) as follows: write

$$\Delta_P = e^{A+B} - \prod_{k=1}^{n} e^{A\Delta s_k} e^{B\Delta s_k} = e^{A+B} - e^{B} \prod_{k=1}^{n} e^{\hat{A}(s_k)\Delta s_k}$$

$$= e^{B} \left\{ \prod_{0}^{1} e^{\hat{A}(s)\,ds} - \prod_{k=1}^{n} e^{\hat{A}(s_k)\Delta s_k} \right\}. \tag{3.9}$$

The product $\prod_{k=1}^{n} e^{\hat{A}(s_k)\Delta s_k}$ is the product integral over $[0,1]$ of the point-value approximant $\hat{A}_P$ corresponding to the function $\hat{A}$ and the partition P [see Definitions 1.1.2 and 1.1.3, and Eq. (1.8.4)]. Also by the Corollary to Theorem 1.5.1 we have

$$\left\| \prod_{0}^{1} e^{\hat{A}(s)\,ds} - \prod_{0}^{1} e^{\hat{A}_P(s)\,ds} \right\| \leqslant e^{\|\hat{A}\|_1 + \|\hat{A}_P\|_1} \|\hat{A} - \hat{A}_P\|_1, \tag{3.10}$$

where $\| \ \|_1$ is the L^1-norm on the interval $[0,1]$. Now

$$\|\hat{A}(s)\| = \|e^{-Bs} A e^{Bs}\| \leqslant \|A\| e^{2\|B\|} \equiv C \qquad (s \in [0,1]) \tag{3.11}$$

and integrating over $[0,1]$ we find

$$\|\hat{A}\|_1 \leqslant C. \tag{3.12}$$

The same inequality holds for $\|\hat{A}_P\|_1$. Also

$$\|\hat{A} - \hat{A}_P\|_1 = \sum_{k=1}^{n} \int_{s_{k-1}}^{s_k} \|\hat{A}(s) - \hat{A}(s_k)\|\,ds. \tag{3.13}$$

ISBN 0-201-13509-4

We have

$$\hat{A}(s)-\hat{A}(s_k)=e^{-Bs}Ae^{Bs}-e^{-Bs_k}Ae^{Bs_k}$$
$$=\int_{s_k}^{s}\frac{d}{dt}(e^{-Bt}Ae^{Bt})\,ds=-\int_{s_k}^{s}e^{-Bt}[B,A]e^{Bt}\,dt, \quad (3.14)$$

where $[B,A]=BA-AB$. Thus by a simple estimate

$$\|\hat{A}(s)-\hat{A}(s_k)\|\leqslant e^{2\|B\|}\|[B,A]\|(s-s_k)\leqslant e^{2\|B\|}\|[B,A]\|\,\mu(P). \quad (3.15)$$

Equation (3.13) then immediately yields

$$\|A-\hat{A}_P\|_1\leqslant e^{2\|B\|}\|[B,A]\|\,\mu(P). \quad (3.16)$$

Combining (3.9) with the above estimates, we find

$$\|\Delta_P\|\leqslant e^{3\|B\|}e^{2C}\|[B,A]\|\,\mu(P). \quad (3.17)$$

This estimate preserves the information that $\Delta_P=0$ if $[A,B]=0$. In the case of the Lie product formula, the estimate (3.17) shows that the difference between e^{A+B} and $(e^{A/n}e^{B/n})^n$ is $O(1/n)$.

Formula (3.8) can be written in a natural notation as

$$e^{A+B}=\prod_0^1 e^{Ads}e^{Bds}. \quad (3.18)$$

The derivation of (3.8) suggests an alternative formulation of the sum rule, namely

$$\prod_a^b e^{\{A(s)+B(s)\}ds}=\prod_a^b e^{A(s)ds}e^{B(s)ds}, \quad (3.19)$$

where the right-hand side is defined in the obvious manner: letting $P=\{s_0,s_1,\ldots,s_n\}$ denote a partition of $[a,b]$, we set

$$\prod_a^b e^{A(s)ds}e^{B(s)ds}=\lim_{\mu(P)\to 0}\prod_{k=1}^{n} e^{A(s_k)\Delta s_k}e^{B(s_k)\Delta s_k}. \quad (3.20)$$

If A and B are continuous on $[a,b]$, then the limit indicated in (3.20) exists and (3.19) is valid. We omit the proof, which is not difficult.

ISBN 0-201-13509-4

4.4 The Hille-Yosida Theorem

In this section we prove that portion of the Hille-Yosida theorem [MR2], which gives sufficient conditions for an operator A to be the infinitesimal generator of a contraction semigroup. (The conditions are also necessary, but we will not prove this.) The proof is a slight modification of a proof due to Kato [TK2]. We give this proof (i) because it exploits the basic idea of the product integral, (ii) because it illustrates the desirability in some situations of dealing with product integrals of the form $\prod_a^b (I-A(s)\,ds)^{-1}$ (see Section 1.8), (iii) because it applies product integral ideas directly to an unbounded operator, and (iv) because it illustrates the sort of combinatorial argument characteristic of earlier treatments of product integration. We shall need the following basic fact: Let $B_1,\dots,B_n$ and $C_1,\dots,C_n$ all denote bounded linear operators from a Banach space X to itself. Then

$$\prod_{k=1}^{n} B_k - \prod_{k=1}^{n} C_k = \sum_{l=1}^{n} \left\{ \prod_{k=l+1}^{n} B_k \right\} (B_l - C_l) \left\{ \prod_{k=1}^{l-1} C_k \right\}. \tag{4.1}$$

In (4.1) we agree that $\prod_{n+1}^{n} C_k = \prod_{1}^{0} C_k = I$. Formula (4.1) holds because the right-hand side is a telescoping sum and after cancellations only the terms on the left-hand side remain. This is best seen by writing out a few terms of the sum on the right-hand side.

Let X be a Banach space. Let A be a linear operator from its domain $D_A \subset X$ to X, and having the following properties:

(i) A is closed and D_A is dense in X.

(ii) For each $\lambda > 0$, λ is in the resolvent set of A and

$$\|(\lambda - A)^{-1}\| \leq \lambda^{-1}. \tag{4.2}$$

We will show that A is the infinitesimal generator of a contraction semigroup T^t. In other words, we will prove the existence of a family $\{T^t \mid t \geq 0\}$ of bounded linear operators from X to itself such that

(a) $\|T^t\| \leq 1, \quad t \geq 0,$

(b) $T^t T^s = T^{t+s}, \quad s,t \geq 0,$

(c) $s-\lim_{t\to 0} T^t = T^0 = I,$

(d) for $\varphi \in D_A$, we have (The derivative is in the norm topology on X.)

$$\frac{dT^t\varphi}{dt} = AT^t\varphi. \tag{4.3}$$

The operator T^t is also written as e^{At} because of Eq. (4.3). If A were an $n \times n$ matrix, then because $(1-z)^{-1}$ is a P-function (Section 1.8) we would have for any a, and $t \geq 0$

$$\prod_{a}^{a+t} (I - A\,ds)^{-1} = \prod_{a}^{a+t} e^{A\,ds} = e^{\int_a^{a+t} A\,ds} = e^{At}. \tag{4.4}$$

ISBN 0-201-13509-4

Guided by this equation, we will try to define T^t by the formula (The reason for the "a" will be clear later.)

$$T^t = \prod_a^{a+t} (I - A\,ds)^{-1} \tag{4.5}$$

when A satisfies conditions (i) and (ii) above. We intend to restrict our attention to those portions of the argument which deal specifically with product integration. We shall list here some facts which will be used and which are easily proved using properties (i) and (ii) of A and familiar manipulations with operators. First, clearly (ii) is equivalent to

$$\|(I - A\alpha)^{-1}\| \leqslant 1, \qquad \alpha > 0, \tag{4.6}$$

where we write $A\alpha$ instead of αA to correspond to the notation $(I - A\,ds)^{-1}$. Second, we have

$$\lim_{\alpha \to 0} (I - A\alpha)^{-1}\varphi = \varphi \qquad (\varphi \in X) \tag{4.7}$$

and

$$A(I - A\alpha)^{-1}\varphi = (I - A\alpha)^{-1}A\varphi \qquad (\varphi \in D_A). \tag{4.8}$$

From (4.7) it follows that

$$\lim_{\alpha \to 0} (I - A\alpha)^{-2}\varphi = \varphi \qquad (\varphi \in X). \tag{4.9}$$

Since the domain D_{A^2} of the operator A^2 is the range of the operator $(I - A\alpha)^{-2}$, it follows that D_{A^2} is dense in X. Also if $\varphi \in D_A$, then writing $\varphi_\alpha = (I - A\alpha)^{-1}\varphi$, we have $\varphi_\alpha \in D_{A^2}$, and using (4.7) and (4.8) we find

$$\lim_{\alpha \to 0} \varphi_\alpha = \varphi; \qquad \lim_{\alpha \to 0} A\varphi_\alpha = A\varphi. \tag{4.10}$$

We now prove some lemmas on products involving factors of the form $(I - A\alpha)^{-1}$.

Definition 4.1. *Let $P = \{s_0, s_1, \ldots, s_n\}$ be a partition of the interval $J =$ $[c,d]$. Let $\Delta s_k = s_k - s_{k-1}$. Then*

$$\Pi_P(J) = \Pi_P(c,d) = \prod_{k=1}^{n} (I - A\Delta s_k)^{-1}. \tag{4.11}$$

Remark. We note that the factors in $\Pi_P(c,d)$ commute. Also, the value of $\Pi_P(c,d)$ depends only on the length of $[c,d]$ and the Δs_k: if $a \in \mathbb{R}$ and $P + a = \{s_0 + a, s_1 + a, \ldots, s_n + a\}$, then

$$\Pi_{P+a}(c + a, d + a) = \Pi_P(c,d). \tag{4.12}$$

ISBN 0-201-13509-4

By (4.6) we have clearly

$$\|\Pi_P(c,d)\| \leqslant 1. \tag{4.13}$$

LEMMA 1. *Let P be a partition of $[c,d]$, with notation as above. Let $\varphi \in D_A$. Then*

$$\|(\Pi_P(c,d)-I)\varphi\| \leqslant (d-c)\|A\varphi\|. \tag{4.14}$$

Proof. Using (4.1) we have

$$\begin{aligned} I-\Pi_P(c,d) &= \prod_{k=1}^{n} I - \prod_{k=1}^{n} (I-A\Delta s_k)^{-1} \\ &= \sum_{l=1}^{n} \left(I-(I-A\Delta s_l)^{-1}\right) \prod_{k=1}^{l-1} (I-A\Delta s_k)^{-1} \\ &= \sum_{l=1}^{n} -A\Delta s_l \prod_{k=1}^{l} (I-A\Delta s_k)^{-1}. \end{aligned} \tag{4.15}$$

For $\varphi \in D_A$, using (4.8) and (4.6), we find

$$\|(\Pi_P(c,d)-I)\varphi\| \leqslant \sum_{l=1}^{n} \|A\varphi\|\Delta s_l = (d-c)\|A\varphi\|, \tag{4.16}$$

completing the proof. ■

LEMMA 2. *Let $\alpha, \beta > 0$ and let $\varphi \in D_{A^2}$. Then*

$$\|\{(I-A(\alpha+\beta))^{-1}-(I-A\alpha)^{-1}(I-A\beta)^{-1}\}\varphi\| \leqslant \alpha\beta\,\|A^2\varphi\|. \tag{4.17}$$

Proof. We have

$$\begin{aligned} &(I-A(\alpha+\beta))^{-1}-(I-A\alpha)^{-1}(I-A\beta)^{-1} \\ &\qquad = (I-A(\alpha+\beta))^{-1}\{(I-A\alpha)(I-A\beta)-(I-A(\alpha+\beta))\} \\ &\qquad\quad \times (I-A\alpha)^{-1}(I-A\beta)^{-1} \\ &\qquad = (I-A(\alpha+\beta))^{-1}\alpha\beta A^2(I-A\alpha)^{-1}(I-A\beta)^{-1}. \end{aligned} \tag{4.18}$$

The conclusion of the lemma now follows immediately from (4.8) and (4.6).

LEMMA 3. *Let J be an interval of length L and let $P=\{s_0,s_1,\ldots,s_n\}$ be a partition of J. Let $\varphi \in D_{A^2}$. Then*

$$\|\{(I-AL)^{-1}-\Pi_P(J)\}\varphi\| \leqslant L^2\|A^2\varphi\|. \tag{4.19}$$

ISBN 0-201-13509-4

Proof. If P has exactly three points (hence two subintervals, whose lengths may be called α and β), then the conclusion of Lemma 3 is immediate from Lemma 2. Assume for induction that Lemma 3 holds with $n=m-1$ and let $P=\{s_0,s_1,\ldots,s_m\}$ be a partition of J. Then $Q=\{s_0,s_1,\ldots,s_{m-1}\}$ is a partition of $[s_0,s_{m-1}]$. If $\varphi\in D_{A^2}$, then

$$\begin{aligned}\{(I-AL)^{-1}-\Pi_P(J)\}\varphi \\ &=\{(I-AL)^{-1}-(I-A(s_{m-1}-s_0))^{-1}(1-A\Delta s_m)^{-1}\}\varphi \\ &\quad+(I-A\Delta s_m)^{-1}\{(I-A(s_{m-1}-s_0))^{-1}-\Pi_Q(s_0,s_{m-1})\}\varphi.\end{aligned} \tag{4.20}$$

Applying Lemma 2 to the first term and the induction hypothesis to the second, we find

$$\begin{aligned}\|\{(I-AL)^{-1}-\Pi_P(J)\}\varphi\| &\leqslant (s_{m-1}-s_0)\Delta s_m\|A^2\varphi\|+(s_{m-1}-s_0)^2\|A^2\varphi\| \\ &=(s_{m-1}-s_0)L\|A^2\varphi\|\leqslant L^2\|A^2\varphi\|.\end{aligned} \tag{4.21}$$

This proves the lemma.

LEMMA 4. *Let P be a partition of $[a,b]$ with mesh $\mu(P)$ and let P' be a refinement of P. Let $\varphi\in D_{A^2}$. Then*

$$\|\{\Pi_P(a,b)-\Pi_{P'}(a,b)\}\varphi\|\leqslant\mu(P)(b-a)\|A^2\varphi\|. \tag{4.22}$$

Proof. Let $J_1,J_2,\ldots,J_n$ be the intervals of P, and let L_k be the length of J_k. The refinement P' is obtained from P by adding points to these intervals. Let $P_0=P$ and for $k\geqslant 1$ let P_k be the partition in which the extra points of P' have been added to the first k intervals of P. Then P_k is a refinement of P_{k-1} and $P_n=P'$. Also

$$\|\{\Pi_P(a,b)-\Pi_{P'}(a,b)\}\varphi\|\leqslant\sum_{k=1}^{n}\|\{\Pi_{P_k}(a,b)-\Pi_{P_{k-1}}(a,b)\}\varphi\|. \tag{4.23}$$

Lemma 3 and (4.6) immediately imply

$$\|\{\Pi_{P_k}(a,b)-\Pi_{P_{k-1}}(a,b)\}\varphi\|\leqslant L_k^2\|A^2\varphi\|. \tag{4.24}$$

Thus

$$\begin{aligned}\|\{\Pi_P(a,b)-\Pi_{P'}(a,b)\}\varphi\| &\leqslant\sum_{k=1}^{n}L_k^2\|A^2\varphi\|\leqslant\mu(P)\sum_{k=1}^{n}L_k\|A^2\varphi\| \\ &=\mu(P)(b-a)\|A^2\varphi\|,\end{aligned} \tag{4.25}$$

so we are done. ■

We can now prove existence of T^t:

ISBN 0-201-13509-4

THEOREM 4.1. *Let $t>0$ and let $a\in\mathbb{R}$. Let P denote a partition of $[a,a+t]$. Then the strong limit*

$$T^t = \lim_{\mu(P)\to 0} \Pi_P(a,a+t) \equiv \prod_a^{a+t} (I-A\,ds)^{-1} \tag{4.26}$$

exists. Let $T^\circ = I$. Then we have

$$\|T^t\| \leqslant 1; \qquad T^{t+s} = T^t T^s \qquad (t \geqslant 0); \qquad \lim_{t\downarrow 0} T^t = I \tag{4.27}$$

Proof. If $\varphi \in D_{A^2}$, then convergence of $\Pi_P(a,a+t)\varphi$ as $\mu(P)\to 0$ is immediate from Lemma 4 [one estimates the difference between $\Pi_{P_1}(a,a+t)\varphi$ and $\Pi_{P_2}(a,a+t)\varphi$ by considering the common refinement of P_1 and P_2). Since $\|\Pi_P(a,a+t)\| \leqslant 1$ and D_{A^2} is dense in X, convergence of $\Pi_P(a,a+t)\varphi$ for $\varphi \in X$ follows by a standard argument, and the inequality $\|T^t\| \leqslant 1$ is immediate. The function T^t is (as the notation indicates) independent of a because of (4.12). Now for $s,t>0$ we have

$$T^{t+s} = \lim_{\mu(P)\to 0} \Pi_P(a,a+t+s). \tag{4.28}$$

But choosing P to have a division point at $a+s$, we find

$$\Pi_P(a,a+t+s) = \Pi_{P_2}(a+s,a+t+s)\Pi_{P_1}(a,a+s), \tag{4.29}$$

where P_1 and P_2 are the partitions of $[a,a+s]$ and $[a+s,a+t+s]$ corresponding to P. Letting $\mu(P)\to 0$ in (4.29), we obtain $T^{t+s} = T^t T^s$. Recall $T^0 = I$. By Lemma 1, we have for $\varphi \in D_A$

$$\|(T^t - I)\varphi\| = \lim_{\mu(P)\to 0} \|(\Pi_P(a,a+t) - I)\varphi\| \leqslant t\|A\varphi\|. \tag{4.30}$$

Thus T^t converges to I on D_A as $t\to 0$, and convergence on all of X follows because D_A is dense and $\|T^t\| \leqslant 1$. ■

THEOREM 4.2. *If $\varphi \in D_A$, then*

$$\frac{d}{dt} T^t\varphi = A T^t \varphi. \tag{4.31}$$

Proof. First note that obviously

$$(I-A)^{-1} T^t = T^t (I-A)^{-1}, \tag{4.32}$$

so that T^t maps D_A to itself and

$$A T^t = T^t A. \tag{4.33}$$

ISBN 0-201-13509-4

Let $\varphi \in D_{A^2}$. We have for $h>0$

$$\left\| \left\{ \frac{T^{t+h}-T^t}{h} - AT^t \right\} \varphi \right\| \leqslant \left\| \frac{T^h - I}{h} - A\varphi \right\|$$

$$\leqslant \left\| \frac{T^h-(I-Ah)^{-1}}{h}\varphi \right\| + \left\| \left\{ \frac{(I-Ah)^{-1}-I}{h} - A \right\} \varphi \right\|. \tag{4.34}$$

Lemma 3 applied to the interval $[0,h]$ shows that

$$\left\| \frac{T^h-(I-Ah)^{-1}}{h}\varphi \right\| = \lim_{\mu(P)\to 0} \left\| \frac{\Pi_P(0,h)-(I-Ah)^{-1}}{h}\varphi \right\| \leqslant h\|A^2\varphi\|. \tag{4.35}$$

Also

$$\left(\frac{(I-Ah)^{-1}-I}{h} - A \right)\varphi = \{(I-Ah)^{-1}-I\}A\varphi = h(I-Ah)^{-1}A^2\varphi. \tag{4.36}$$

The last two equations prove that

$$\left\| \left\{ \frac{T^{t+h}-T^t}{h} - AT^t \right\} \varphi \right\| \leqslant 2h\|A^2\varphi\| \underset{h\to 0}{\to} 0. \tag{4.37}$$

This proves right differentiability (with the correct derivative) for $\varphi \in D_{A^2}$. If $\varphi \in D_A$, let $\varphi_\alpha = (I-A\alpha)^{-1}\varphi$. Then

$$\left\| \left\{ \frac{T^{t+h}-T^t}{h} - AT^t \right\} \varphi \right\| \leqslant \left\| \left\{ \frac{T^h-I}{h} - A \right\} \varphi \right\|$$

$$\leqslant \left\| \left\{ \frac{T^h-I}{h} - A \right\} (\varphi_\alpha - \varphi) \right\| + \left\| \left\{ \frac{T^h-I}{h} - A \right\} \varphi_\alpha \right\|. \tag{4.38}$$

Using (4.30) on the first term on the right-hand side and (4.37) on the second, we find

$$\left\| \left\{ \frac{T^{t+h}-T^t}{h} - AT^t \right\} \varphi \right\| \leqslant 2\|A(\varphi_\alpha - \varphi)\| + 2h\|A^2\varphi_\alpha\|. \tag{4.39}$$

Given $\varepsilon>0$, by (4.10) we can choose a fixed α so small that $2\|A(\varphi_\alpha-\varphi)\| < \varepsilon/2$. For this α, we can then choose h so small that $2h\|A^2\varphi_\alpha\| < \varepsilon/2$. This

ISBN 0-201-13509-4

proves right differentiability for $\varphi \in D_A$. But left differentiability follows in a trivial way (for $t>0$) from right differentiability and the semigroup property $T^{t+s}=T^tT^s$. Thus the theorem is proved.

4.5 An Example Involving Unbounded Operators with Variable Domain

In Section 3.6, we treated the equation

$$\frac{dU(x,c)\varphi}{dx}=iB(x)U(x,c)\varphi, \qquad U(c,c)=I, \tag{5.1}$$

for a case in which the operators $B(x)$ were (not necessarily bounded) self-adjoint operators on a Hilbert space $\mathcal{H}$. The operators $B(x)$ had a common domain D, and φ belonged to D. In some cases, it is desirable to study equations of the form (5.1) with operators $B(x)$ whose domains depend on x. [The operators $B(x)$ may even act in different Hilbert spaces.] This makes the study of (5.1) much more difficult. Still, there are cases in which progress can be made. One line of attack which has been useful [TK3] is the following: formally (5.1) is solved by

$$U(x,c)=\prod_c^x e^{iB(s)\,ds}. \tag{5.2}$$

The problem is to give meaning to the right-hand side of (5.2). Still arguing formally, suppose that $T(x)$ is a nonsingular operator-valued function which is differentiable in an appropriate sense. The similarity rule (3.5.34) yields

$$\prod_c^x e^{iB(s)\,ds}=T^{-1}(x)\prod_c^x e^{\{(LT)(s)+iT(s)B(s)T^{-1}(s)\}\,ds}T(c). \tag{5.3}$$

If T is chosen properly, it may happen that the product integral on the right-hand side of (5.3) can be given a sense (by Theorem 3.6.1, for example). If so, then we can regard (5.3) as the equation which defines $\prod_c^x e^{iB(s)\,ds}$. We can then ask whether $U(x,c)$ of (5.2) satisfies the desired equation (5.1). Formally, there is every reason to hope that it does. We now give an example to show that the approach described can actually work. The example to be given is somewhat unusual in that the operators $B(t)$ act in different Hilbert spaces. More typically [TK3] one studies families $\{B(t)\}$ all densely defined in the same space. Our example is chosen to illustrate the flexibility of the suggested method of solution. The equation to be discussed arises in conjunction with the problem of "continual measurement" in quantum mechanics. We refer the reader to [CF2] for a discussion of the physical questions involved. Because of the physical

ISBN 0-201-13509-4

significance of the equation to be studied, we use the variable t (for time) to parametrize the operators B.

Let $[t_0, t_1]$ denote an interval in $\mathbb{R}$. Let a and b denote twice continuously differentiable functions from $[t_0, t_1]$ to $\mathbb{R}$ with the property that

$$a(t) < b(t) \qquad (t \in [t_0, t_1]). \tag{5.4}$$

For each $t \in [t_0, t_1]$ we thus have a nontrivial interval $[a(t), b(t)] \subset \mathbb{R}$, whose length we denote by $l(t)$:

$$l(t) = b(t) - a(t), \qquad t \in [t_0, t_1]. \tag{5.5}$$

For each $t \in [t_0, t_1]$ we introduce the Hilbert space $\mathcal{H}_t$ consisting of all functions square-integrable over $[a(t), b(t)]$ with respect to Lebesgue measure:

$$\mathcal{H}_t = L^2(a(t), b(t)). \tag{5.6}$$

We shall write x to denote the variable on which the functions in $\mathcal{H}_t$ depend. We now introduce the family of self-adjoint operators $B(t)$: for $t \in [t_0, t_1]$, the operator $B(t)$ is "the operator d^2/dx^2 on the $\mathcal{H}_t$ with zero boundary conditions at the endpoints $a(t)$ and $b(t)$." More precisely, $B(t)$ is defined as follows:

DEFINITION 5.1.

(i) A function $f \in \mathcal{H}_t$ belongs to the domain D_t of $B(t)$ if and only if f has the following representation:

$$f(x) = \int_{a(t)}^{x} \left\{ \int_{a(t)}^{x_1} g(x_2)\, dx_2 \right\} dx_1, \tag{5.7}$$

where $g \in \mathcal{H}_t$ is such that, if $f(x)$ is given by (5.7), then $f(b(t)) = 0$. [Of course (5.7) implies $f(a(t)) = 0$.]

(ii) For $f \in D_t$ given by (5.7), $B(t)f$ is defined by

$$B(t)f = g. \tag{5.8}$$

With the above definition, $B(t)$ is self-adjoint by a standard argument.

The Hilbert spaces $\mathcal{H}_t$ can all be viewed in an obvious way as subspaces of $L^2(\mathbb{R})$. We denote by J_t the natural map of $\mathcal{H}_t$ into $L^2(\mathbb{R})$: if $\varphi \in \mathcal{H}_t$, then

$$(J_t\varphi)(x) = \begin{cases} \varphi(x), & x \in [a(t), b(t)], \\ 0, & x \notin [a(t), b(t)]. \end{cases}$$

Obviously J_t is one-to-one, and the range of J_t consists of all elements in $L^2(\mathbb{R})$ with support in $[a(t), b(t)]$.

ISBN 0-201-13509-4

We remark that the operators $B(t)$ defined above have domains which vary with t. Technically the various $B(t)$ act in different Hilbert spaces so that their domains have only the zero function in common. However, even if one views each domain D_t as a subset of $L^2(\mathbb{R})$ by considering J_tD_t instead of D_t, the sets J_tD_t and J_sD_s may have only the zero function in common. (This will happen if $[a(t),b(t)]\cap[a(s),b(s)]=\varnothing$.)

We now consider the following initial-value problem: suppose that φ is a given function in $\mathcal{H}_{t_0}$. For $t_0 \leqslant t \leqslant t_1$ we seek a function $\hat{\varphi}_t \in L^2(\mathbb{R})$ with the following properties:

(i) $\hat{\varphi}_t$ has support in $[a(t),b(t)]$, so that $\hat{\varphi}_t$ can be viewed as an element φ_t of $\mathcal{H}_t$. (That is, $\hat{\varphi}_t = J_t\varphi_t$.) Furthermore, $\varphi_{t_0}=\varphi$.

(ii) If $\varphi \in D_{t_0}$, then $\hat{\varphi}_t$ is differentiable in the sense of the norm topology on $L^2(\mathbb{R})$, $\varphi_t \in D_t$, and

$$\frac{d\hat{\varphi}_t}{dt} = iJ_tB(t)\varphi_t.$$

We will show that this initial-value problem has a solution. In fact, we will prove more than is required: we will demonstrate the existence of a family $\{U(t,s) \mid t,s\in[t_0,t_1]\}$ of unitary operators with the following properties:

(i) $U(t,s)$ maps $\mathcal{H}_s$ onto $\mathcal{H}_t$. $U(s,s)=I$.

(ii) For $r,s,t\in[t_0,t_1]$ we have

$$U(t,s)\,U(s,r)=U(t,r). \tag{5.9}$$

(iii) If φ belongs to the domain D_{t_0} of $B(t_0)$, then $U(t,t_0)\varphi$ belongs to the domain D_t of $B(t)$ and

$$\frac{d}{dt}J_tU(t,t_0)\varphi = iJ_tB(t)\,U(t,t_0)\varphi. \tag{5.10}$$

To prove existence of $U(t,s)$, we proceed as outlined at the start of this section. First, for each $t\in[t_0,t_1]$ let $\sigma_t:[0,1]\to[a(t),b(t)]$ be the coordinate transformation

$$\sigma_t(\xi) = (b(t)-a(t))\xi + a(t) = l(t)\xi + a(t) \qquad (\xi\in[0,1]). \tag{5.11}$$

We set

$$\mathcal{H} = L^2(0,1) \tag{5.12}$$

and define a map $T_t: \mathcal{H}_t \to \mathcal{H}$ by

$$(T_tf)(\xi) = l(t)^{1/2}f(\sigma_t(\xi)). \tag{5.13}$$

ISBN 0-201-13509-4

It is easily verified that the map T_t is a unitary map of $\mathcal{H}_t$ onto $\mathcal{H}$. We now study the function $LT+iTBT^{-1}$ occurring on the right-hand side of (5.3). First, since T_t is unitary, the operator $T_tB(t)T_t^{-1}$ is, for each $t\in[t_0,t_1]$, a self-adjoint operator on $\mathcal{H}$, and it is easy to see that $T_tB(t)T_t^{-1}$ is the operator "$(1/l(t)^2)d^2/dx^2$ with zero boundary conditions at 0 and 1." For each fixed t, this operator is a constant multiple of "d^2/dx^2 with zero boundary conditions at 0 and 1," so the domain D of this operator is *independent of* t. Next, we analyze LT. In this analysis we shall use a dot to denote differentiation with respect to t and a prime to denote differentiation with respect to ξ or x. Thus

$$(LT)(t)=\dot{T}_tT_t^{-1}. \tag{5.14}$$

Now assuming $f\in\mathcal{H}_t$ to be differentiable we have

$$\begin{aligned}(\dot{T}_tf)(\xi)&\equiv\frac{d}{dt}(T_tf)(\xi)=\frac{d}{dt}\left\{l(t)^{1/2}f(\sigma_t(\xi))\right\}\\&=\frac{1}{2}\frac{\dot{l}(t)}{l(t)^{1/2}}f(\sigma_t(\xi))+l(t)^{1/2}\dot{\sigma}_t(\xi)f'(\sigma_t(\xi))\\&=\frac{1}{2}\frac{\dot{l}(t)}{l(t)^{1/2}}f(\sigma_t(\xi))+l(t)^{1/2}\left\{\dot{l}(t)\xi+\dot{a}(t)\right\}f'(\sigma_t(\xi)).\end{aligned} \tag{5.15}$$

Setting $f=T_t^{-1}g$, we have

$$f(x)=l(t)^{-1/2}g(\sigma_t^{-1}(x))=l(t)^{-1/2}g\left(\frac{x-a(t)}{l(t)}\right) \tag{5.16}$$

and hence

$$f'(x)=l(t)^{-3/2}g'\left(\frac{x-a(t)}{l(t)}\right)=l(t)^{-3/2}g'(\sigma_t^{-1}(x)). \tag{5.17}$$

Inserting (5.16) and (5.17) into (5.15) we find

$$(\dot{T}_tT_t^{-1}g)(\xi)=\frac{\dot{l}(t)}{l(t)}\left\{\frac{1}{2}g(\xi)+\xi g'(\xi)\right\}+\frac{\dot{a}(t)}{l(t)}g'(\xi). \tag{5.18}$$

Thus $\dot{T}_tT_t^{-1}$ is the operator

$$\frac{\dot{l}(t)}{l(t)}\left\{\frac{1}{2}+\xi\frac{d}{d\xi}\right\}+\frac{\dot{a}(t)}{l(t)}\frac{d}{d\xi}$$

acting on $\mathcal{H}=L^2(0,1)$. It is necessary to be a little careful about the

ISBN 0-201-13509-4

domain of this operator, but this is not a problem for the following reason: we wish to product integrate the expression

$$iT_tB(t)T_t^{-1}+\dot{T}_tT_t^{-1}=i\{T_tB(t)T_t^{-1}-i\dot{T}_tT_t^{-1}\}\equiv iC(t). \tag{5.19}$$

Now the domain D of $T_tB(t)T_t^{-1}$ is known explicitly and is independent of t. It is easy to verify that all functions in D are differentiable (so that $\dot{T}_tT_t^{-1}$ can be applied to them) and that the operator $i\dot{T}_tT_t^{-1}$ is symmetric on D. Furthermore, a simple estimate shows that $-i\dot{T}_tT_t^{-1}$ is relatively bounded [MR1] with respect to the self-adjoint operator $T_tB(t)T_t^{-1}$, with relative bound less than 1, and thus, by a theorem of Kato [MR1], $C(t)$ is self-adjoint on the domain D for each $t\in[t_0,t_1]$. Writing out all operators explicitly, we thus have to deal with

$$C(t)=\frac{1}{l(t)^2}\frac{d^2}{d\xi^2}+\frac{\dot{l}(t)}{l(t)}\left\{\frac{1}{2}+\xi\frac{d}{d\xi}\right\}+\frac{\dot{a}(t)}{l(t)}\frac{d}{d\xi} \tag{5.20}$$

on the domain D. Now because we have assumed that the functions $a(t)$ and $b(t)$ are *twice* continuously differentiable, it follows immediately that $C(t)$ is strongly C^1 on D. Thus by the last remark in Chapter 3, it follows that $C(t)$ satisfies the hypotheses of Theorem 3.6.1, so that $\prod_s^t e^{iC(u)du}$ is defined and unitary for $s,t\in[t_0,t_1]$. We now put

$$U(t,s)=T_t^{-1}\prod_s^t e^{iC(u)du}T_s. \tag{5.21}$$

By the way T_t and T_s are defined, it is immediate that $U(t,s)$ is a unitary map of $\mathcal{H}_s$ onto $\mathcal{H}_t$. Clearly $U(s,s)=I$. The multiplication property (5.9) follows from the form of $U(t,s)$ and the fact that it holds for the product integral in (5.21). Finally, suppose that φ belongs to the domain D_{t_0} of $B(t_0)$. Then $T_{t_0}\varphi$ belongs to the domain D in $\mathcal{H}$, so $\prod_{t_0}^t e^{iC(u)du}T_{t_0}\varphi$ is differentiable with respect to t by Theorem 3.6.1. Furthermore, T_t^{-1} maps D to D_t, so $U(t,t_0)\varphi\in D_t$. The same argument used to prove the product rule for differentiation in calculus shows that $J_tU(t,t_0)\varphi$ is differentiable with respect to t and (5.10) is satisfied.

Notes to Chapter 4

The advantages of the systematic use of the product integral is evident in the methods of Chapter 4. The use of the sum and similarity formulas to write a product integral as the product of two product integrals one of which has an invertible limit as x tends to infinity is quite mechanical and straightforward, but, nevertheless, powerful. A translation of these methods into differential equation

ISBN 0-201-13509-4

language without mention of product integration is of course possible (see [JD5]) but only at the price of considerable awkwardness and circumlocution. (The reader should compare our presentation with that in books on differential equations which derive results on asymptotics of solutions.) We believe, for example, that we would not have discovered the "delicate convergence theorem" (Theorem 1.6.3) and its applications outside of the framework of product integration.

Some results on the possibility of existence of positive energy bound states for the Schrödinger equation are proved in [BS]; in the case of the Wigner and von Neumann example where a bound state occurs for $k=1$, it is shown that no bound state occurs for $k>4$. (Our results show that a bound state could only occur for $k=1$.)

Various results on asymptotics of solutions of differential equations, some of which seem to be applications of special cases of our "delicate convergence theorem", are proved by Atkinson [FA] and Wintner [AW]. However, Atkinson did not seem to be aware of our theorem in any general form, and Wintner only proves a special case of our theorem. The nonexistence of bound states for $k\neq 1$ in the Wigner-von Neumann examples is implied by Atkinson's results.

The Lie product formula in the form (3.1) with $A,B\in\mathbb{C}_{n\times n}$ can be proved in a quite elementary way by noting that if we put

$$F(t)=e^{tA}e^{tB},$$

then

$$F(t/n)=I+\frac{t}{n}(A+B)+o(t/n)$$

so

$$F(t/n)^n=\left(I+\frac{t}{n}(A+B)+o(t/n)\right)^n=\left(I+\frac{t}{n}(A+B)\right)^n+o(1)$$

by a simple calculation. This is equivalent to (3.1). The result (3.8) is more troublesome to prove directly.

The theory of semigroups of operators began with Stone's analysis of the structure of one-parameter unitary groups in Hilbert space [MS]. An extensive treatment of the situation in the Banach space case may be found in [EH]. In the present monograph we have viewed product integration as basically distinct from the theory of exponentiation of operators; indeed, excepting the discussion of Section 4 of the present chapter, we are always concerned with the construction of the product integral of an operator valued function $A(s)$ *given* the notion of the exponential $e^{A(s)\Delta s}$ (or a suitable replacement).

The example of Section 5 arose in the study of "continual position measurements" in a sense defined by Feynman. If $\psi\in\mathcal{H}_s$, then $\|U(t,s)\psi\|_{\mathcal{H}_t}$ represents the probability that a "continual observation" during the time interval $[s,t]$ to determine whether the particle described by ψ is in $[a(\xi),b(\xi)]$ for each $\xi\in[s,t]$ yields an affirmative answer. See [CF2] for further details.

Applications to scattering theory of our methods of determining the asymptotics of solutions of the Schrödinger equation are developed in [JD2].

ISBN 0-201-13509-4

CHAPTER 5

Product Integration of Measures

5.1. Introduction

In this chapter we shall extend some of the theory developed so far by introducing the notion of the product integral of a *measure*. This extension is a quite natural one from several points of view. On the one hand, from a purely integration-theoretic viewpoint, one might envision the possibility of product integrating the most general objects of ordinary integration theory, especially in the light of the development so far. Of course, it would be expected that such a theory would be closely related to the theory of (linear) differential equations with measures as coefficients. Such equations have occasionally been considered ([VG], [DP]) at least in special cases, but usually not in any systematic manner. They have some importance, however; for example, one problem in (quantum mechanical) scattering theory concerns the description of self-adjoint operators H with a part unitarily equivalent to $H_0 = -\Delta = -\sum_{i=1}^{n} \partial^2/\partial x_i^2$ operating in $L^2(\mathbb{R}^n)$, and some of these operators are of the form $H = H_0 + \mu$, with μ a measure. The scattering theory associated with such operators may be developed by using the product integral to study the asymptotic form of solutions of the equation $H\psi = k^2\psi$. (See [JD6] for a development along these lines.) From a slightly different but related point of view, investigations of the structure of operator-valued functions $M(x,y)$ satisfying the "propagator identities"

$$M(x,y)M(y,z) = M(x,z), \qquad M(x,x) = I$$

led H. S. Wall and his students to the notion of product integration of various types of measures. (See the notes at the end of this chapter.)

ENCYCLOPEDIA OF MATHEMATICS and Its Applications, Gian-Carlo Rota (ed.). Vol. 10: John D. Dollard and Charles N. Friedman, Product Integration with Applications to Differential Equations

ISBN 0-201-13509-4

The fact that the product integral of an operator-valued *function* is expressible as a time-ordered exponential expansion should reinforce the belief that product integration of measures is quite feasible, since that expansion continues to make perfectly good sense if the integrand is replaced by a more general operator-valued *measure*. As might be expected, the theory to be discussed is more complicated than, say, the theory of Chapter 1, and some surprises are involved. It is possible to discuss the new theory in the Banach space context, but in order to avoid unnecessary complications we will confine the discussion to $n\times n$ matrices as in Chapter 1. This discussion will suffice to indicate the problems which arise and the results which are obtainable. Because the development of the theory is substantially parallel to the development in Chapter 1, our treatment will be abbreviated at certain points. We shall give full detail only when new features appear or extensive modification is required in proofs.

To begin with, we must settle a small point about the objects to be product integrated. We will need

DEFINITION 1.1. *Let μ be a positive Borel measure on an interval $[a,b)$ of real numbers. The space $L^1([a,b);\mu)$ is the set of all (equivalence classes of) functions $A:[a,b)\to\mathbb{C}_{n\times n}$ satisfying*

$$\|A\|_1\equiv\int_{[a,b)}\|A(s)\|\,\mu(ds)<\infty. \tag{1.1}$$

We plan to define $\prod_a^b e^{A(s)\mu(ds)}$ for $A\in L^1([a,b);\mu)$. This procedure has a close formal similarity with the procedure followed in previous chapters, and is convenient for this reason. However, there are certain formal drawbacks to restricting oneself to the notation $A(s)\mu(ds)$ for an object to be product integrated. In proving the analogue of the sum rule, for instance, one will presumably wish to product integrate an expression of the form $A(s)\mu(ds)+B(s)\nu(ds)$, which is not of the form $C(s)\lambda(ds)$. Of course, by the use of the Radon-Nikodym Theorem, setting $\lambda=\mu+\nu$ we could write

$$A(s)\mu(ds)+B(s)\nu(ds)=\left\{A(s)\frac{d\mu}{d\lambda}(s)+B(s)\frac{d\nu}{d\lambda}(s)\right\}\lambda(ds),$$

thus avoiding the difficulty. However, the necessity of continually rewriting objects in this way would make the theory cumbersome to deal with. Our solution to this notational difficulty is to introduce the concept of a matrix-valued measure:

DEFINITION 1.2. *An* n×n complex matrix-valued Borel measure *on an interval $[a,b)$ of real numbers is a function $\hat{\nu}$ from the Borel sets of $[a,b)$ to $\mathbb{C}_{n\times n}$ which is countably additive on disjoint families.*

ISBN 0-201-13509-4

An $n\times n$ matrix-valued measure $\hat{\mu}$ on $[a,b)$ can be considered in an obvious way as an $n\times n$ matrix whose (i,j) entry μ_{ij} is a complex-valued measure. An important fact about such measures [WR] is that they are always of bounded total variation, i.e., if $|\mu_{ij}|$ denotes the total variation of μ_{ij}, then $|\mu_{ij}|$ is a *finite* measure on $[a,b)$. We define the *total variation* $|\hat{\mu}|$ of $\hat{\mu}$ by

$$|\hat{\mu}|=\sum_{i,j=1}^{n}|\mu_{ij}|. \tag{1.2}$$

Now if μ is a positive Borel measure on $[a,b)$ and $A\in L^1([a,b);\mu)$, then the expression $A(s)\mu(ds)$ can be viewed as a matrix-valued measure $\hat{\mu}$ by the association

$$\hat{\mu}(T)=\int_T A(s)\mu(ds) \tag{1.3}$$

for any Borel set T. Conversely if $\hat{\mu}$ is a matrix-valued measure, then using the Radon-Nikodym Theorem the expression $\hat{\mu}(ds)$ can be replaced by $A(s)|\hat{\mu}|(ds)$, where $A=d\hat{\mu}/d|\hat{\mu}|\in L^1([a,b);|\hat{\mu}|)$. For this reason, one can maintain that the objects to be product integrated are $n\times n$ matrix-valued measures, and the content of the theory is the same as if one decided to product integrate objects of the form $A(s)\mu(ds)$. In the matrix-valued measure language, the product integral appears as $\prod_a^b e^{\hat{\mu}(ds)}$. There is no difficulty in formulating the sum rule because if $\hat{\mu}_1$ and $\hat{\mu}_2$ are matrix-valued measures, then the same is true of $\hat{\mu}_1+\hat{\mu}_2$.

The lesson to be learned from the above discussion is that both the notation involving $A(s)\nu(ds)$ and the notation of matrix-valued measures have advantages. Since it is an entirely elementary matter to convert from one notation to the other, our policy will be to use whichever notation is more convenient in a given context. For example, in proving existence of the product integral we use the notation $A(s)\mu(ds)$ to stress and exploit the similarity with previous existence proofs. In proving the sum rule, of course, we use the notation of matrix-valued measures. Some results are written out in both notations for emphasis.

In the rest of this chapter, all measures are understood to be Borel measures. As above, $n\times n$ matrix-valued measures are denoted using hats ($\hat{\mu}$, $\hat{\nu}$, etc.). The reader should not lose sight of the fact that any such measure is of finite total variation. A measure without a hat (μ, ν, etc.) is always taken to be a positive measure unless explicit mention is made to the contrary. We stress that in using the word "positive" we mean to *exclude* the value $+\infty$, so that any measure μ is *finite* unless otherwise indicated. We shall comment occasionally on the removability of the finiteness restriction on μ.

ISBN 0-201-13509-4

5.2 The Product Integral

A general measure μ can have "pure points," i.e., points $x \in \mathbb{R}$ such that $\mu(\{x\}) \neq 0$. (Such points are also called *atoms*. We follow here the terminology of [MR1].) For this reason, in integrating or product integrating over an interval it is necessary to specify whether or not the endpoints are included in the integration region. In this chapter our policy will be the following: *in integrating or product integrating over an interval from a smaller number y to a larger number x, we will choose the interval to be closed at y and open at x.* To avoid later notational clutter, it is convenient to make the following definition:

DEFINITION 2.1. *Let f be a function, let μ be a measure, and let $x,y \in \mathbb{R}$. Assuming all indicated integrals make sense, the symbol $\int_y^x f(s)\mu(ds)$ is defined as follows:*

If $y<x$, then

$$\int_y^x f(s)\mu(ds) = \int_{[y,x)} f(s)\mu(ds). \tag{2.1}$$

If $y>x$, then

$$\int_y^x f(s)\mu(ds) = -\int_x^y f(s)\mu(ds) = -\int_{[x,y)} f(s)\mu(ds). \tag{2.2}$$

Finally

$$\int_y^y f(s)\mu(ds) = 0. \tag{2.3}$$

With this definition, $\int_y^x f(s)\mu(ds)$ has all the usual formal properties, for example,

$$\int_x^z f(s)\mu(ds) + \int_y^x f(s)\mu(ds) = \int_y^z f(s)\mu(ds). \tag{2.4}$$

Until further notice let μ denote a fixed (finite, positive Borel) measure on $[a,b)$. In discussing product integration with respect to μ we will, as in previous chapters, be concerned with partitions of $[a,b)$ and step-functions on $[a,b)$. To fix notation we make

DEFINITION 2.2. *A* partition *P of $[a,b)$ is a finite collection $P = \{s_0, s_1, \ldots, s_n\}$ with $a = s_0 < s_1 < \cdots < s_n = b$. The* intervals *of P are the disjoint sets $I_k = [s_{k-1}, s_k)$, $k = 1, \ldots, n$.*

DEFINITION 2.3. *A function $A : [a,b) \to \mathbb{C}_{n\times n}$ is a* step-function *if and only if there is a partition P of $[a,b)$ such that A is constant on each interval I_k of P. The value of A on I_k is denoted A_k.*

ISBN 0-201-13509-4

Definition 2.3 is slightly more restrictive than the usual definition of step-function. For instance if $c \in [a,b)$, then the function A equal to I at c and zero elsewhere is not a step-function. However, our only interest in step-functions is the fact that they are dense in $L^1([a,b);\mu)$, and as will be seen this is true of the step-functions defined in Definition 2.3.

DEFINITION 2.4. *Let $A \in L^1([a,b);\mu)$ and let $P = \{s_0, s_1, \ldots, s_n\}$ be a partition of $[a,b)$. The* mean-value approximant $\bar{A}_P$ *corresponding to A and P is the step-function which on the subinterval $I_k = [s_{k-1}, s_k)$ of P takes the value*

$$\bar{A}_{P_k} = \begin{cases} \dfrac{1}{\mu(I_k)} \displaystyle\int_{s_{k-1}}^{s_k} A(s)\mu(ds), & \text{if } \mu(I_k) \neq 0 \\ 0, & \text{if } \mu(I_k) = 0. \end{cases} \tag{2.5}$$

As in the corresponding argument in Chapter 3, it is not difficult to prove that any $A \in L^1([a,b);\mu)$ is the L^1-limit of $\bar{A}_P$ as the mesh of P approaches zero, i.e.,

$$\lim_{\text{mesh}(P) \to 0} \|A - \bar{A}_P\|_1 = 0. \tag{2.6}$$

Thus the step-functions are dense in $L^1([a,b);\mu)$.

We now proceed to the construction of the product integral. We begin by defining for each step-function B a function $E_B(x)$ which will turn out to be the product integral of B over $[a,x)$ with respect to μ.

DEFINITION 2.5. *Let $B:[a,b) \to \mathbb{C}_{n\times n}$ be a step-function, with notations as in Definitions 2.2 and 2.3. We define $E_B:[a,b] \to \mathbb{C}_{n\times n}$ as follows:*

$$\begin{aligned} E_B(a) &= I && (\text{recall } a = s_0) \\ E_B(x) &= e^{B_1 \mu([s_0,x))}, && x \in (s_0, s_1] \\ &= e^{B_2 \mu([s_1,x))} e^{B_1 \mu(I_1)}, && x \in (s_1, s_2] \\ &\;\vdots \\ &= e^{B_n \mu([s_{n-1},x))} \cdots e^{B_2 \mu(I_2)} e^{B_1 \mu(I_1)}, && x \in (s_{n-1}, s_n]. \end{aligned} \tag{2.7}$$

The function E_B above is plainly analogous to the function E_B of Definition 1.1.3, and it has a number of similar properties. We shall need the following:

LEMMA 2.1.

(i) E_B is left-continuous on $[a,b]$. If $x \in [a,b]$ and x is not a pure point of μ then E_B is continuous at x.

(ii) E_B satisfies the bound

$$\|E_B(x)\| \leqslant e^{\int_a^x \|B(s)\| \mu(ds)} \leqslant e^{\|B\|_1}. \tag{2.8}$$

ISBN 0-201-13509-4

Proof. Part (i) is a simple consequence of the following facts: (a) since μ is a measure we have

$$\mu([c,d)) = \lim_{\varepsilon\downarrow 0} \mu([c,d-\varepsilon)) \tag{2.9}$$

and (b) if d is not a pure point on μ, then also

$$\mu([c,d)) = \lim_{\varepsilon\downarrow 0} \mu([c,d+\varepsilon)). \tag{2.10}$$

Part (ii) is handled just as in the proof of Lemma 1.1.1. Thus we are done. ■

The reader will note that we have not included an integral equation for E_B in Lemma 2.1, in contrast to what was done in Lemma 1.1.1. This is because in the present context integral equations present certain difficulties which are best deferred until later. We can now prove existence of the product integral:

THEOREM 2.1. *Let $A \in L^1([a,b);\mu)$ and let $\{A_n\}$ be any sequence of step-functions converging to A in the L^1 sense. Then the sequence $\{E_{A_n}(x)\}$ converges uniformly on $[a,b]$ to a matrix denoted $\prod_a^x e^{A(s)\mu(ds)}$ and called* the product integral of A over $[a,x)$ with respect to μ.

Proof. Let B and C be step-functions on $[a,b)$. By taking a common refinement if necessary, we may assume that B and C have the same partition $P = \{s_0, s_1, \ldots, s_n\}$. Using the telescoping sum formula of equation (4.4.1), we then have

$$\begin{aligned} E_B(b) - E_C(b) &= \prod_{k=1}^{n} e^{B_k\mu(I_k)} - \prod_{k=1}^{n} e^{C_k\mu(I_k)} \\ &= \sum_{l=1}^{n} \prod_{k=l+1}^{n} e^{B_k\mu(I_k)}\{e^{B_l\mu(I_l)} - e^{C_l\mu(I_l)}\} \prod_{k=1}^{l-1} e^{C_k\mu(I_k)}. \end{aligned} \tag{2.11}$$

Using an estimate similar to (2.8) on the products in (2.11), we obtain

$$\|E_B(b) - E_C(b)\| \leqslant e^{\|B\|_1 + \|C\|_1} \sum_{l=1}^{n} \|e^{B_l\mu(I_l)} - e^{C_l\mu(I_l)}\|. \tag{2.12}$$

The elementary estimate

$$\|e^M - e^N\| \leqslant \|M-N\| e^{\|M\| + \|N\|} \qquad (M,N \in \mathbb{C}_{n\times n}) \tag{2.13}$$

ISBN 0-201-13509-4

then yields

$$\|E_B(b)-E_C(b)\| \leqslant e^{2\{\|B\|_1+\|C\|_1\}} \sum_{l=1}^{n} \|B_l - C_l\| \, \mu(I_l)$$
$$= e^{2\{\|B\|_1+\|C\|_1\}} \|B-C\|_1. \tag{2.14}$$

Now Eq. (2.14) actually holds with b replaced by any $x \in [a,b]$; if we had gone through the above argument with b replaced by x, we would have obtained the same result except that on the right-hand side the L^1 norms would have been norms in the space $L^1([a,x);\mu)$. These are dominated by the corresponding norms in $L^1([a,b);\mu)$, so (2.14) holds with b replaced by x.

The proof of the theorem is now immediate: we have for $x \in [a,b]$

$$\|E_{A_m}(x) - E_{A_n}(x)\| \leqslant e^{2\{\|A_m\|_1+\|A_n\|_1\}} \|A_m - A_n\|_1. \tag{2.15}$$

Since $\{A_n\}$ converges to A in the L^1 sense, Eq. (2.15) shows that $\{E_{A_n}\}$ is uniformly Cauchy, hence uniformly convergent. If $\{B_n\}$ is another sequence of step-functions converging to A in L^1, then estimating $\|E_{B_n}(x) - E_{A_n}(x)\|$ by the above method, we find that E_{B_n} has the same limit as E_{A_n}. This completes the proof. ■

We pause briefly in our development to show how the product integral $\prod_a^b e^{\hat{\mu}(ds)}$ of a matrix-valued measure $\hat{\mu}$ could have been defined directly. Namely, write as before

$$\hat{\mu}(dx) = A(x) |\hat{\mu}|(dx) \tag{2.16}$$

with $A \in L^1([a,b);|\mu|)$. Letting P denote a partition of $[a,b)$ and letting $\bar{A}_P$ be the corresponding mean-value approximant computed using $|\mu|$, Theorem 2.1 implies that

$$\prod_a^b e^{A(s)|\mu|(ds)} = \lim_{\text{mesh}(P)\to 0} E_{\bar{A}_P}(x)$$
$$= \lim_{\text{mesh}(P)\to 0} \prod_{k=1}^{n} e^{\bar{A}_{P_k} |\mu|(I_k)}. \tag{2.17}$$

But the definition (2.5) of A_{P_k} and Eq. (2.16) show that

$$\bar{A}_{P_k} |\mu|(I_k) = \int_{s_{k-1}}^{s_k} \hat{\mu}(ds) \equiv \hat{\mu}(I_k). \tag{2.18}$$

ISBN 0-201-13509-4

Thus we could have defined directly

$$\prod_a^b e^{\hat{\mu}(ds)} = \lim_{\text{mesh}(P)\to 0} \prod_{k=1}^{n} e^{\hat{\mu}(I_k)}.$$

(Of course $\prod_a^x e^{\hat{\mu}(ds)}$ is defined by partitioning $[a,x)$.) This discussion should serve to emphasize the equivalence of product integrating expressions of the form $A(s)\mu(ds)$ and ones of the form $\hat{\mu}(ds)$.

One other remark is in order: if ν is a positive Borel measure on $[a,b)$ which is not finite, then if $A \in L^1([a,b);\nu)$ it is possible to define $\prod_a^x e^{A(s)\nu(ds)}$ at least for $x \in [a,b)$, for the following reason: ν is *regular*, so the restriction of ν to $[a,c)$ is a finite measure for each $c \in [a,b)$, so that Theorem 2.1 guarantees existence of $\prod_a^x e^{A(s)\nu(ds)}$ for $x \in [a,c]$. Our later study of improper integration will show that $\prod_a^b e^{A(s)\nu(ds)}$ can also be defined in the above situation as $\lim_{x\uparrow b} \prod_a^x e^{A(s)\nu(ds)}$. We now proceed with the development of the theory with the finite measure μ.

By applying Theorem 2.1 to the interval $[y,x)$, we can say that if $A \in L^1([a,b),\mu)$ then $\prod_y^x e^{A(s)\mu(ds)}$ has been defined for all $x,y \in [a,b]$ with $y \leqslant x$. As might be suspected and as is formally stated in the next theorem, $\prod_y^x e^{A(s)\mu(ds)}$ is *nonsingular*. Assuming this fact, we make

DEFINITION 2.6. *Let $A \in L^1([a,b);\mu)$ and let $x,y \in [a,b]$ with $y > x$. We define*

$$\prod_y^x e^{A(s)\mu(ds)} = \left(\prod_x^y e^{A(s)\mu(ds)} \right)^{-1}. \tag{2.19}$$

We rapidly summarize some useful properties of the product integral in the next theorem:

THEOREM 2.2. *Let $A \in L^1([a,b);\mu)$. Then*

(*i*) *if A is a step-function, we have*

$$\prod_a^x e^{A(s)\mu(ds)} = E_A(x). \tag{2.20}$$

(*ii*) *For $x,y \in [a,b]$, $\prod_y^x e^{A(s)\mu(ds)}$ is nonsingular and*

$$\det\left(\prod_y^x e^{A(s)\mu(ds)} \right) = e^{\int_y^x \operatorname{tr} A(s)\mu(ds)}. \tag{2.21}$$

(*iii*) *If the values $A(s)$ and $A(s')$ commute for all $s,s' \in [a,b)$ [except possibly for $s,s' \in Z$, where $\mu(Z)=0$], then for $x,y \in [a,b)$ we have*

$$\prod_y^x e^{A(s)\mu(ds)} = e^{\int_y^x A(s)\mu(ds)}. \tag{2.22}$$

ISBN 0-201-13509-4

(iv) For $x,y,z \in [a,b]$ we have

$$\prod_x^z e^{A(s)\mu(ds)} \prod_y^x e^{A(s)\mu(ds)} = \prod_y^z e^{A(s)\mu(ds)}. \tag{2.23}$$

(v) For $x,y \in [a,b]$, $\prod_y^x e^{A(s)\mu(ds)}$ is left-continuous in x and in y. If x_0 is not a pure point of μ, then $\prod_y^x e^{A(s)\mu(ds)}$ is continuous in x at x_0, and the analogous statement holds with x_0 and x replaced by y_0 and y.

(vi) For $x,y \in [a,b]$ we have

$$\left\| \prod_y^x e^{A(s)\mu(ds)} \right\| \leqslant e^{|\int_y^x \|A(s)\| \mu(ds)|} \leqslant e^{\|A\|_1}. \tag{2.24}$$

Proof. Statements (i) through (iv) and (vi) all have direct analogues in Chapter 1 and are proved in the same way. We leave the details to the reader. In statement (v), the assertions about dependence on x can be obtained from the corresponding property (i) of E_B in Lemma 2.1, and the uniformity of the convergence asserted in Theorem 2.1. The assertions about dependence on y (which will play almost no role in our development) can be proved by first studying the case in which A is a step-function. Then $\prod_y^x e^{A(s)\mu(ds)}$ can be written out explicitly and the assertions about y-dependence can be verified using properties of measures like (2.9) and (2.10). For general $A \in L^1([a,b);\mu)$ one then views $\prod_y^x e^{A(s)\mu(ds)}$ as the limit (uniform in y) of $\prod_y^x e^{A_n(s)\mu(ds)}$, where $\{A_n\}$ is a sequence of step-functions converging to A in the L^1 sense. Again we leave details to the reader. ■

5.3 Integral and Differential Equations

If

$$P(x) = \prod_a^x e^{A(s)\mu(ds)}, \tag{3.1}$$

then, on the basis of previous results, it might be expected that

$$P(x) = I + \int_a^x A(s)P(s)\mu(ds). \quad (\textit{False in general.}) \tag{3.2}$$

In fact, this equation does hold *provided that μ has no pure points*. However, (3.2) need not hold in general. Since this is somewhat surprising, we give an example in which (3.2) is false: let $A(x) = I$ for $x \in [a,b)$, let $c \in (a,b)$, and let μ be a unit point mass at $c: \mu(\{c\}) = 1$ and $\mu([a,b) \setminus \{c\}) = 0$. Since the values of A commute, we have

$$P(x) = e^{\int_a^x I\mu(ds)} = \begin{cases} I, & \text{if } x \leqslant c, \\ eI & \text{if } x > c. \end{cases} \tag{3.3}$$

ISBN 0-201-13509-4

On the other hand, using this expression for $P(x)$ we find

$$I+\int_a^x A(s)P(s)\mu(ds)=\begin{cases} I, & \text{if } x\leqslant c,\\ 2I & \text{if } x>c.\end{cases} \tag{3.4}$$

Thus (3.2) is false for $x>c$.

We will now exhibit the correct integral equation for a product integral involving a measure. The reason for the form of this equation is best seen in the proof of its correctness, so no preliminary heuristic motivation will be given. We will, however, need some definitions:

DEFINITION 3.1. *The entire functions* $\varphi:\mathbb{C}\to\mathbb{C}$ *and* $\check{\varphi}:\mathbb{C}\to\mathbb{C}$ *are defined by*

$$\varphi(z)=\sum_{n=0}^{\infty}\frac{z^n}{(n+1)!}=\begin{cases}\dfrac{e^z-1}{z} & (z\neq 0)\\ 1 & (z=0),\end{cases} \tag{3.5}$$
$$\check{\varphi}(z)=\varphi(-z)=e^{-z}\varphi(z).$$

If $B\in\mathbb{C}_{n\times n}$, then $\varphi(B)$ and $\check{\varphi}(B)$ are defined in the usual way via the power series for $\varphi(z)$ and $\check{\varphi}(z)$. Because $z\varphi(z)=e^z-1$ and $z\check{\varphi}(z)=1-e^{-z}$ for *all* $z\in\mathbb{C}$, we have

$$\varphi(B)B=B\varphi(B)=e^B-I,$$
$$\check{\varphi}(B)B=B\check{\varphi}(B)=I-e^{-B} \qquad (B\in\mathbb{C}_{n\times n}). \tag{3.6}$$

We also note that if $\|B\|\leqslant M$, then

$$\|\varphi(B)\|\leqslant\sum_{n=0}^{\infty}\frac{\|B\|^n}{(n+1)!}\leqslant\sum_{n=0}^{\infty}\frac{M^n}{(n+1)!}=\varphi(M),$$
$$\|\check{\varphi}(B)\|=\|\varphi(-B)\|\leqslant\varphi(M). \tag{3.7}$$

DEFINITION 3.2. *Let μ be a (positive finite Borel) measure on $[a,b)$. For each $x\in[a,b)$ we define $\mu(x)$ to be the μ-measure of the set $\{x\}$:*

$$\mu(x)\equiv\mu(\{x\}). \tag{3.8}$$

We make the same convention for a matrix-valued measure $\hat{\mu}$. We note that $\mu(x)=0$ except for an at most countable set, the set of pure points of μ. If $A\in L^1([a,b);\mu)$, then the function $s\to A(s)\mu(s)$ is well defined for all $s\in[a,b)$. [If $\mu(s)=0$, then the values of $A(s)$ may be indeterminate but $A(s)\mu(s)=0$.] We note that for any $x\in[a,b)$ we have

$$\|A(x)\mu(x)\|\leqslant\int_a^b\|A(s)\|\,\mu(ds)=\|A\|_1. \tag{3.9}$$

ISBN 0-201-13509-4

Combining (3.9) and (3.7), we have

$$\begin{aligned}&\varphi(A(x)\mu(x))\leqslant\varphi(\|A\|_1),\\&\check{\varphi}(A(x)\mu(x))\leqslant\varphi(\|A\|_1),\qquad(x\in[a,b)).\end{aligned}\tag{3.10}$$

DEFINITION 3.3. *Let $A\in L^1([a,b);\mu)$. The functions A_φ and $A_{\check{\varphi}}$ in $L^1([a,b);\mu)$ are defined by*

$$\begin{aligned}&A_\varphi(x)=\varphi(A(x)\mu(x))A(x),\\&A_{\check{\varphi}}(x)=\check{\varphi}(A(x)\mu(x))A(x),\qquad(x\in[a,b)).\end{aligned}\tag{3.11}$$

By (3.10) we have

$$\|A_\varphi(x)\|\leqslant\varphi(\|A\|_1)\|A(x)\|\tag{3.12}$$

and correspondingly

$$\|A_\varphi\|_1\leqslant\varphi(\|A\|_1)\|A\|_1=e^{\|A\|_1}-1.\tag{3.13}$$

The same argument shows that $\|A_{\check{\varphi}}\|_1\leqslant e^{\|A\|_1}-1$. We remark that if x is not a pure point of μ, then $A(x)\varphi(x)=0$, so that $A_\varphi(x)=\varphi(0)A(x)=A(x)$, and similarly $A_{\check{\varphi}}(x)=A(x)$. In particular, if μ has no pure points, then $A_\varphi=A_{\check{\varphi}}=A$.

We can now state

THEOREM 3.1. *Let $A\in L^1([a,b);\mu)$. Let $x,y\in[a,b]$ and let*

$$P(x,y)=\prod_y^x e^{A(s)\mu(ds)}.\tag{3.14}$$

Then $P(x,y)$ is a solution of both of the equations

$$P(x,y)=I+\int_y^x A_\varphi(s)P(s,y)\mu(ds)\tag{3.15}$$

and

$$P(x,y)=I+\int_y^x P(x,s)A_{\check{\varphi}}(s)\mu(ds).\tag{3.16}$$

Remark. As mentioned above, if μ has no pure points, then $A_\varphi=A$. In this case (3.15) (with $y=a$) reduces to the "expected" equation (3.2).

Proof of Theorem 3.1. We will prove (3.15) with $y=a$. The result for general y then follows as in Theorem 3.5.3. The proof of (3.16) will be left

ISBN 0-201-13509-4

to the reader. Thus we wish to establish that

$$P(x,a)=I+\int_a^x A_\varphi(s)P(s,a)\mu(ds). \tag{3.17}$$

First suppose that A is a constant function, whose constant value we also denote by A. Since the values of A commute we have

$$P(x,a)=e^{\int_a^x A\mu(ds)}=e^{A\mu([a,x))}. \tag{3.18}$$

Let $Q=\{s_0,s_1,\ldots,s_n\}$ be a partition of $[a,x)$ with $I_k=[s_{k-1},s_k), k=1,\ldots,n$. Then using (3.18) and (3.6) we have

$$\begin{aligned} P(x,a)-I &= \sum_{k=1}^{n}\{P(s_k,a)-P(s_{k-1},a)\}=\sum_{k=1}^{n}\{e^{A\mu(I_k)}-I\}P(s_{k-1},a) \\ &= \sum_{k=1}^{n}\varphi(A\mu(I_k))A\mu(I_k)P(s_{k-1},a)\equiv\int_a^x \psi_Q(s)\mu(ds), \end{aligned} \tag{3.19}$$

where $\psi_Q(s)$ is the step-function taking the value $\varphi(A\mu(I_k))AP(s_{k-1},a)$ on the interval I_k. Now let $\{Q_n\}$ be a sequence of partitions with mesh tending to zero. Because φ is continuous and $P(s,a)$ is left continuous, $\psi_{Q_n}(s)$ then approaches pointwise the function $\varphi(A\mu(s))AP(s,a)=A_\varphi(s)P(s,a)$. A trivial estimate [see (3.7)] shows that

$$\|\psi_{Q_n}(s)\|\leqslant\varphi(\|A\|\,\mu([a,b)))\|A\|e^{\|A\|\,\mu([a,b))} \tag{3.20}$$

so by Lebesgue's bounded convergence theorem

$$\lim_{n\to\infty}\int_a^x \psi_{Q_n}(s)\mu(ds)=\int_a^x A_\varphi(s)P(s,a)\mu(ds), \tag{3.21}$$

and (3.17) is established.

Next suppose that A is a step-function with corresponding partition Q. Then (3.17) can be established by using the result just proved on each subinterval of Q and piecing together the results [this requires use of the multiplicative property of product integrals, Theorem 2.2, part (iv)].

Finally suppose that A is a general function in $L^1([a,b);\mu)$. Let $\{A_n\}$ be a sequence of step-functions converging to A in the L^1 sense. Let

$$P_n(x,a)=\prod_a^x e^{A_n(s)\mu(ds)}. \tag{3.22}$$

Theorem 2.1 implies that $P_n(x,a)=E_{A_n}(x)$ converges uniformly to $P(x,a)$ on $[a,b]$. Furthermore, since A_n is a step-function we have

$$P_n(x,a)=I+\int_a^x \varphi(A_n(s)\mu(s))A_n(s)P_n(s,a)\mu(ds). \tag{3.23}$$

ISBN 0-201-13509-4

Now for $x \in [a,b)$ we have [see (3.9)]

$$\|\{A(x)-A_n(x)\}\,\mu(x)\| \leqslant \|A-A_n\|_1. \tag{3.24}$$

Thus $A_n(x)\mu(x)$ converges uniformly to $A(x)\mu(x)$ as $n\to\infty$. Since also $\|A_n(x)\mu(x)\| \leqslant \|A_n\|_1$ and φ is uniformly continuous on bounded sets, $\varphi(A_n(x)\mu(x))$ converges uniformly to $\varphi(A(x)\mu(x))$ as $n\to\infty$. This fact together with uniform convergence of P_n to P and L^1 convergence of A_n to A allows us to take the limit as $n\to\infty$ under the integral sign in (3.23) to obtain (3.17). This completes the proof. ■

COROLLARY 1. *The product integral $P(x,y)$ of* (3.14) *is a function of bounded variation on $[a,b]$ (as a function of x or of y).*

Proof. We prove only bounded variation in x. If $\{s_0,s_1,\ldots,s_n\}$ is a partition of $[a,b]$, then using (3.15) we have

$$\sum_{k=1}^{n} \|P(s_k,y)-P(s_{k-1},y)\| = \sum_{k=1}^{n} \left\| \int_{s_{k-1}}^{s_k} A_\varphi(s)P(s,y)\mu(ds) \right\|$$

$$\leqslant e^{\|A\|_1} \int_a^b \|A_\varphi(s)\|\,\mu(ds) < \infty, \tag{3.25}$$

where we have used the bound (2.24) for $\|P(s,y)\|$. This completes the proof. ■

COROLLARY 2. *Considered as a function of x, the product integral $P(x,y)$ of* (3.15) *has a right-hand limit at each point $x\in[a,b)$ and this right-hand limit is nonsingular. The same statement holds considering $P(x,y)$ as a function of y on $[a,b)$.*

Proof. We discuss only dependence on x. Existence of the right-hand limit

$$P(x+,y) = \lim_{\varepsilon\downarrow 0} P(x+\varepsilon,y) \tag{3.26}$$

follows from the fact that $P(x,y)$ is of bounded variation in x. Nonsingularity of $P(x+,y)$ follows from the formula

$$\det P(x+,y) = \lim_{\varepsilon\downarrow 0} e^{\int_y^{x+\varepsilon} \operatorname{tr} A(s)\mu(ds)} = e^{\lim_{\varepsilon\downarrow 0} \int_y^{x+\varepsilon} \operatorname{tr} A(s)\mu(ds)} \neq 0. \tag{3.27}$$

For convenience, we restate Eq. (3.17) in the language of matrix-valued measures: if $\hat{\mu}$ is a matrix-valued measure on $[a,b)$ then setting $P(x,a) = \prod_a^x e^{\hat{\mu}(ds)}$, we have

$$P(x,a) = I + \int_a^x \hat{\mu}_\varphi(ds)P(s,a), \tag{3.28}$$

ISBN 0-201-13509-4

where

$$\hat{\mu}_{\varphi}(ds)=\varphi(\hat{\mu}(s))\hat{\mu}(ds). \tag{3.29}$$

■

THEOREM 3.2. *Let $A\in L^1([a,b);\mu)$ and let $y\in[a,b]$. Then the solutions of Eq.* (3.15) *and* (3.16) *are unique and are given by the time-ordered exponential series*

$$P(x,y)=I+\int_y^x A_{\varphi}(s_1)\mu(ds_1)+\int_y^x A_{\varphi}(s_1)\left\{\int_y^{s_1} A_{\varphi}(s_2)\mu(ds_2)\right\}\mu(ds_1)+\cdots \tag{3.30}$$

and

$$P(x,y)=I+\int_y^x A_{\check{\varphi}}(s_1)\mu(ds_1)+\int_y^x\left\{\int_{s_1}^x A_{\check{\varphi}}(s_2)\mu(ds_2)\right\}A_{\check{\varphi}}(s_1)\mu(ds_1)+\cdots \tag{3.31}$$

which converge uniformly on $[a,b]$. Thus the product integral can be represented as a time-ordered exponential.

Proof. We discuss Eq. (3.30) for the case $y=a$. As in Theorem 3.1, the other cases are similar and are left to the reader. Thus we will prove that (3.30) with $y=a$ gives the unique solution of (3.17). We first note that if $P(x,a)$ is any solution of (3.17), then the integrand $A_{\varphi}P(\cdot,a)\mu$ must be a matrix-valued measure in order for the indicated integral to have a sense for all $x\in[a,b]$. But if ν is a matrix-valued measure on $[a,b)$, then $\int_a^x\nu(ds)$ is a bounded function on $[a,b]$, and it follows that $P(x,a)$ is bounded, say

$$\|P(x,a)\|\leqslant M \qquad (x\in[a,b]). \tag{3.32}$$

Iterating (3.17) (with $y=a$) n times we obtain

$$P(x,a)=I+\sum_{k=1}^{n-1} J_k(x,a)+R_n(x,a) \tag{3.33}$$

where

$$J_k(x,a)=\int_a^x A_{\varphi}(s_1)\left\{\int_a^{s_1}\cdots\left\{\int_a^{s_{k-1}} A_{\varphi}(s_k)\mu(ds_k)\right\}\cdots\right\}\mu(ds_1) \tag{3.34}$$

and

$$R_n(x,a)=\int_a^x A_{\varphi}(s_1)\left\{\int_a^{s_1}\cdots\left\{\int_a^{s_{n-1}} A_{\varphi}(s_n)P(s_n,a)\mu(ds_n)\right\}\cdots\right\}\mu(ds_1). \tag{3.35}$$

ISBN 0-201-13509-4

To prove (3.30) with $y=a$, we need only show that $R_n(x,a)$ approaches zero uniformly on $[a,b]$ as $n\to\infty$. But using (3.32) we have

$$\|R_n(x,a)\| \leqslant M\int_a^x \|A_\varphi(s_1)\|\left\{\int_a^{s_1}\cdots\left\{\int_a^{s_{n-1}}\|A_\varphi(s_n)\|\,\mu(ds_n)\right\}\cdots\right\}\mu(ds_1)$$

$$\leqslant M\frac{\left\{\int_a^x \|A_\varphi(s)\|\,\mu(ds)\right\}^n}{n!} \tag{3.36}$$

where the last inequality holds because the expression $\{\int_a^x\|A_\varphi(s)\|\,\mu(ds)\}^n$ can be viewed as the sum of (i) $n!$ equal n-fold integrals in each of which the variables occur in a fixed order $s_{i_1}<s_{i_2}<\cdots<s_{i_n}$ and (ii) an n-fold integral over the set of $(s_1,\ldots,s_n)$ defined by the condition that at least two of the variables are equal. From (3.36) we have immediately

$$\|R_n(x,a)\| \leqslant M\frac{\|A_\varphi\|_1^n}{n!} \qquad (x\in[a,b]) \tag{3.37}$$

and this proves the desired uniform convergence of R_n to zero when $n\to\infty$.

To prove uniqueness of the solution of (3.30) (with $y=a$) we need only note that if $P(x,a)$ and $Q(x,a)$ satisfy this equation, then their difference satisfies the corresponding *homogeneous* equation. Iterating this equation n times, we find that only the term corresponding to $R_n(x,a)$ is present. Proceeding as above we find for any n

$$\|P(x,a)-Q(x,a)\| \leqslant M'\frac{\|A_\varphi\|_1^n}{n!}, \tag{3.38}$$

proving that $P(x,a)=Q(x,a)$ for all $x\in[a,b]$. This completes the proof. ■

It will be useful in the sequel to have a differential equation for the product integral $P(x,y)$ of Eq. (3.14). Because in general $P(x,y)$ is not literally differentiable, the differential equation will have to be valid in a generalized sense, which we take to be the sense of distribution theory. Before deriving the differential equation we make some preliminary remarks and state a lemma which will be useful now and in the later development.

If $f:[a,b]\to\mathbb{C}$ is a left-continuous function of bounded variation, then f generates a unique Lebesgue-Stieltjes measure df on $[a,b)$: df is the measure which to the subinterval $[c,d)$ of $[a,b)$ assigns the value $f(d)-f(c)$. Naturally the measure df has bounded total variation. The same construction yields a matrix-valued measure if f has values in $\mathbb{C}_{n\times n}$. In agreement with the convention of Definition 3.2, we denote by $df(x)$ the measure assigned to the set $\{x\}$ by df. If f is continuous at x, then $df(x)=0$. Now if f and g are both left-continuous and of bounded

ISBN 0-201-13509-4

variation (with values in $\mathbb{C}$ or $\mathbb{C}_{n\times n}$), then the same holds for fg. The lemma to be proved relates $d(fg)$ to df and dg.

LEMMA 3.1. *Let f and g be left-continuous functions of bounded variation on an interval $[a,b)$ with values in $\mathbb{C}$ (or $\mathbb{C}_{n\times n}$). Then*

$$d(fg)=(df)g+f\,dg+\rho_{fg} \tag{3.39}$$

where ρ_{fg} is the measure defined by

$$\rho_{fg}(S)=\sum_{x\in S} df(x)\,dg(x). \tag{3.40}$$

Proof. First note that in the sum over the (possibly uncountable) set S of (3.40), only countably many terms are nonzero and the sum is absolutely convergent, because both f and g have bounded variation. Next, if either f or g is continuous, then $\rho_{fg}=0$ and (3.39) is a special case of a result given in [FR], p. 118. For general f and g, we may decompose both f and g into the sum of a continuous function and a left-continuous saltus function [FR]. Using the known validity of (3.39) when one of the functions is continuous, it is seen that the only problem is to check (3.39) when both f and g are left-continuous saltus functions, which we now assume. In this case, df and dg are pure point measures. Let $[c,d)\subset[a,b)$. We will verify that

$$f(d)g(d)-f(c)g(c)=\int_c^d (df)g+\int_c^d f\,dg+\rho_{fg}([c,d)) \tag{3.41}$$

and this will establish (3.39). Let $F=\{s\in[c,d)\,|\,df(s)\neq 0\}$ and let G be the corresponding set for dg. Assuming for simplicity that $f(c)=g(c)=0$ we have for $x\in[c,d)$

$$\begin{aligned} f(x)&=\int_c^x df(s)=\sum_{\substack{s\in F\\ s<x}} df(s),\\ g(x)&=\sum_{\substack{s\in G\\ s<x}} dg(s). \end{aligned} \tag{3.42}$$

Thus

$$\int_c^d (df)g=\sum_{s\in F} df(s)\sum_{\substack{t\in G\\ t<s}} dg(t) \tag{3.43}$$

ISBN 0-201-13509-4

and

$$\int_c^d f\,dg = \sum_{t\in G}\sum_{\substack{s\in F\\ s<t}} df(s)\,dg(t). \tag{3.44}$$

Finally,

$$\rho_{fg}([c,d)) = \sum_{\substack{s\in F\\ t\in G\\ t=s}} df(s)\,dg(t). \tag{3.45}$$

We now add the last three equations to find

$$\int_c^d (df)g + \int_c^d f\,dg + \rho_{fg}([c,d)) = \sum_{\substack{s\in F\\ t\in G}} df(s)\,dg(t)$$

$$= \sum_{s\in F} df(s) \sum_{t\in G} dg(t) = f(d)g(d) = f(d)g(d) - f(c)g(c), \tag{3.46}$$

where rearrangement of the sum in (3.46) is permissible because it converges absolutely. If $f(c)$ and $g(c)$ are nonzero, the argument is no harder but there are more terms in the equations. The proof is now complete. ■

We call attention to one consequence of the lemma:

COROLLARY. (Integration by parts formula.) *If f and g are left-continuous functions of bounded variation from $[a,b)$ to $\mathbb{C}$ or $\mathbb{C}_{n\times n}$, and if f and g have no common points of discontinuity, then for any $[c,d)\subset[a,b)$ we have*

$$\int_c^d (df)g + \int_c^d f\,dg = f(d)g(d) - f(c)g(c). \tag{3.47}$$

Proof. Under the hypotheses of the corollary we have $\rho_{fg}=0$, so (3.47) follows directly from (3.39). ■

We can now state

LEMMA 3.2. *Let $A\in L^1([a,b);\mu)$ and for $x,y\in[a,b]$ let $P(x,y)=\prod_y^x e^{A(s)\mu(ds)}$. Then the following equation holds in the sense of distribution theory:*

$$\frac{dP(x,y)}{dx} = A_\varphi(x)P(x,y)\mu. \tag{3.48}$$

Proof. Write $P(x,y)$ as $P_y(x)$ to emphasize its dependence on x. P_y is left-continuous and of bounded variation on $[a,b)$. Also, according to

ISBN 0-201-13509-4

(3.15), the Lebesgue-Stieltjes measure dP_y associated with P_y is

$$dP_y = A_\varphi P_y \mu. \tag{3.49}$$

Let ψ be a C^∞ complex-valued function with compact support in (a,b). Since P_y and ψ have no common points of discontinuity we can use the corollary of Lemma 3.1 and (3.49) to conclude that

$$\int_a^b \psi(s) A_\varphi(s) P(s,y) \mu(ds) = -\int_a^b P(s,y)\psi'(s)\,ds. \tag{3.50}$$

But (3.50) is precisely the statement that (3.48) holds in the sense of distribution theory. Thus we are done. ■

We note that in the terminology of matrix-valued measures, Eq. (3.48) appears as follows: if $P(x,y) = \prod_y^x e^{\hat{\mu}(ds)}$, then in the sense of distribution theory we have

$$\frac{dP(x,y)}{dx} = \hat{\mu}_\varphi P(x,y), \tag{3.51}$$

where $\hat{\mu}_\varphi$ is defined in (3.29).

The results proved above on integral equations imply that if $\hat{\mu}$ is a matrix-valued measure on $[a,b)$, then the integral equation

$$P(x,y) = I + \int_y^x \hat{\mu}(ds) P(s,y) \tag{3.52}$$

has a product integral solution provided that $\hat{\mu}$ can be written in the form

$$\hat{\mu} = \hat{\nu}_\varphi \tag{3.53}$$

for some matrix-valued measure $\hat{\nu}$. The solution is then uniquely given by

$$P(x,y) = \prod_y^x e^{\hat{\nu}(ds)}. \tag{3.54}$$

We note that if $\hat{\mu}$ has no pure points, then $\hat{\mu} = \hat{\mu}_\varphi$ so that (3.52) is satisfied by $P(x,y) = \prod_y^x e^{\hat{\mu}(ds)}$. If $\hat{\mu}$ has pure points, the situation is of course more complicated. If (3.53) is to hold, we must have for each pure point x of μ the equation

$$\hat{\mu}(x) = \hat{\nu}_\varphi(x) = \varphi(\hat{\nu}(x))\hat{\nu}(x) = e^{\hat{\nu}(x)} - I. \tag{3.55}$$

Thus a necessary condition for (3.53) to hold is that $\hat{\mu}(x) + I$ should be

ISBN 0-201-13509-4

nonsingular at each pure point x of $\hat{\mu}$. It is not difficult to prove that this condition is also sufficient:

LEMMA 3.3. *Let $\hat{\mu}$ be a matrix-valued measure on $[a,b)$ such that $\hat{\mu}(x)+I$ is nonsingular for each pure point x of $\hat{\mu}$. Then there is a matrix-valued measure $\hat{\nu}$ on $[a,b)$ such that* (3.53) *holds.*

Proof. We specify the continuous and the pure-point part of $\hat{\nu}$ separately. The continuous part $\hat{\nu}_c$ is taken to be the same as the continuous part $\hat{\mu}_c$ of μ. To specify the pure-point part $\hat{\nu}_{pp}$, let $S=\{x_1,x_2,\dots\}$ be an enumeration of the pure points of $\hat{\mu}$. Then

$$\sum_{n=1}^{\infty} \|\hat{\mu}(x_n)\| < \infty, \tag{3.56}$$

so there is an N so large that $\|\hat{\mu}(x_n)\| \leqslant 1/2$ for $n>N$. We now define $\hat{\nu}_{pp}(x)$ as follows: if $x \notin S$, then $\hat{\nu}_{pp}(x)=0$. For $n=1,\dots,N$ we choose $\hat{\nu}_{pp}(x_n)$ to be any matrix such that

$$I+\hat{\mu}(x_n)=e^{\hat{\nu}_{pp}(x_n)}. \tag{3.57}$$

[Such a matrix exists because $I+\hat{\mu}(x_n)$ is nonsingular. See Appendix I.] For $n>N$, we define

$$\hat{\nu}_{pp}(x_n)=\log(I+\hat{\mu}(x_n)) \equiv \sum_{k=1}^{\infty} (-1)^{k+1} \frac{\hat{\mu}(x_n)^k}{k}. \tag{3.58}$$

The series in (3.58) clearly converges, and $\hat{\nu}_{pp}(x_n)$ as defined in (3.58) satisfies (3.57). Furthermore, for $n>N$ we have

$$\|\hat{\nu}_{pp}(x_n)\| \leqslant \|\hat{\mu}(x_n)\| \sum_{k=0}^{\infty} \left(\frac{1}{2}\right)^k = 2\|\hat{\mu}(x_n)\|. \tag{3.59}$$

This inequality shows that

$$\sum_{n=1}^{\infty} \|\hat{\nu}_{pp}(x_n)\| < \infty. \tag{3.60}$$

Thus $\hat{\nu} \equiv \hat{\nu}_c + \hat{\nu}_{pp}$ is a matrix-valued measure on $[a,b)$, and Eq. (3.53) holds by construction, completing the proof. ■

Now Eq. (3.53) is not a necessary condition for the integral equation (3.52) to have a solution of *some* sort. In fact, the proof of Theorem 3.2

ISBN 0-201-13509-4

shows that (3.52) has a unique time-ordered exponential solution for *any* matrix-valued measure $\hat{\mu}$. However, if (3.52) has a *product integral* solution, then (3.53) must be satisfied. In fact, if $P(x,y)$ is a product integral and satisfies (3.52), then at any pure point x of μ we have

$$P(x+,y)-P(x,y)=\lim_{\varepsilon\downarrow 0}\int_x^{x+\varepsilon}\hat{\mu}(ds)P(s,y)=\hat{\mu}(x)P(x,y). \tag{3.61}$$

Since $P(x,y)$ is nonsingular, (3.61) implies

$$\hat{\mu}(x)+I=P(x+,y)P(y,x). \tag{3.62}$$

Since $P(x+,y)$ is also invertible, (3.62) shows that $\hat{\mu}(x)+I$ is invertible, so there is a $\hat{\nu}$ satisfying (3.53). Thus (3.53) is a necessary and sufficient condition for (3.52) to have a product integral solution. We can use the above type of analysis to give a characterization of product integrals, which we state as

THEOREM 3.3. *Let $Q:[a,b]\to\mathbb{C}_{n\times n}$ be a function. Then there is a matrix-valued measure $\hat{\nu}$ on $[a,b)$ such that*

$$Q(x)=\prod_y^x e^{\hat{\nu}(ds)}Q(y) \tag{3.63}$$

for $x,y\in[a,b]$ if and only if the following conditions are satisfied:

(i) *$Q(x)$ is left-continuous and of bounded variation on $[a,b]$.*

(ii) *$Q(x)$ is nonsingular for all $x\in[a,b]$ and the right-hand limit $Q(x+)$ is nonsingular for all $x\in[a,b)$.*

Proof. The necessity of conditions (i) and (ii) has been established in our previous work. To prove sufficiency, suppose that (i) and (ii) hold. Then $\|Q^{-1}(x)\|$ is left-continuous on $[a,b]$ and has a right-hand limit at each $x\in[a,b)$, so that $\|Q^{-1}(x)\|$ is bounded. Now for $x,y\in[a,b]$ we have

$$Q(x)=Q(y)+\int_y^x dQ=Q(y)+\int_y^x\hat{\mu}(ds)Q(s), \tag{3.64}$$

where

$$\hat{\mu}=(dQ)Q^{-1}. \tag{3.65}$$

As indicated by the notation, $\hat{\mu}$ is a complex-valued measure on $[a,b)$ (this follows from the boundedness of $\|Q^{-1}\|$). Multiply (3.64) on the right by $Q^{-1}(y)$ and write $Z(x)=Q(x)Q^{-1}(y)$. Then

$$Z(x)=I+\int_y^x\hat{\mu}(ds)Z(s). \tag{3.66}$$

ISBN 0-201-13509-4

Furthermore, $\hat{\mu}$ can be represented as in (3.53) because by (3.65) the pure points of $\hat{\mu}$ are the points of discontinuity of Q, and at any such point x we have

$$\hat{\mu}(x)=\{Q(x+)-Q(x)\}Q^{-1}(x)=Q(x+)Q^{-1}(x)-I. \tag{3.67}$$

Thus $\hat{\mu}(x)+I$ is nonsingular and a $\hat{\nu}$ can be found such that (3.53) holds. But then the unique solution of (3.66) is given by

$$Z(x)=\prod_y^x e^{\hat{\nu}(ds)}. \tag{3.68}$$

This last equation is identical with (3.63), and the theorem is proved. ■

The following corollary is an immediate consequence of the theorem and its proof:

COROLLARY. (The fundamental theorem of product integration.) *Let Q satisfy the hypotheses of Theorem* 3.3. *Let LQ denote any matrix-valued measure on $[a,b)$ such that*

$$(LQ)_\varphi=dQ\cdot Q^{-1}. \tag{3.69}$$

Then for $x,y\in[a,b]$ we have

$$\prod_y^x e^{LQ(ds)}=Q(x)Q^{-1}(y). \tag{3.70}$$

5.4 Further Properties of Product Integrals

In this section we obtain some results which are of general use and which will prepare us for the study of improper product integration in the next section.

THEOREM 4.1. *Let $A\in L^1([a,b);\mu)$ and let $x,y\in[a,b]$. Then*

$$\left\|\prod_y^x e^{A(s)\mu(ds)}\right\|\leqslant e^{\left|\int_y^x\|A_\varphi(s)\|\,\mu(ds)\right|} \tag{4.1}$$

and

$$\left\|\prod_y^x e^{A(s)\mu(ds)}-I\right\|\leqslant e^{\left|\int_y^x\|A_\varphi(s)\|\,\mu(ds)\right|}-1. \tag{4.2}$$

ISBN 0-201-13509-4

Remark. The inequality (4.1) differs from the earlier bound (2.24) because A_φ instead of A appears in the integral on the right-hand side.

Proof of Theorem 4.1. Both (4.1) and (4.2) follow directly from the time-ordered exponential expansion (3.30) for $P(x,y)=\prod_y^x e^{A(s)\mu(ds)}$, because as in the proof of Theorem 3.2 we can estimate the nth term on the right-hand side of (3.30) by $(1/n!)|\int_y^x \|A_\varphi(s)\|\,\mu(ds)|^n$. This completes the argument. We pause to remark that the bound (4.1) is suggestive. The right-hand side of (4.1) is finite if $A_\varphi \in L^1([a,b);\mu)$, and this is a *less* restricitive condition than the condition $A\in L^1([a,b);\mu)$. {For example, suppose that μ is a pure point measure concentrated on the countable set $S=\{x_1,x_2,\dots\}\subset[a,b)$ and assigning weight $1/n^2$ to x_n. Let A be the function such that $A(x_n)=2\pi i n^3 I$ and $A(x)=0$ for $x\notin S$. Then $A_\varphi=0\in L^1([a,b);\mu)$ but $A\notin L^1([a,b);\mu)$.} The bound (4.1) suggests the possibility that $\prod_y^x e^{A(s)\mu(ds)}$ could be defined under the condition $A_\varphi\in L^1([a,b);\mu)$. This is in fact the case, although we shall not pursue the point here because situations in which A_φ belongs to $L^1([a,b);\mu)$ but A does not are somewhat pathological, as indicated by the example given above. We note one consequence of (4.2): since $\|A_\varphi(s)\|\leqslant\varphi(\|A\|_1)\|A(s)\|$, (4.2) implies that

$$\left\|\prod_y^x e^{A(s)\mu(ds)}-I\right\|\leqslant e^{\varphi(\|A\|_1)|\int_y^x\|A(s)\|\,\mu(ds)|}-1. \tag{4.3}$$

■

We state the next theorem in the language of matrix-valued measures:

THEOREM 4.2. (The sum rule.) *Let $\hat\mu$ and $\hat\nu$ be $n\times n$ matrix-valued measures on $[a,b)$ having no common pure points. Let y be fixed in $[a,b]$ and for $x\in[a,b]$ define*

$$P(x)=\prod_y^x e^{\hat\mu(ds)}. \tag{4.4}$$

Then

$$\prod_y^x e^{\hat\mu(ds)+\hat\nu(ds)}=P(x)\prod_y^x e^{P^{-1}(s)\hat\nu(ds)P(s)}. \tag{4.5}$$

Remark. The expressions $\hat\mu+\hat\nu$ and $P^{-1}\hat\nu P$ are both in an obvious way matrix-valued measures on $[a,b)$, so the product integrals occurring in (4.5) are well defined.

Proof of Theorem 4.2. Write

$$Q(x)=\prod_y^x e^{P^{-1}(s)\hat\nu(ds)P(s)}. \tag{4.6}$$

ISBN 0-201-13509-4

The function P and Q are both left-continuous and of bounded variation on $[a,b)$, and they have no common points of discontinuity because $\hat{\mu}$ and $\hat{\nu}$ have no common pure points. The Corollary of Lemma 3.1 then implies that

$$P(x)Q(x)=P(y)Q(y)+\int_y^x P\,dQ+\int_y^x (dP)Q. \tag{4.7}$$

Now $P(y)Q(y)=I$, and because of the integral equations satisfied by P and Q we have

$$dP=\hat{\mu}_\varphi P \tag{4.8}$$

and

$$dQ=(P^{-1}\hat{\nu}P)_\varphi Q. \tag{4.9}$$

The definition (3.5) of φ makes it clear that for $B\in\mathbb{C}_{n\times n}$ we have $\varphi(P^{-1}BP)=P^{-1}\varphi(B)P$. Thus

$$dQ=\varphi([P^{-1}\hat{\nu}P](s))(P^{-1}\hat{\nu}P)Q=P^{-1}\varphi(\hat{\nu}(s))\hat{\nu}PQ=P^{-1}\hat{\nu}_\varphi PQ. \tag{4.10}$$

From (4.7) we therefore obtain

$$(PQ)(x)=I+\int_y^x \{\hat{\mu}_\varphi(ds)+\hat{\nu}_\varphi(ds)\}(PQ)(s). \tag{4.11}$$

Letting $W(x)$ denote the left-hand side of (4.5), we have

$$W(x)=I+\int_y^x \{\hat{\mu}+\hat{\nu}\}_\varphi(ds)\,W(s). \tag{4.12}$$

By our uniqueness theorem for integral equations, we will have $W=PQ$ if we can show that

$$\{\hat{\mu}+\hat{\nu}\}_\varphi=\hat{\mu}_\varphi+\hat{\nu}_\varphi. \tag{4.13}$$

Let S and T be the sets of pure points of $\hat{\mu}$ and $\hat{\nu}$, respectively, and let $U=[a,b)\setminus S\cup T$. Let χ_R denote the characteristic function of the set R. Now $S\cap T=\varnothing$. Also S and T are both countable. Since $\hat{\nu}$ has no pure points in S, we thus have $\hat{\nu}(S)=0$. Similarly, $\hat{\mu}(T)=0$. Thus $\hat{\nu}\chi_S=\hat{\mu}\chi_T=0$.

ISBN 0-201-13509-4

Combining these facts and the fact that $\varphi(0)=I$, we have

$$\begin{aligned}\{\hat{\mu}+\hat{\nu}\}_{\varphi} &= \varphi([\hat{\mu}+\hat{\nu}](s))(\hat{\mu}+\hat{\nu})(\chi_U+\chi_S+\chi_T)\\ &= (\hat{\mu}+\hat{\nu})\chi_U+\varphi(\hat{\mu}(s))\hat{\mu}\chi_S+\varphi(\hat{\nu}(s))\hat{\nu}\chi_T\\ &= \{\varphi(\hat{\mu}(s))\hat{\mu}+\varphi(\hat{\nu}(s))\hat{\nu}\}\chi_U\\ &\quad +\varphi(\hat{\mu}(s))\hat{\mu}(\chi_S+\chi_T)+\varphi(\hat{\nu}(s))\hat{\nu}(\chi_S+\chi_T)\\ &= \hat{\mu}_{\varphi}+\hat{\nu}_{\varphi},\end{aligned} \tag{4.14}$$

as desired. This proves the theorem. ■

COROLLARY 1. *Let $\hat{\lambda}$ be an $n\times n$ matrix-valued measure on $[a,b)$ and fix $y\in[a,b]$. Let*

$$W(x,y)=\prod_y^x e^{\hat{\lambda}(ds)}. \tag{4.15}$$

Then $W(x,y)$ can be written as

$$W(x,y)=P(x)Q(x) \tag{4.16}$$

where P is a saltus function and Q is continuous on $[a,b]$.

Proof. It suffices to write

$$\hat{\lambda}=\hat{\mu}+\hat{\nu} \tag{4.17}$$

where $\hat{\mu}$ is the pure point part of $\hat{\lambda}$ and $\hat{\nu}$ is its continuous part. Define P and Q as in Theorem 4.1. Using the integral equation for P, it is easy to show that P is a saltus function. Also, the measure $P^{-1}\hat{\nu}P$ has no pure points, so Q is continuous. This completes the proof. It should be obvious that we could equally well have chosen P to be continuous and Q to be a saltus function. ■

COROLLARY 2. (The similarity rule). *Let $\hat{\mu}$ be a matrix-valued measure on $[a,b)$. Let Q be a function satisfying the hypotheses of Theorem 3.3 and such that the measure LQ defined in the corollary to Theorem 3.3 has no pure points in common with $\hat{\mu}$. Then*

$$\prod_y^x e^{\hat{\mu}(ds)}=Q^{-1}(x)\prod_y^x e^{LQ(ds)+Q(s)\hat{\mu}(ds)Q^{-1}(s)}Q(y). \tag{4.18}$$

Proof. Apply the sum rule to the product integral on the right-hand side of (4.18) and use the fact that $\prod_y^x e^{LQ(ds)}=Q(x)Q^{-1}(y)$. The right-hand side of (4.18) then becomes $Q^{-1}(y)\prod_y^x e^{Q(y)\hat{\mu}(ds)Q^{-1}(y)}Q(y)$. But this agrees with the left-hand side of (4.18) because, by an obvious argument, if M is a

ISBN 0-201-13509-4

constant nonsingular matrix, then

$$\prod_y^x e^{M\hat{\mu}(ds)M^{-1}} = M \prod_y^x e^{\hat{\mu}(ds)} M^{-1}. \tag{4.19}$$

This completes the proof. ■

5.5 Improper Product Integration

We adopt the usual notion of improper product integrals:

DEFINITION. *Suppose that μ is a measure on $[a,\infty)$ such that $\mu([a,b))<\infty$ for each $b\in[a,\infty)$. Suppose that $A:[a,\infty)\to\mathbb{C}_{n\times n}$ is a function belonging to $L^1([a,b);\mu)$ for each $b\in[a,\infty)$. We define the* improper product integral *$\prod_a^\infty e^{A(s)\mu(ds)}$ by*

$$\prod_a^\infty e^{A(s)\mu(ds)} = \lim_{x\to\infty} \prod_a^x e^{A(s)\mu(ds)} \tag{5.1}$$

provided that the indicated limit exists.

We can prove theorems on improper product integration as in Chapter 1. The basic result is

THEOREM 5.1. *Suppose that $A\in L^1([a,\infty);\mu)$. Then $\prod_a^\infty e^{A(s)\mu(ds)}$ exists and is nonsingular.*

Proof. Writing

$$\|A\|_1 = \int_a^\infty \|A(s)\|\,\mu(ds) \tag{5.2}$$

we have the estimate (4.3) for all $x,y\in[a,\infty)$. The proof of Theorem 5.1 is a verbatim repetition of the proof of the analogous Theorem 1.6.1, except that the estimate (4.3) is used instead of the estimate (1.4.2). We leave the verification to the reader. ■

In the language of matrix-valued measures, Theorem 5.1 states that if $\hat{\mu}$ is a matrix-valued measure (necessarily of finite total variation) on $[a,\infty)$, then $\prod_a^\infty e^{\hat{\mu}(ds)}$ exists and is nonsingular. We can prove a corollary analogous to the corollary of Theorem 1.6.1:

COROLLARY. *Suppose $\hat{\mu}$ and $\hat{\nu}$ are functions on the bounded Borel sets in $[a,\infty)$ which are measures when restricted to $[a,b)$ for each $b\in[a,\infty)$. Further suppose that $\prod_a^\infty e^{\hat{\mu}(ds)}$ exists and is nonsingular and that $\hat{\nu}$ has finite total variation on $[a,\infty)$. Finally, suppose that $\hat{\mu}$ and $\hat{\nu}$ have no common pure points. Then $\prod_a^\infty e^{\hat{\mu}(ds)+\hat{\nu}(ds)}$ exists and is nonsingular.*

ISBN 0-201-13509-4

Proof. The proof is such a simple modification of the proof of the corollary of Theorem 1.6.1 that we again leave verification to the reader. ■

It is possible in the present context to prove an analogue of the "delicate" convergence result given in Theorem 1.6.3. To do this, we recall that the function $\check{\varphi}:\mathbb{C}\to\mathbb{C}$ is defined by

$$\check{\varphi}(z)=\varphi(-z)=\varphi(z)e^{-z} \tag{5.3}$$

and if $A\in L^1([a,b);\mu)$, we have set

$$A_{\check{\varphi}}(s)=\check{\varphi}(A(s)\mu(s))A(s)=A_{\varphi}(s)e^{-\mu(s)A(s)}. \tag{5.4}$$

THEOREM 5.2. *Suppose that* $A:[a,\infty)\to\mathbb{C}_{n\times n}$ *is a function such that the improper integral*

$$\operatorname{imp}\int_a^{\infty}A_{\check{\varphi}}(s)\mu(ds)\equiv\lim_{b\to\infty}\int_a^b A_{\check{\varphi}}(s)\mu(ds) \tag{5.5}$$

exists and is nonsingular. For $x\in[a,\infty)$, *let*

$$H(x)=\operatorname{imp}\int_x^{\infty}A_{\check{\varphi}}(s)\mu(ds). \tag{5.6}$$

Then if $HA_{\varphi}\in L^1([a,\infty),\mu)$ *the product integral* $\prod_a^{\infty}e^{A(s)\mu(ds)}$ *exists and is nonsingular.*

Remark. If μ has no pure points then $A_{\check{\varphi}}=A_{\varphi}=A$, and the theorem more closely resembles Theorem 1.6.3.

Proof of Theorem 5.2. Following the proof of Theorem 1.6.3 we begin by noting that we can choose c so large that $\|H(x)\|\leqslant 1/2$ for $x\geqslant c$. Then $\|(I+H(x))^{-1}\|\leqslant 2$ for $x\geqslant c$ and by the hypothesis of the theorem we have $HA_{\varphi}(I+H)^{-1}\in L^1((c,\infty);\mu)$. It follows that

$$\sum_{\substack{s\geqslant c\\ s\text{ a pure point of }\mu}}\|H(s)A_{\varphi}(s)(I+H(s))^{-1}\|\,\mu(s)<\infty. \tag{5.7}$$

Thus we can choose a $b\geqslant c$ so large that for $x\geqslant b$ we have simultaneously

$$\begin{gathered}\|H(x)\|\leqslant 1/2,\\ \|H(x)A_{\varphi}(x)(I+H(x))^{-1}\|\,\mu(x)\leqslant 1/2.\end{gathered} \tag{5.8}$$

As in Theorem 1.6.3, we will prove existence and nonsingularity of

ISBN 0-201-13509-4

$\prod_b^x e^{A(s)\mu(ds)}$. For $x \geqslant b$ let

$$P(x) = \prod_b^x e^{A(s)\mu(ds)}(I + H(b))^{-1}. \tag{5.9}$$

Note that $\prod_b^x e^{A(s)\mu(ds)}$ has a nonsingular limit as $x \to \infty$ if and only if the same holds for $P(x)$. Let

$$Q(x) = (I + H(x))P(x). \tag{5.10}$$

By Lemma 3.1 we have

$$dQ = (dH)P + (I + H)\,dP + \rho_{HP} \tag{5.11}$$

where for any set S

$$\rho_{HP}(S) = \sum_{x \in S} dH(x)\,dP(x).$$

Now

$$dH = -A_{\breve{\varphi}}\mu \tag{5.12}$$

and

$$dP = A_{\varphi}\mu P. \tag{5.13}$$

We note from (5.12) and (5.13) that the pure points of dH and dP are the same and are pure points of μ. We have

$$dQ = -A_{\breve{\varphi}}\mu P + (I + H)A_{\varphi}\mu P + \rho_{HP} = (A_{\varphi} - A_{\breve{\varphi}})\mu P + HA_{\varphi}\mu P + \rho_{HP}. \tag{5.14}$$

Now for any set S

$$\rho_{HP}(S) = \sum_{x \in S} \left[-A_{\breve{\varphi}}(x)A_{\varphi}(x)\mu(x)\right]\mu(x)P(x). \tag{5.15}$$

But

$$\begin{aligned} A_{\varphi}(x) - A_{\breve{\varphi}}(x) - A_{\breve{\varphi}}(x)A_{\varphi}(x)\mu(x) &= A_{\varphi}(x) - A_{\breve{\varphi}}(x)(I + A_{\varphi}(x)\mu(x)) \\ &= A_{\varphi}(x) - A_{\breve{\varphi}}(x)e^{A(x)\mu(x)} = 0. \end{aligned} \tag{5.16}$$

Equation (5.16) shows that the pure point part of $(A_{\varphi} - A_{\breve{\varphi}})\mu P$ is precisely cancelled by ρ. And the continuous part of $(A_{\varphi} - A_{\breve{\varphi}})\mu P$ is zero, so we obtain from (5.14) the equation

$$dQ = HA_{\varphi}\mu P = HA_{\varphi}\mu(I + H)^{-1}Q. \tag{5.17}$$

ISBN 0-201-13509-4

Thus (noting that $Q(b)=I$)

$$Q(x)=I+\int_b^x H(s)A_\varphi(s)(I+H(s))^{-1}\mu(ds)Q(s). \tag{5.18}$$

The function $HA_\varphi(I+H)^{-1}$ is in $L^1([b,\infty);\mu)$ by our choice of b. If we can show that

$$HA_\varphi(I+H)^{-1}=B_\varphi \tag{5.19}$$

for some $B\in L^1([b,\infty);\mu)$, then we will have $Q(x)=\prod_b^x e^{B(s)\mu(ds)}$ and Theorem 5.1 will imply that $\lim_{x\to\infty}Q(x)$ exists and is nonsingular. Since $H(x)\to 0$ it will then follow that $\lim_{x\to\infty}P(x)$ exists and is nonsingular, proving the theorem. Now by the second inequality in (5.8) it follows that $H(x)A_\varphi(x)(I+H(x))^{-1}\mu(x)+I$ is nonsingular at each pure point, and we can find $B(x)$ at such points by the formula

$$B(x)\mu(x)=\log\big(I+H(x)A_\varphi(x)(I+H(x))^{-1}\mu(x)\big), \tag{5.20}$$

where the right-hand side of (5.20) is defined by the series

$$\log(1+C)=\sum_{k=1}^{\infty}(-1)^{k+1}\frac{C^k}{k}. \tag{5.21}$$

As in the similar estimate (3.59) in the proof of Lemma 3.3 we have

$$\|B(x)\mu(x)\|\leqslant 2\|H(x)A_\varphi(x)(I+H(x))^{-1}\mu(x)\| \tag{5.22}$$

so that by (5.7)

$$\sum_{\substack{x\geqslant b\\ x \text{ a pure point of } \mu}} \|B(x)\|\,\mu(x)<\infty.$$

Now defining B to be $HA(I+H)^{-1}$ at points of continuity of μ and defining B by (5.20) at pure points of μ, we have $B\in L^1([b,\infty))$ and (5.19) holds. Thus we are done. ■

5.6 The Schrödinger Equation

As an application of the theory just presented, we discuss in this section the Schrödinger equation with the potential V replaced by a measure μ. This replacement is of some physical interest (see, e.g., [DP]). We will discuss the Schrödinger equation on the real line rather than on $(0,\infty)$ as before. This choice will result in greater notational simplicity in the example to be given, but has no other significance.

ISBN 0-201-13509-4

DEFINITION 6.1. *A* local real-valued measure on $\mathbb{R}$ *is a set function* μ *such that for any finite interval* $[a,b)\subset\mathbb{R}$ *we have*

(i) $\mu(B)$ *is defined for any Borel set* $B\subset[a,b)$ *and*

(ii) *the restriction of* μ *to Borel subsets of* $[a,b)$ *is a real-valued measure (hence necessarily of bounded total variation).*

Note that if μ is a local real-valued measure on $\mathbb{R}$, then $\mu([a,b))$ may approach $\pm\infty$ as $a\to-\infty$ or as $b\to\infty$. We now consider the Schrödinger equation

$$y''=(\mu-E)y \tag{6.1}$$

on the real line, where μ is a local real-valued measure on $\mathbb{R}$, E is a nonzero real number, and (6.1) is meant in the sense of distribution theory. Converting as usual to matrix notation by setting $Y=\binom{y}{y'}$, Eq. (6.1) becomes

$$Y'(x)=\hat{\mu}Y(x) \tag{6.2}$$

where $\hat{\mu}$ is the matrix-valued (local) measure

$$\hat{\mu}=\begin{pmatrix}0 & \lambda\\ \mu-E\lambda & 0\end{pmatrix} \tag{6.3}$$

with λ representing Lebesgue measure. In order to solve (6.2) we consider as usual the initial value problem

$$P'(x)=\hat{\mu}P(x),\qquad P(a)=I \tag{6.4}$$

with $a\in\mathbb{R}$. According to our previous work Eq. (6.4) has a product-integral solution provided that $\hat{\mu}$ can be written as $\hat{\nu}_\varphi$. The solution is then $P(x)=\prod_a^x e^{\hat{\nu}(ds)}$. Now in the present situation we have actually

$$\hat{\mu}=\hat{\mu}_\varphi. \tag{6.5}$$

To see this, it is only necessary to check that at each pure point x of $\hat{\mu}$ we have

$$\hat{\mu}(x)=\hat{\mu}_\varphi(x)=\varphi(\hat{\mu}(x))\hat{\mu}(x)=e^{\hat{\mu}(x)}-I. \tag{6.6}$$

Now since λ has no pure points, we have

$$\hat{\mu}(x)=\begin{pmatrix}0 & 0\\ \mu(x) & 0\end{pmatrix}. \tag{6.7}$$

Clearly $\hat{\mu}(x)^n=0$ for $n\geqslant 2$, and this shows that (6.6) is indeed correct.

ISBN 0-201-13509-4

Because of (6.5) we see that a solution of the initial-value problem (6.4) is given by

$$P(x)=\prod_a^x e^{\hat{\mu}(ds)}. \tag{6.8}$$

A solution of (6.2) with initial condition $Y(a)=Y_0$ is then given by

$$Y(x)=\prod_a^x e^{\hat{\mu}(ds)}Y_0. \tag{6.9}$$

To study the behavior of $P(x)$ it is convenient to transform $P(x)$ as in Chapter 1: let

$$k=\begin{cases} \sqrt{E}\,, & \text{if } E>0 \\ i\sqrt{|E|} & \text{if } E<0 \end{cases} \tag{6.10}$$

where the square roots in (6.10) are positive. Let

$$M=\begin{pmatrix} 1 & 1 \\ ik & -ik \end{pmatrix}. \tag{6.11}$$

Then by the simplest instance of the similarity rule [Eq. (4.19)] we have

$$P(x)=M\prod_a^x e^{M^{-1}\hat{\mu}(ds)M}M^{-1}\equiv MQ(x)M^{-1}. \tag{6.12}$$

We write

$$\hat{\nu}=M^{-1}\hat{\mu}M=\begin{pmatrix} ik & 0 \\ 0 & -ik \end{pmatrix}\lambda+\frac{i}{2k}\begin{pmatrix} -1 & -1 \\ 1 & 1 \end{pmatrix}\mu\equiv\hat{\nu}_1+\hat{\nu}_2. \tag{6.13}$$

By inspection $\hat{\nu}_1$ and $\hat{\nu}_2$ have no common pure points, so we can use the sum rule to write

$$Q(x)=\prod_a^x e^{\hat{\nu}_1(ds)+\hat{\nu}_2(ds)}=Q_1(x)Q_2(x) \tag{6.14}$$

where [using the fact that $\left(\begin{smallmatrix} ik & 0 \\ 0 & -ik \end{smallmatrix}\right)$ is a function of s whose values commute] we have

$$Q_1(x)=\prod_a^x e^{\hat{\nu}_1(ds)}=e^{\int_a^x\hat{\nu}_1(s)}=\begin{pmatrix} e^{ik(x-a)} & 0 \\ 0 & e^{-ik(x-a)} \end{pmatrix} \tag{6.15}$$

and $Q_2(x)$ is the product integral over $[a,x)$ of the measure

$$\hat{\sigma}=Q_1^{-1}(x)\hat{\nu}_2Q_1(x)=\frac{i}{2k}\begin{pmatrix} -1 & -e^{-2ik(x-a)} \\ e^{2ik(x-a)} & 1 \end{pmatrix}\mu. \tag{6.16}$$

ISBN 0-201-13509-4

To reinforce understanding of our solution of the Schrödinger equation, we consider the following elementary example: suppose that μ is a constant multiple of the "δ-function" at 0, i.e., $\mu(0)=\alpha\in\mathbb{R}$ and $\mu(\mathbb{R}\setminus\{0\})=0$. We then note that the values of the function multiplying μ on the right-hand side of (6.16) commute for μ-almost every x. Taking $a=0$, we have thus

$$Q_2(x)=\prod_0^x e^{\hat{\sigma}(ds)}=e^{\int_0^x\hat{\sigma}(ds)}=\begin{cases} I, & \text{if } x\leqslant 0 \\ e^{\hat{\sigma}(0)}, & \text{if } x>0. \end{cases} \tag{6.17}$$

Now

$$\hat{\sigma}(0)=\frac{i\alpha}{2k}\begin{pmatrix} -1 & -1 \\ 1 & 1 \end{pmatrix} \tag{6.18}$$

so that

$$e^{\hat{\sigma}(0)}=I+\frac{i\alpha}{2k}\begin{pmatrix} -1 & -1 \\ 1 & 1 \end{pmatrix}. \tag{6.19}$$

Letting $H(x)$ denote the Heaviside function

$$H(x)=\begin{cases} 0, & \text{if } x\leqslant 0, \\ 1, & \text{if } x>0, \end{cases} \tag{6.20}$$

we thus have

$$Q_2(x)=I+\frac{i\alpha}{2k}H(x)\begin{pmatrix} -1 & -1 \\ 1 & 1 \end{pmatrix}. \tag{6.21}$$

Taking account of (6.12), (6.14), and (6.15) we have after an elementary computation

$$P(x)=\begin{bmatrix} \cos kx+\alpha H(x)\dfrac{\sin kx}{k} & \dfrac{\sin kx}{k} \\ -k\sin kx+\alpha H(x)\cos kx & \cos kx \end{bmatrix}. \tag{6.22}$$

We note the following:

(i) If $E<0$, then k is pure imaginary. In this case $\cos kx$ and $\sin kx$ are of course defined by the usual power series.

(ii) The entries in the top row of P, which give two linearly independent solutions of (6.1), are continuous functions, despite the presence of $H(x)$ in the $(1,1)$ entry.

(iii) The entries in the bottom row of P are the distributional derivatives of the entries in the top row.

If $E>0$, then k is real, and it is easy to see that no linear combination of the entries of the top row of P can be square-integrable. Thus no solution

ISBN 0-201-13509-4

of (6.1) is square-integrable. Now suppose that $E<0$. Let us determine whether (6.1) has a square-integrable solution: writing

$$Y(x)=\begin{pmatrix} y(x) \\ y'(x) \end{pmatrix}=P(x)\begin{pmatrix} c \\ d \end{pmatrix} \tag{6.23}$$

and setting $k=i\kappa$ with $\kappa>0$, we have

$$\begin{aligned} y(x)&=c\left(\cos i\kappa x+\alpha H(x)\frac{\sin i\kappa x}{i\kappa}\right)+d\frac{\sin i\kappa x}{i\kappa} \\ &=\left\{\frac{c}{2}-\frac{d}{2\kappa}-\frac{c\alpha}{2\kappa}H(x)\right\}e^{-\kappa x}+\left\{\frac{c}{2}+\frac{d}{2\kappa}+\frac{c\alpha}{2\kappa}H(x)\right\}e^{\kappa x}. \end{aligned} \tag{6.24}$$

The function $y(x)$ will be square-integrable if and only if the coefficient of $e^{-\kappa x}$ vanishes for $x<0$ and the coefficient of $e^{\kappa x}$ vanishes for $x>0$, i.e., if and only if

$$\begin{aligned} &\frac{c}{2}-\frac{d}{2\kappa}=0, \\ &\frac{c}{2}+\frac{d}{2\kappa}=-\frac{c\alpha}{2\kappa}. \end{aligned} \tag{6.25}$$

Equations (6.25) have a nontrivial solution if and only if $\alpha<0$ and $\kappa=-\alpha/2$. Taking $c=1$, $d=\kappa$, we then obtain

$$y(x)=H(x)e^{\alpha x/2}+\{1-H(x)\}e^{-\alpha x/2}=e^{\alpha|x|/2}. \tag{6.26}$$

The energy E for this solution is $E=\kappa^2=\alpha^2/4$. Thus for $\alpha<0$ there is a nontrivial square-integrable solution of (6.1) for precisely one value of E, and for $\alpha\geqslant 0$ there are no nontrivial square-integrable solutions.

Using the representations (6.9), (6.12), and (6.14), one can now proceed to study the solutions of (6.1) by the same methods employed to study solutions of the "ordinary" Schrödinger equation in Chapters 1 and 4. In particular, using the results on improper product integration in Section 5 of the present chapter, one can obtain information about the asymptotic behavior of solutions of (6.1) as $x\to\pm\infty$. (We note that the use of Theorem 5.2 is relatively simple in this context because $\mu_\varphi=\mu_{\check{\varphi}}=\mu$.) For an extensive development along these lines with applications to scattering theory, see [JD6].

5.7 The Equation $y''+p(dx)y'+q(dx)y=0$

As a further example of the theory developed so far, we consider the second-order homogeneous linear equation

$$y''+p(dx)y'+q(dx)y=0, \tag{7.1}$$

ISBN 0-201-13509-4

where p and q are real or complex measures on the real line $\mathbb{R}$. As usual, we write (7.1) in the equivalent matrix form

$$Y' = \hat{\mu} Y \tag{7.2}$$

with

$$Y = \begin{pmatrix} y \\ y' \end{pmatrix}, \qquad \hat{\mu} = \begin{pmatrix} 0 & \lambda \\ -q & -p \end{pmatrix}. \tag{7.3}$$

[In (7.3) λ denotes Lebesgue measure on $\mathbb{R}$.] A solution of (7.2) is then given by

$$Y(x) = \prod_{x_0}^{x} e^{\hat{\nu}(ds)} \tag{7.4}$$

with

$$\hat{\nu}_{\varphi} = \hat{\mu} \tag{7.5}$$

provided a $\hat{\nu}$ satisfying (7.5) exists. This will be the case if and only if $\hat{\mu}(x)+I$ is invertible for each pure point x of $\hat{\mu}$, and then $\hat{\nu}$ must satisfy

$$\hat{\mu}(x) = e^{\hat{\nu}(x)} - I. \tag{7.6}$$

Since

$$\hat{\mu}(x) + I = \begin{pmatrix} 1 & 0 \\ -q(x) & 1-p(x) \end{pmatrix}, \tag{7.7}$$

[note that $\lambda(x)=0$, since λ has no pure points]. It follows that (7.2) has a product integral solution if and only if

$$p(x) \neq 1 \tag{7.8}$$

for all x.

An elementary calculation shows that

$$e^{\begin{pmatrix} 0 & 0 \\ -\alpha & -\beta \end{pmatrix}} - I = \begin{bmatrix} 0 & 0 \\ \dfrac{\alpha}{\beta}(e^{-\beta}-1) & e^{-\beta}-1 \end{bmatrix}. \tag{7.9}$$

(If $\beta=0$, the $(2,1)$ entry is to be interpreted as $-\alpha$.) Hence, if (7.8) holds, we may take

$$\hat{\nu} = \begin{pmatrix} 0 & \lambda \\ -\tilde{q} & -\tilde{p} \end{pmatrix}, \tag{7.10}$$

ISBN 0-201-13509-4

where $\tilde{p}$ and $\tilde{q}$ are measures with the same continuous parts as p and q, respectively, and

$$e^{-\tilde{p}(x)}-1=-p(x), \qquad (\tilde{q}(x)/\tilde{p}(x))(e^{-\tilde{p}(x)}-1)=-q(x) \qquad (7.11)$$

or

$$\tilde{p}(x)=-\log(1-p(x)), \qquad \tilde{q}(x)=-q(x)\log(1-p(x))/p(x). \quad (7.12)$$

We may take $\hat{\nu}=\hat{\mu}$ provided that the choice $\tilde{p}(x)=p(x)$ and $\tilde{q}(x)=q(x)$ satisfies (7.11). Because of the first equation in (7.11), this will hold only if $p(x)=0$ for each x. The second equation in (7.11) then becomes $-q(x)=-q(x)$. Thus the value of $q(x)$ is immaterial, and we find that the choice $\hat{\nu}=\hat{\mu}$ is permissible if and only if p has no pure points.

It is interesting to see explicitly why (7.2) has no product integral solution if (7.8) is violated. As an example, consider the equation

$$y''+\delta y'=0 \qquad (7.13)$$

where δ is a unit point mass at $x=0$. This has no product-integral solution since $\delta(0)=1$. A fundamental set of left-continuous solutions of (7.13) is given by $y_1(x), y_2(x)$ where

$$y_1(x)\equiv 1 \qquad y_2(x)=\left\{\begin{matrix} 1+x, & x\leqslant 0 \\ 1, & x>0 \end{matrix}\right\} \qquad (7.14)$$

and we take y_1' and y_2' to be left-continuous. We have

$$y_1(-1)=1, \quad y_1'(-1)=0; \qquad y_2(-1)=0, \quad y_2'(-1)=1. \qquad (7.15)$$

However, the matrix

$$\begin{pmatrix} y_1(x) & y_2(x) \\ y_1'(x) & y_2'(x) \end{pmatrix} = \left\{\begin{matrix} \begin{pmatrix} 1 & 1+x \\ 0 & 1 \end{pmatrix}, & x\leqslant 0 \\ \begin{pmatrix} 1 & 1 \\ 0 & 0 \end{pmatrix}, & x>0 \end{matrix}\right\}$$

is singular for $x>0$ and so cannot be of the form $\prod_{-1}^{x} e^{\hat{\nu}(ds)}$.

Notes to Chapter 5

Product integration of measures has been considered by several authors in varying degrees of generality. H. S. Wall [HW] developed product integral solutions of the integral equation

$$M(x,y)=I+\int_x^y dF(s)\cdot M(s,y),$$

ISBN 0-201-13509-4

where $F(s)$ is a continuous matrix-valued function of bounded variation. The continuity of F implies that the measure dF has no pure points. J. S. MacNerney [JM5] considered product integrals of the form

$$\prod(I+dF)$$

where F is continuous and of bounded variation, and taking values in the ring of continuous transformations of a linear space; MacNerney also extended this work in several directions. The work of Wall and MacNerney has been carried on by various authors and generalized to include product integration of functions and measures with values in a normed ring or normed abelian group. (See Chapter 6 and the notes to the references.) G. Schmidt [GS] has considered product integrals of the form

$$\prod_a^b e^{A(s)\mu(ds)}$$

where μ is a positive Borel measure and $A(s)$ is Banach space operator valued and "strongly integrable." (See the notes to Chapter 3.) His proof of existence of the product integral is rather complicated; existence of solutions to a related integral equation is first proved, and the existence of the product integral is then deduced. His proof is quite similar to our existence proof for the strong product integral in Chapter 3. For the context we have considered, our existence proof is much simpler and direct. Our proof also goes through with essentially no change for the case of a Banach space operator valued function $A(x)$ which is Bochner integrable with respect to a positive Borel measure μ.

In Section 3, we discussed the solution of the integral equation

$$P(x,y)=I+\int_y^x \hat{\mu}(ds)P(s,y)$$

and showed that a necessary and sufficient condition for a *product integral solution* is that $\hat{\mu}=\hat{\nu}_{\varphi}$ for some measure $\hat{\nu}$. A similar integral equation has been studied by Hildebrandt [TH2]; however, in his equation the integration is over a *closed* rather than *half-open* interval. Hildebrandt constructs solutions to such equations by using a *product integral which is altered at pure points of the measure involved.* His discussion and ours are similar in many respects, but differ in various technical details.

The Schrödinger equation with "potentials" which are measures is considered in [VG] and [DP]. Product integration plays no part in either of these papers.

ISBN 0-201-13509-4

CHAPTER 6

Complements; other Work and further Results on Product Integration

In the present chapter, we shall discuss additional results and generalizations of the theory that we have presented so far. A large part of our discussion will focus on the work of other authors. Some of this work, e.g., product integration of nonlinear operator-valued functions, has not been developed in the previous chapters and will only be discussed briefly here. This is partly due to the fact that for these results, a really systematic and complete presentation is simply not available at the present time. On the other hand, what we have presented above has been somewhat determined by our own predispositions; a treatise of reasonable length covering all aspects of product integration in detail is hardly feasible, and we have therefore selected material which is of the most interest to us and which at the same time is not readily accessible in the modern literature. Nevertheless, the results we shall discuss in this chapter are of considerable significance and importance and certainly deserve some attention in any general discussion of product integration.

If we consider the theory of product integration of matrix-valued functions presented in Chapter 1 as a starting point, then there are several possible directions for generalization. For example, a first type of generalization might focus on reducing to a minimum the *regularity* assumptions on the function $A(x)$ being product integrated with regard to its dependence on x, while not changing essentially the nature of the values of this function. A second type of generalization might include theories of product integration of functions $A(x)$ *whose values are of a more general type*, i.e., unbounded operators, nonlinear operators, functions on more general

ENCYCLOPEDIA OF MATHEMATICS and Its Applications, Gian-Carlo Rota (ed.). Vol. 10: John D. Dollard and Charles N. Friedman, Product Integration with Applications to Differential Equations

ISBN 0-201-13509-4

kinds of spaces, etc. For example, the theory of product integration of functions in L_s^1 developed in Chapter 3 is primarily a generalization of the first type. Of course the functions being integrated have values which are bounded operators in a Banach space rather than matrices, but the real import of the theory derives from the possibility of product-integrating functions $A(x)$ which are merely *strongly integrable*; a product integration theory for *Bochner integrable* functions $A(x)$ is, as noted in Chapter 3, an immediate and elementary generalization of the theory of Chapter 1, but is far less useful than the L_s^1 theory. We view the theory of product integration of measures as another generalization of this type, a measure being a "less regular" version of the object $A(s)ds$ (which is an absolutely continuous measure). On the other hand, the theory of the last section of Chapter 3 concerning product integration of *unbounded* operator valued functions is a generalization primarily of the second type.

In the present monograph we have been concerned mainly with product integration of functions whose values are *bounded linear operators* in various spaces. In this context the use of product integration yields some quite detailed and useful results concerning solutions of differential equations (e.g., the applications in Chapter 4). Many other authors have utilized product integration of functions whose values are unbounded operators, nonlinear operators, elements of an abstract normed ring, etc. In these works usually the major results concern *existence* of the product integral and sometimes verification that the product integral solves a differential or integral equation properly interpreted, such verification often requiring additional assumptions beyond those necessary for existence of the product integral. We shall discuss some of these results, our intention being to indicate the kinds of work that have been done rather than to give any detailed exposition of theories.

A large amount of work has been done concerning the product integration of functions $A(x)$ whose values are nonlinear operators in Banach spaces. One of the first works of this type is the paper [GB] of Birkhoff. Birkhoff defines and studies the product integral of the *infinitesimal transformation* associated with a one-parameter family of transformations or *displacements*. His results include existence theorems for systems of ordinary differential equations (linear or nonlinear) and some existence results for "Lebesgue product integrals." The paper is difficult to read and contains formulas and terminology whose precise meanings must be supplied by the reader. The product integral considered is essentially of the form

$$\prod (I + dsA(s)) \tag{6.1}$$

defined as a limit of Riemann products (products of operators denoting composition). Various later works have treated product integration of families $A(x)$ of nonlinear Banach space operators, usually utilizing a

ISBN 0-201-13509-4

product integral of the form

$$\prod (I - dsA(s))^{-1} \tag{6.2}$$

defined as a limit of Riemann products. For example, Webb [GW1] uses (6.2) to solve the evolution equation

$$g'(t) \;= Ag(t)_{0 \leq t < \infty} \,(A \text{ independent of } t), \tag{6.3}$$

where A is a continuous function from a Banach space S to itself and g is a continuously differentiable function from $[0, \infty)$ to S; it is hypothesized that A is *dissipative* on S, i.e., $\|(I - \varepsilon A)u - (I - \varepsilon A)v\| \geq \|u - v\|$ for $u, v \in S$ and $\varepsilon \geq 0$, and that Range $(I - \varepsilon A) = S$ for $\varepsilon \geq 0$. These hypotheses are satisfied if, for example, $S = \mathbb{R}^1$ and A is a continuous nonincreasing function from $\mathbb{R}^1$ to $\mathbb{R}^1$, e.g., $A\xi = -\xi^3$. In a later paper [GW2] of Webb the case of a t-dependent family $A(t)$ is considered. The proofs that (6.1) or (6.2) provide a solution to (6.3) are extremely technical and not at all transparent in even the most elementary cases. For example, for the case $A\xi = -\xi^3$ mentioned above, Eq. (6.3) is

$$g'(t) = -(g(t))^3, \tag{6.4}$$

which has the solution

$$g(t) = (2t + \text{const.})^{-1/2}. \tag{6.5}$$

The product integral solution of (6.4) with $g(0) = 1$, is given by

$$\begin{aligned} g(t) &= \prod_0^t (I - dsA)^{-1} \cdot 1 = \lim_{n \to \infty} \left(\left(I - \frac{t}{n} A \right)^{-1} \right)^n \cdot 1 \\ &= \lim_{n \to \infty} \underbrace{\Phi_{t/n}(\cdots(\Phi_{t/n}(\Phi_{t/n}(1))))}_{n\text{-fold composition}}, \end{aligned} \tag{6.6}$$

where for $h > 0$, x real, $\Phi_h(x)$ denotes the unique real solution of $h(\Phi_h(x))^3 + \Phi_h(x) - x = 0$. It is certainly not easy to see directly that the right-hand side of (6.6) is equal to $(2t + 1)^{-1/2}$, and it would seem rather hopeless to try to derive specific properties of $g(t)$ from (6.6). Evidently, these difficulties are magnified in less trivial examples. The results on product integrals that we have proved in the previous chapters, e.g., the sum and similarity formulas, convergence theorems, etc., and which may be applied to obtain precise information about solutions to differential equations (as in Chapter 4) are simply not available in the nonlinear case. For these reasons the subject of nonlinear product integration has a completely different flavor.

ISBN 0-201-13509-4

Other works on nonlinear product integration include [MC1] and [MC2] in which (6.2) is used to solve (in a generalized sense) the initial value problem

$$\begin{aligned} &\frac{du}{dt}+A(t)u\ni 0, \qquad s\leqslant t\leqslant T \\ &u(s)=x, \end{aligned} \tag{6.7}$$

where the $A(t)$ are nonlinear operators (possibly multivalued) in a Banach space. For some of the cases considered, the existence of the product integral is proved, but the fact that $u(t)=\prod_s^t(I-d\xi A(\xi))^{-1}\cdot x$ satisfies (6.7) is only proved under the hypothesis that (6.7) *has* a solution.

Several authors have considered product integrals of functions whose values belong to a normed ring or abelian group. Masani [PM1] considers Riemann product integrals of the form $\prod(I+A(s)ds)$ where the $A(s)$ have values in a normed ring. MacNerney [JM5] and [JM9] studies the correspondence between two classes of interval functions, $V(a,b)$, $W(a,b)$, from a linearly ordered set to a complete normed ring given by

$$\begin{aligned} V(a,b)&=\sum_a^b\left[W-I\right]=\lim\sum\left[W(t_p,t_{p-1})-I\right], \\ W(a,b)&=\prod_a^b\left[I+V\right]=\lim\prod\left[I+V(t_p,t_{p-1})\right], \end{aligned} \tag{6.8}$$

which is an abstract version of the correspondence between product integrals and integral equations. This work has been continued by various authors (see the notes to the references). Hamilton [JH1] develops product integration of functions with values in the Lie algebra $\mathcal{G}$ of a Lie group G, the product integral taking values in G. McKean [HM] has used product integration of exponentials of differentials of (skew) Brownian motions on the Lie algebra of the group $SO(3)$ to construct Brownian motions on $SO(3)$. Other applications of product integration are cited in the notes to the references.

It might be of some interest to study product integration in a setting in which "product" is replaced by an arbitrary binary operation. There are various possible formulations of problems of this type and we shall consider only one example. Suppose $\mathfrak{x}$ is a topological space with a continuous binary operation $*:\mathfrak{x}\times\mathfrak{x}\to\mathfrak{x}$. Suppose $F(t,s)$ is a mapping with values in $\mathfrak{x}$ defined for $a\leqslant t\leqslant b$ and s sufficiently close to 0. We could then define the $*$*-integral* of F to be

$$\mathop{*}_a^b F(t,dt)=\lim_{n\to\infty}F_n*(\cdots F_3*(F_2*F_1)\cdots), \tag{6.9}$$

ISBN 0-201-13509-4

where

$$F_k = F\left(a + \frac{k(b-a)}{n}, \frac{b-a}{n}\right)$$

provided the indicated limit exists. In the case that $\mathfrak{x}$ is the space of $n \times n$ matrices, $F(t,s) = e^{sA(t)}$, $I + sA(t)$, or $(I - sA(t))^{-1}$, where $A(t)$ is a continuous function with values in $\mathfrak{x}$ and $*$ is matrix multiplication, then (6.9) is equivalent to the definition of $\prod_a^b e^{A(t)dt}$ given in Chapter 1. We remark that if $F(t,s) = F(s)$ is constant in t and is matrix-valued with $F(s)|_{s=0} = I$, $dF(s)/ds|_{s=0} = A$, then the limit in (6.9) exists and equals $e^{(b-a)A}$. This remains true if the values of F are bounded operators on a Banach space and A is a bounded operator. The case that A is unbounded is much more subtle. This type of problem is discussed in [PC], [JEM], [HT]. See also the example on the Hille-Yosida Theorem in Chapter 4. The case that $\mathfrak{x}$ is a suitable space of functions and $*$ is convolution of functions may be analyzed in view of the connection between multiplication and convolution provided by the Fourier transform. If $\mathfrak{x}$ is a space of functions, $*$ is *composition* of functions, and $F(t,s) = I + sA(t)$ or $(I - sA(t))^{-1}$, then (6.9) has been much studied, as indicated above. In its most abstract setting (6.9) encompasses almost all forms of integration (ordinary integration as well as product integration) and constructions involving "continuous" operations, so criteria for existence of the indicated limit might be of quite general interest and application. As far as we know, no general study of (6.9) has been attempted.

ISBN 0-201-13509-4

APPENDIX I

APPENDIX I

Matrices

A.I.1 Elementary Definitions

In this appendix we present some matrix theory which we use in the text. No attempt at completeness is intended; we discuss only those subjects pertinent to our development of product integration. Some of the more difficult results of the theory are merely stated without proof; in these cases we give references to the literature where proofs may be found.

We denote the complex number field by $\mathbb{C}$; the elements of $\mathbb{C}$ are numbers of the form $z=a+ib$, a,b real numbers. For $z=a+ib\in\mathbb{C}$, the *complex conjugate* of z, is $\bar{z}=a-ib$ and the *modulus* or *absolute value* of z is $|z|=\sqrt{z\bar{z}}=\sqrt{a^2+b^2}$ where $\sqrt{\ }$ denotes the nonnegative square root.

For n a positive integer we denote by $\mathbb{C}_n$ the n dimensional complex vector space of all $n\times 1$ column vectors,

$$\begin{bmatrix} z_1 \\ \vdots \\ z_n \end{bmatrix},$$

where $z_i\in\mathbb{C}$, $i=1,\dots,n$. The *Euclidian norm*, $\|\ \|$, on $\mathbb{C}_n$ is defined by

$$\left\|\begin{bmatrix} z_1 \\ \vdots \\ z_n \end{bmatrix}\right\| = \sqrt{|z_1|^2+\cdots+|z_n|^2}\ .$$

ENCYCLOPEDIA OF MATHEMATICS and Its Applications, Gian-Carlo Rota (ed.). Vol. 10: John D. Dollard and Charles N. Friedman, Product Integration with Applications to Differential Equations

ISBN 0-201-13509-4

An $n \times n$ *complex matrix*, A, is a square array,

$$\begin{bmatrix} a_{11} & \cdots & a_{1n} \\ \vdots & & \vdots \\ a_{n1} & \cdots & a_{nn} \end{bmatrix},$$

of complex numbers with n rows and n columns. *The* (i,j) *entry* of such a matrix A is the element in the ith row and jth column; sometimes we denote the (i,j) entry of A by $[A]_{ij}$. The set of all $n \times n$ complex matrices is denoted by $\mathbb{C}_{n \times n}$. $\mathbb{C}_{n \times n}$ forms an algebra under the operations

$$\begin{aligned} [A+B]_{ij} &= [A]_{ij} + [B]_{ij}, \\ [AB]_{ij} &= \sum_{k=1}^{n} [A]_{ik}[B]_{kj}, \\ [\lambda A]_{ij} &= \lambda [A]_{ij}, \qquad \text{for } \lambda \in \mathbb{C}. \end{aligned} \tag{1.1}$$

If $A \in \mathbb{C}_{n \times n}$ and

$$Z = \begin{bmatrix} z_1 \\ \vdots \\ z_n \end{bmatrix} \in \mathbb{C}_n,$$

we define AZ to be the element of $\mathbb{C}_n$ whose ith entry is $\sum_{k=1}^{n}[A]_{ik}z_k$. In this way the elements of $\mathbb{C}_{n \times n}$ are realized as *linear transformations* (operators) on $\mathbb{C}_n$. (Conversely all linear operators on $\mathbb{C}_n$ are of this form. See [TK2].) We define a norm, $\| \ \|$, on $\mathbb{C}_{n \times n}$ by

$$\|A\| = \max_{\substack{Z \in \mathbb{C}_n \\ \|Z\| \leqslant 1}} \|AZ\| \tag{1.2}$$

[In (1.2) $\|AZ\|$ and $\|Z\|$ are norms of elements of $\mathbb{C}_n$, while $\|A\|$ is the norm of an element of $\mathbb{C}_{n \times n}$.] It is not easy to give an explicit formula for $\|A\|$ in terms of the entries of the matrix A. However, it is not difficult to show that

$$\max_{i,j} |[A]_{ij}| \leqslant \|A\| \leqslant n^2 \max_{i,j} |[A]_{ij}|. \tag{1.3}$$

For $A, B \in \mathbb{C}_{n \times n}$, $\lambda \in \mathbb{C}$, we have

$$\begin{aligned} \|A+B\| &\leqslant \|A\| + \|B\|, \\ \|AB\| &\leqslant \|A\| \|B\|, \\ \|\lambda A\| &= |\lambda| \|A\|. \end{aligned} \tag{1.4}$$

ISBN 0-201-13509-4

These properties follow easily from the definition (1.2) and elementary properties of the norm on $\mathbb{C}_n$.

We denote by I the element of $\mathbb{C}_{n\times n}$ for which

$$[I]_{ij}=\left\{\begin{matrix} 1 \text{ if } i=j \\ 0 \text{ if } i\neq j \end{matrix}\right\},$$

and by 0 the element of $\mathbb{C}_{n\times n}$ all of whose entries are 0. These are, respectively, the multiplicative and additive identities in the algebra $\mathbb{C}_{n\times n}$. If $A,B\in\mathbb{C}_{n\times n}$ and $AB=BA=I$, then we say that A is the inverse of B and write $A=B^{-1}$; obviously then, it is also true that $B=A^{-1}$. A matrix with an inverse is called *nonsingular*. It is a fact (requiring some effort to prove) that either of the equations $AB=I$ or $BA=I$ implies the other. (See [TK2].) If S is a nonsingular matrix and A,B are matrices with $A=S^{-1}BS$, then we say that A and B are *similar*; the relation of similarity is an *equivalence relation* on $\mathbb{C}_{n\times n}$.

If $A\in\mathbb{C}_{n\times n}$, the conjugate of A, denoted $\bar{A}$, is by definition the matrix with entries $[\bar{A}]_{ij}=\overline{[A]_{ij}}$ and the *transpose* of A, denoted A^t, is the matrix with entries $[A^t]_{ij}=[A]_{ji}$. The matrix $\overline{A^t}=(\bar{A})^t$ is called the *adjoint* of A and denoted A^*. We have $[A^*]_{ij}=\overline{[A]_{ji}}$. If $A^*=A$, A is said to be *self-adjoint*, and if $A^*=-A$, A is called *skew-adjoint*.

A.I.2 Calculus of $\mathbb{C}_{n\times n}$-Valued Functions

Throughout this section we shall consider functions, from a fixed interval $[a,b]=\{t|a\leqslant t\leqslant b\}$ of the real numbers, with values in $\mathbb{C}_{n\times n}$. If $A(t)$ is such a function, we may think of $A(t)$ as a matrix whose entries are complex-valued functions of the variable t. The set of all such functions will be denoted $\mathfrak{F}([a,b],\mathbb{C}_{n\times n})$ or merely by $\mathfrak{F}$ since $[a,b]$ and n will be fixed throughout our discussion.

Suppose $A(t)\in\mathfrak{F}$, and $t_0\in[a,b]$. We say $A(t)$ is *continuous at* t_0 if $\lim_{t\to t_0}\|A(t)-A(t_0)\|=0$. Using (1.3) it follows easily that $A(t)$ is continuous at t_0 if and only if each entry function of $A(t)$ is continuous at t_0. $A(t)$ is said to be *continuous* if it is continuous at each $t\in[a,b]$. From (1.4) it follows easily that if $A(t),B(t)\in\mathfrak{F}$ are continuous at a point t_0, then so are $A(t)+B(t)$ and $A(t)B(t)$; this also follows from the fact that continuity is equivalent to "entrywise" continuity. We remark that if $B\in\mathbb{C}_{n\times n}$, we consider $\lim_{t\to t_0}A(t)=B$ to mean $\lim_{t\to t_0}\|A(t)-B\|=0$. With this convention, the condition defining continuity of $A(t)$ at t_0 becomes $\lim_{t\to t_0}A(t)=A(t_0)$.

If $A(t)\in\mathfrak{F}$, $t_0\in[a,b]$, we say $A(t)$ is *differentiable at* t_0 if there is a matrix $B\in\mathbb{C}_{n\times n}$ such that

$$\lim_{t\to t_0}\frac{A(t)-A(t_0)}{t-t_0}=B.$$

ISBN 0-201-13509-4

In this case we call B the derivative of $A(t)$ at t_0 and write $B=A'(t_0)=dA/dt|_{t=t_0}$. $A(t)$ is said to be *differentiable* if it is differentiable at each $t\in[a,b]$. (One-sided derivatives are meant at a and b.) It follows easily from (1.3) that $A'(t_0)$ exists if and only if each entry of $A(t)$ is differentiable at t_0; in this case $A'(t_0)$ is the matrix whose entries are $(d/dt)([A(t)]_{ij})|_{t=t_0}$. If $A(t)$ and $B(t)$ are differentiable, then so are $A(t)+B(t)$ and $A(t)B(t)$ and

$$\begin{aligned}(A(t)+B(t))' &= A'(t)+B'(t),\\ (A(t)B(t))' &= A'(t)B(t)+A(t)B'(t).\end{aligned} \tag{2.1}$$

The proofs of these facts are essentially the same as the familiar calculus proofs for the case $n=1$.

We next discuss integration of matrix-valued functions. We assume that the notions of the Riemann and Lebesgue integral are known for the case of scalar functions (real or complex valued). Let $A(t)\in\mathcal{F}$. We say that $A(t)$ is Riemann (Lebesgue) integrable if each entry function of $A(t)$ is Riemann (Lebesgue) integrable; if this is the case then we define $\int_a^b A(t)\,dt$ to be the matrix with entries $[\int_a^b A(t)\,dt]_{ij}=\int_a^b [A(t)]_{ij}\,dt$. From (1.2) and (1.3) it follows that a function $A(t)$ with measurable entries is Lebesgue integrable if and only if $\|A(t)\|$ is integrable. The Lebesgue integrable functions are denoted by $L^1([a,b],\mathbb{C}_{n\times n})$ or sometimes just $L^1(a,b)$ or L^1. Most of the familiar properties and theorems concerning integration of scalar functions extend immediately to the case of functions in $\mathcal{F}$. For example, integration is a linear function from $\mathcal{F}$ to $\mathbb{C}_{n\times n}$, i.e., if $A(t),B(t)\in\mathcal{F}$ and $\lambda,\mu\in\mathbb{C}$, then $\int_a^b(\lambda A(t)+\mu B(t))\,dt=\lambda\int_a^b A(t)\,dt+\mu\int_a^b B(t)\,dt$. The analogue of the fundamental theorem of calculus holds for matrix-valued functions. Namely, if $A(t)\in\mathcal{F}$ is continuous, and $a\leqslant t\leqslant b$, then $(d/dt)\int_a^t A(s)\,ds=A(t)$.

A.I.3 The Canonical Form of a Matrix

In this section we discuss the Jordan canonical form of a matrix. We shall merely state the pertinent results without proof; for proofs and further discussion, see [TK2] or [EC]. We begin with some definitions. Suppose $\lambda\in\mathbb{C}$ and k is a positive integer. We define *the elementary Jordan block associated with* λ *and* k to be the $k\times k$ matrix $J(\lambda,k)$ with entries given by

$$[J(\lambda,k)]_{ij}=\left\{\begin{array}{ll}\lambda, & \text{if } i=j\\ 1, & \text{if } j=i+1\\ 0, & \text{in all other cases}\end{array}\right\}, \tag{3.1}$$

ISBN 0-201-13509-4

for example,

$$J(\lambda,3)=\begin{bmatrix}\lambda & 1 & 0\\ 0 & \lambda & 1\\ 0 & 0 & \lambda\end{bmatrix}.$$

Next suppose that for $s=1,\dots,l$ we have matrices A_s of size $k_s\times k_s$ where $\sum_{s=1}^{l}k_s=n$. We define an $n\times n$ matrix $\amalg(A_1,\dots,A_l)$ by the prescription

$$\begin{aligned}
\left[\amalg(A_1,\dots,A_l)\right]_{ij}&=\left[A_1\right]_{ij}, &&\text{if } 1\leqslant i,j\leqslant k_1,\\
\left[\amalg(A_1,\dots,A_l)\right]_{k_1+i,k_1+j}&=\left[A_2\right]_{ij}, &&\text{if } 1\leqslant i,j\leqslant k_2,\\
&\ \vdots\\
\left[\amalg(A_1,\dots,A_l)\right]_{k_1+\cdots+k_{l-1}+i,k_1+\cdots+k_{l-1}+j}&=\left[A_l\right]_{ij}, &&\text{if } 1\leqslant i,j\leqslant k_l.
\end{aligned}\tag{3.2}$$

All other entries of $\amalg(A_1,\dots,A_l)=0$. In an obvious notation we have

$$\amalg(A_1,\dots,A_l)=\begin{bmatrix}A_1 & & & & & \\ & A_2 & & & & \\ & & \cdot & & & \\ & & & \cdot & & \\ & & & & \cdot & \\ & & & & & A_l\end{bmatrix}.\tag{3.3}$$

For example, if $\lambda,\mu\in\mathbb{C}$, then

$$\amalg(J(\lambda,2),J(\mu,3))=\begin{bmatrix}\lambda & 1 & 0 & 0 & 0\\ 0 & \lambda & 0 & 0 & 0\\ 0 & 0 & \mu & 1 & 0\\ 0 & 0 & 0 & \mu & 1\\ 0 & 0 & 0 & 0 & \mu\end{bmatrix}.\tag{3.4}$$

If $A\in\mathbb{C}_{n\times n}$ and there are complex numbers $\lambda_1,\dots,\lambda_l\in\mathbb{C}$ and positive integers $k_1,\dots,k_l$ with $A=\amalg(J(\lambda_1,k_1),J(\lambda_2,k_2),\dots,J(\lambda_l,k_l))$, then A is said to be in (Jordan) *Canonical form*. Note that it is not necessary for the $\lambda_1,\dots,\lambda_l$ to be distinct.

THEOREM 3.1. *Suppose $A\in\mathbb{C}_{n\times n}$. Then there is an invertible matrix $S\in\mathbb{C}_{n\times n}$ such that $S^{-1}AS$ is in canonical form. The canonical form of a matrix is unique up to rearrangement of the $J(\lambda_i,k_i)$. Two $n\times n$ matrices are similar if and only if their canonical forms are the same up to rearrangement of the λ_i.*

For the proof, see [TK2]. A consequence of Theorem 3.1 is the

ISBN 0-201-13509-4

Corollary 1. *Let $A \in \mathbb{C}_{n\times n}$. Then we can decompose $\mathbb{C}_n$ as a direct sum $\mathbb{C}_n = V_1 \oplus \cdots \oplus V_l$ where in each V_i there is a basis $\{f_{i,1},\ldots,f_{i,k_i}\}$ with*

$$\begin{aligned} Af_{i,1} &= \lambda_i f_{i,1} + f_{i,2}, \\ Af_{i,2} &= \lambda_i f_{i,2} + f_{i,3}, \\ &\vdots \\ Af_{i,k_i} &= \lambda_i f_{i,k_i}. \end{aligned}$$

Proof. This follows easily from the fact that if $f_1,\ldots,f_k$ are vectors in $\mathbb{C}_k$ given by

$$f_j = \begin{bmatrix} 0 \\ 0 \\ \vdots \\ 0 \\ 1 \\ 0 \\ \vdots \\ 0 \end{bmatrix} \leftarrow (k-j)\text{th entry}, \qquad j = 0,\ldots,k-1.$$

Then

$$\begin{aligned} J(\lambda,k) f_1 &= \lambda f_1 + f_2, \\ J(\lambda,k) f_2 &= \lambda f_2 + f_3, \\ \vdots &= \vdots \\ J(\lambda,k) f_k &= \lambda f_k. \end{aligned}$$

■

Another consequence of Theorem 3.1 is the

Corollary 2. *Let $A \in \mathbb{C}_{n\times n}$. Then A is similar to a matrix of the form $D+N$ where D is* diagonal *(i.e., $[D]_{ij} = 0$ if $i \neq j$) and N is* nilpotent *(i.e., $N^m = 0$ for some integer m).*

Proof. This follows from the theorem and the fact that any *upper triangular* matrix [one whose (i,j) entry is zero unless $j > i$] is nilpotent. ■

A.I.4 The Spectrum of a Matrix

Let $A \in \mathbb{C}_{n\times n}$. We recall the definition of the determinant of A and the trace of A:

$$\det A = \sum_{\tau} \operatorname{sgn}(\tau) [A]_{1\tau(1)} [A]_{2\tau(2)} \cdots [A]_{n\tau(n)}, \tag{4.1}$$

$$\operatorname{tr} A = \sum_{i=1}^{n} [A]_{ii}. \tag{4.2}$$

ISBN 0-201-13509-4

In (4.1) the sum is over all permutations τ of n elements and $\operatorname{sgn}(\tau)$ is $+1$ if τ is an even permutation, -1 if τ is an odd permutation. An inductive definition of the determinant can also be given:

$$\begin{aligned}\det A &= [A]_{11} \quad \text{if } A \text{ is } 1\times 1\\ \det A &= \sum_{j=1}^{n} (-1)^{i+j}[A]_{ij} \det\alpha_{ij},\end{aligned} \tag{4.1.1}$$

where α_{ij} is the $(n-1)\times(n-1)$ matrix obtained from A by deleting the ith row and the jth column. We list some properties of det and tr. If $A,B,S \in \mathbb{C}_{n\times n}$, S nonsingular:

$$\begin{aligned}\det(AB) &= \det A \det B,\\ \det(S^{-1}AS) &= \det A,\\ \operatorname{tr}(A+B) &= \operatorname{tr}(A)+\operatorname{tr}(B),\\ \operatorname{tr}(AB) &= \operatorname{tr}(BA),\\ \operatorname{tr}(S^{-1}AS) &= \operatorname{tr}(A).\end{aligned} \tag{4.3}$$

We also note that a matrix is nonsingular if and only if its determinant is not zero.

Now let $A \in \mathbb{C}_{n\times n}$. We define the characteristic polynomial, $\mathrm{Ch}_A(\lambda)$, of A by

$$\mathrm{Ch}_A(\lambda) = \det(A-\lambda I) \tag{4.4}$$

considered as a polynomial in λ with complex coefficients. It is a theorem of linear of algebra (see [TK2]) that a complex number λ is a root of $\mathrm{Ch}_A(\lambda)$, i.e., $\mathrm{Ch}_A(\lambda)=0$, if and only if there is a *nonzero* vector $Z_\lambda \in \mathbb{C}_n$ with $AZ_\lambda = \lambda Z_\lambda$. If this is the case, λ is called an *eigenvalue* and Z_λ a corresponding *eigenvector* of A. The set of eigenvalues of a matrix A is called the *spectrum* of A and denoted $\sigma(A)$. We show how $\det A$, $\operatorname{tr} A$, $\mathrm{Ch}_A(\lambda)$, and $\sigma(A)$ can be evaluated in terms of the canonical form of A. Suppose A is similar to $J=\amalg(J(\lambda_1,k_1),J(\lambda_2,k_2),\ldots,J(\lambda_l,k_l))$. Using (4.3) we find easily

$$\begin{aligned}\operatorname{tr} A &= \operatorname{tr} J = \sum_{i=1}^{l} k_i\lambda_i,\\ \det A &= \det J = \prod_{i=1}^{l} \lambda_i^{k_i},\\ \mathrm{Ch}_A(\lambda) &= \det(J-\lambda I) = \prod_{i=1}^{l} (\lambda_i-\lambda)^{k_i},\\ \sigma(A) &= \{\lambda_1,\ldots,\lambda_l\}.\end{aligned} \tag{4.5}$$

ISBN 0-201-13509-4

Next we define the exponential function on $\mathbb{C}_{n\times n}$. For $A\in\mathbb{C}_{n\times n}$, we put

$$e^A=\sum_{n=0}^{\infty}\frac{A^n}{n!}=I+A+\frac{A^2}{2}+\frac{A^3}{3!}+\cdots. \tag{4.6}$$

It is easily shown that the infinite series in (4.6) converges in $\mathbb{C}_{n\times n}$. An elementary property of the exponential function which follows from the definition is

$$e^{S^{-1}AS}=S^{-1}e^AS \tag{4.7}$$

for $A,S\in\mathbb{C}_{n\times n}$ with S nonsingular. An easy calculation shows that if A is diagonal, then so is e^A and

$$\left[e^A\right]_{ii}=e^{[A]_{ii}} \qquad (A\text{ diagonal}). \tag{4.8}$$

If A and $B\in\mathbb{C}_{n\times n}$, then in general $e^{A+B}\neq e^Ae^B$; however, if A and B commute, i.e., $AB=BA$, then

$$e^{A+B}=e^Ae^B \qquad (\text{if } AB=BA). \tag{4.9}$$

This follows from the definition (4.6) and some rearranging of the series involved. Using these facts, one can compute the function e^A if one knows the canonical form J and the similarity S for which $A=SJS^{-1}$. Indeed, letting $J(\lambda,k)=\lambda I+N$, N nilpotent, we have

$$e^{J(\lambda,k)}=e^{\lambda I}e^N=e^{\lambda}\begin{bmatrix} 1 & 1 & \frac{1}{2} & \frac{1}{3!} & \cdots & \frac{1}{(k-2)!} \\ 0 & 1 & 1 & \frac{1}{2} & \frac{1}{3!} & \cdots \\ \vdots & & & & & \vdots \\ 0 & \cdots & & & 0 & 1 \end{bmatrix}, \tag{4.10}$$

and then

$$e^{\amalg(J(\lambda_1,k_1),\dots,J(\lambda_l,k_l))}=\amalg\left(e^{J(\lambda_1,k_1)},\dots,e^{J(\lambda_l,k_l)}\right). \tag{4.11}$$

Hence, if $A=SJS^{-1}$, $e^A=Se^JS^{-1}$ may be computed explicitly.

The matrix e^A is always nonsingular since $e^Ae^{-A}=I$. Conversely if B is a nonsingular matrix, then there is a matrix A (not unique) with $B=e^A$. We sketch briefly how A may be computed. Suppose that $B=\amalg(J(\lambda_1,k_1),\dots,J(\lambda_l,k_l))$. It suffices to find matrices $A_1,\dots,A_l$ with $J(\lambda_i,k_i)=e^{A_i}$, $i=1,\dots,l$, for then $A=\amalg(A_1,\dots,A_l)$ is a matrix of the desired type.

ISBN 0-201-13509-4

Note that $\lambda_i \neq 0$ since B is nonsingular so that $\det B \neq 0$. (See (4.5).) Now $J(\lambda_i, k_i) = \lambda_i I + N = \lambda_i(I + \lambda_i^{-1}N)$ where N is nilpotent. Using the series

$$\log(1+z) = \sum_{j=1}^{\infty} (-1)^{j+1} \frac{z^j}{j}, \qquad |z|<1 \tag{4.12}$$

we have formally

$$\log J(\lambda_i, k_i) = (\log \lambda_i) I + \sum_{j=1}^{\infty} (-1)^{j+1} \frac{(\lambda_i^{-1}N)^j}{j}. \tag{4.13}$$

Here $\log \lambda_i$ is any logarithm of λ_i and the infinite series is actually finite because N is nilpotent. It is not difficult to see that setting A_i equal to the right-hand side of (4.13), we have $e^{A_i} = J(\lambda_i, k_i)$. For details of the above argument, see [EC].

Finally we remark that if $A \in \mathbb{C}_{n\times n}$, and t is a real or complex variable, then

$$\frac{d}{dt} e^{tA} = A e^{tA} = e^{tA} A \tag{4.14}$$

as is easily proved by differentiating the series for e^{tA} term by term. Hence if $Z \in \mathbb{C}_n$, setting $u(t) = e^{tA} Z$, $u(t)$ solves the initial value problem

$$\frac{du}{dt} = Au(t), \qquad u(0) = Z. \tag{4.15}$$

A.I.5 Some Additional Results

Let $A \in \mathbb{C}_{n\times n}$. We define a linear mapping, denoted adA, from $\mathbb{C}_{n\times n}$ to itself by

$$adA(B) = AB - BA \qquad \text{for } B \in \mathbb{C}_{n\times n}. \tag{5.1}$$

We sometimes denote $AB - BA$ by $[A,B]$, so that $adA(B) = [A,B]$. We define the norm of the mapping adA in a manner similar to (1.2), namely

$$\|adA\| = \max_{\substack{\|B\| \leqslant 1 \\ B \in \mathbb{C}_{n\times n}}} \|adA(B)\|. \tag{5.2}$$

Since $\|AB - BA\| \leqslant \|AB\| + \|BA\| \leqslant 2\|A\|\|B\|$, we have

$$\|adA\| \leqslant 2\|A\|. \tag{5.3}$$

If λ is a complex number, and $Z_\lambda \in \mathbb{C}_{n\times n}$ is a nonzero matrix with

ISBN 0-201-13509-4

$adA(Z_\lambda)=\lambda Z_\lambda$, then λ is called an *eigenvalue* and Z_λ an associated *eigenmatrix* of adA. The set of eigenvalues of adA is called the *spectrum* of adA and denoted $\sigma(adA)$. In Chapter 2 of the text we have shown that

$$\sigma(adA)=\sigma(A)-\sigma(A).$$

We have seen in (4.14) that if $A\in\mathbb{C}_{n\times n}$, e^{tA} is differentiable with derivative Ae^{tA}. We now consider a more general situation. Suppose $C(t)$ is a differentiable $\mathbb{C}_{n\times n}$-valued function with $C'(t)$ continuous. If $n=1$, $(d/dt)e^{C(t)}=C'(t)e^{C(t)}$, but if $n>1$ this is no longer generally true. In fact we have:

THEOREM 5.1.

$$\frac{d}{dt}e^{C(t)}=\Gamma_1(t)\cdot e^{C(t)}=e^{C(t)}\Gamma_2(t), \tag{5.4}$$

where

$$\Gamma_1(t)=C'(t)+\sum_{k=1}^{\infty}\frac{(adC)^k}{(k+1)!}C'(t),$$

$$\Gamma_2(t)=C'(t)+\sum_{k=1}^{\infty}(-1)^k\frac{(adC)^k}{(k+1)!}C'(t),$$

where the notation $(adC)^kC'(t)$ *means* $[C(t),[\cdots[C(t),[C(t),C'(t)]]\cdots]]$ *with* $C(t)$ *appearing* k *times.*

Remark. If $[C(t),C'(t)]=0$, then the theorem implies that $(d/dt)e^{C(t)}=C'(t)e^{C(t)}$.

Proof of the theorem. By the fundamental theorem of calculus, we have for $\delta\neq 0$

$$\frac{e^{C(t+\delta)}-e^{C(t)}}{\delta}=\frac{1}{\delta}\int_0^1\frac{d}{ds}\left(e^{sC(t+\delta)}e^{(1-s)C(t)}\right)ds$$

$$=\int_0^1 e^{sC(t+\delta)}\left(\frac{C(t+\delta)-C(t)}{\delta}\right)e^{(1-s)C(t)}\,ds. \tag{5.5}$$

Letting $\delta\to 0$ we find

$$\frac{d}{dt}e^{C(t)}=\left(\int_0^1 e^{sC(t)}C'(t)e^{-sC(t)}\,ds\right)\cdot e^{C(t)}. \tag{5.6}$$

ISBN 0-201-13509-4

Expanding the integrand in (5.6) in a Taylor series about $s=0$ yields

$$e^{sC(t)}C'(t)e^{-sC(t)}=C'(t)+(adC)C'(t)\cdot s+(adC)^2\cdot C'(t)\frac{s^2}{2!}+\cdots \tag{5.7}$$

and integrating this from 0 to 1 yields the formula

$$\frac{d}{dt}e^{C(t)}=\Gamma_1(t)\cdot e^{C(t)}$$

claimed in the theorem. The formula

$$\frac{d}{dt}e^{C(t)}=e^{C(t)}\cdot\Gamma_2(t)$$

is proved similarly starting with

$$\frac{e^{C(t+\delta)}-e^{C(t)}}{\delta}=\frac{1}{\delta}\int_1^0\frac{d}{ds}\left(e^{(1-s)C(t+\delta)}e^{sC(t)}\right)ds. \tag{5.5'}$$

■

ISBN 0-201-13509-4

References

Notes to the References

In preparing the list of references we have attempted to include all work that has been done on the subject of product integration. Undoubtedly there are omissions, but these are unintentional. We shall group the cited articles and books in several categories as a rough guide for anyone wishing to consult the literature. In spite of the fact that our grouping is somewhat artificial and considerable overlapping occurs, we hope that our remarks will be useful.

We have included several references on differential equations and functional analysis cited in the text that make no use of product integration. These are [FA], [KB], [EC], [JD5], [VG], [TK3], [TK4], [GL], [DP], [MR1], [MR2], [FR], [M-R], [LIS], [BS], [JV], and [JW]. (Product integration is mentioned in an exercise in [EC] but not used in the text.)

Product integration of matrix- (or bounded linear operator) valued functions is developed in [JD3], [JD1], [FG], [GR], [L-S1], [L-S4], [L-S5], and [VV1]—[VV4]. These works are roughly at the level of our Chapter 1. Development and discussion of contour product integrals similar to our development in Chapter 2 may be found in [JD4], [FG], [PM2], [GR], [L-S1], [L-S4], and [VV4].

The use of product integration as a tool in solving evolution equations of the type $Y'(x)=A(x)Y(x)$, where the $A(x)$ are unbounded linear operators, is treated in [JD3], [CF1], [JAG1], [TK1], [TK2], [HT], [KY1], [KY2], [KY3], and [KY4]. [PC], [CF1], [TK2], and [HT] are concerned with the semigroup formula $\lim_{n\to\infty}F(t/n)^n=e^{tA}$ with A unbounded (but not time dependent).

The literature on product integration of nonlinear operator-valued functions may be divided into several groups according to the type of product integral utilized. Product integrals of the type $\prod(I-dsA)^{-1}$, where A is nonlinear but constant occur, in [H-B2], [JC], [JRD2], [J-M]–[SO], [SR], [GW3], and [GW4] where they are used to represent, approximate, or prove existence of nonlinear

ENCYCLOPEDIA OF MATHEMATICS and Its Applications, Gian-Carlo Rota (ed.). Vol. 10: John D. Dollard and Charles N. Friedman, Product Integration with Applications to Differential Equations

ISBN 0-201-13509-4

semigroups. Some general discussion of product formulas of the type $\lim_{n\to\infty} F(t/n)^n$ for nonlinear semigroups may be found in [PC], [H-B1], and [JEM]. As mentioned earlier, we have considered the theory of semigroups to be somewhat separated from the theory of product integration; of course, a semigroup may be viewed as the product integral of a constant function, but we have, in most of our presentation, taken the product integral of a constant function or a step function as our starting point in the construction of the product integral of more general functions. Product integrals of the type $\prod(I+dsA(s))^{-1}$ where $A(s)$ is non-constant are used in [MC1], [MC2], [WF1], [WF3], [JAG2], [AR], [GW2], and [GW3]. Birkhoff ([GB]) makes use of product integrals essentially of type $\prod(I+dsA(s))$, where $A(s)$ is non-constant and nonlinear.

A large number of authors have developed product integration of functions or measures whose values lie in an abelian group or normed ring. One of the first works in this setting is Masani's paper [PM1]. Masani considers product integrals of the form $\prod(I+A(s)ds)$, where $A(s)$ has values in a normed ring, defined as a limit of "Riemann products." In [JH1] Hamilton develops product integration of functions with values in a Lie algebra, the product integral taking values in the corresponding Lie group. In [HW] Wall used product integration to solve the integral equation $M(x,y)=I+\int_x^y dF(s)\cdot M(s,y)$, where F is a continuous matrix-valued function of bounded variation (so dF is a measure without pure points). His work was continued and vastly generalized by MacNerney in [JM1]–[JM9]. For example, in [JM5] and [JM6] MacNerney studies the correspondence between two classes of functions whose domains are linearly ordered sets and whose values lie in a complete normed ring and which are defined, respectively, by an additive and a product integral. (This is discussed briefly in our Chapter 6.) Other articles in this setting whose authors appear to have continued or been influenced by MacNerney's work include [JAC], [RC], [WD], [JRD1], [JRD2], [BH1]–[JVH9], [TH], [DBH], [JK], [DL], [RM], [JN], and [JR]. These papers include applications of product integration to linear and nonlinear equations as well as existence criteria for product integrals in quite abstract settings.

Finally we mention some articles which contain miscellaneous applications of product integration and which do not seem to properly belong to any of the previous categories discussed. In [NA1] and [NA2] product integration figures in applications to stochastic processes, quantum mechanical perturbation theory, and the theory of cosmic radiation. In [JF] a product integral is utilized in analyzing perturbations of a shift operator in Hilbert space. The application of product integration to Nevanlinna theory for matrix-valued functions is developed in [JG1]–[JG6], [PM4], and [VP]. In [JH2], product integration intervenes in the expression of path integrals involving spin. Numerical approximation of matrix-valued Riemann product integrals is developed in [JCH18]. Multiplicative matricial radial limit theorems (using product integrals) are proved in [DH] and [LS]. In [HM] product integration of the exponential of the differential of Brownian motion on the Lie algebra of $SO(3)$ is used to construct Brownian motions on $SO(3)$. Product integration of the curvature tensor to represent the effect of parallel displacement along a closed curve is treated in [L-S3]. Product integration of measures in the sense of our discussion in Chapter 5 is developed in [GS]. In [FS] an abstract integral including product and ordinary integrals is developed.

ISBN 0-201-13509-4

[NA1] Arley, N., "On the theory of stochastic processes and their applications to the theory of cosmic radiation," Wiley, New York, 1943, 1948.

[NA2] Arley, N., and Borschsenius, V., "On the theory of infinite systems of differential equations and their applications to the theory of stochastic processes and the perturbation theory of quantum mechanics," *Acta Math.* **76** (1944), 261–322.

[FA] Atkinson, F. V., "The asymptotic solution of second order differential equations," *Ann. Math. Pura Appl.* **37** (1954), 347–378.

[HB1] Baker, H. F., "Further application of matrix notation to integration problems," *Proc. London Math. Soc.* **34** (1902), 347–360.

[HB2] Baker, H. F., "On the calculation of the finite equations of a continuous group," *Proc. London Math. Soc.* **25** (1903), 332–384.

[HB3] Baker, H. F., "Note on the integration of linear differential equations," *Proc. London Math. Soc., ser. 2,* **2** (1904), 293–296.

[HB4] Baker, H. F., Alternate and continuous groups, *Proc. London Math. Soc., ser. 2,* **3** (1905), 24–47.

[KB] Bichteler, K., *Integration Theory*, Lecture Notes in Mathematics #315, Springer Verlag, Berlin, Heidelberg, New York, 1973.

[GB] Birkhoff, G., "On product integration," J. Math. and Phys. **16** (1938), 104–132.

[CB1] Bitzer, C. W., "Stieltjes-Volterra integral equations," *Ill. J. Math.* **14** (1970), 434–451.

[CB2] Bitzer, C. W., "Convolution, fixed point, and approximations in Stieltjes-Volterra integral equations," *J. Austral. Math. Soc.* **14** (1972), 182–199.

[H-B1] Brezis, H., and Pazy, A., "Convergence and approximation of nonlinear semigroups in Banach spaces," *J. Func. Anal.* **9** (1971), 63–74.

[H-B2] Brezis, H., and Pazy, A., "Semigroups of nonlinear contractions on convex sets," *J. Func. Anal.* **6** (1970), 237–281.

[H-B3] Brezis, H., *Opérateurs Maximaux Monotones et Semi-groupes de Contractions dans les Espaces de Hilbert*, North-Holland, Amsterdam, 1973.

[JC] Chambers, J., and Oharu, S., "Semigroups of local lipschitzians in a Banach space," *Pac. J. Math.* **39** (1971), 89–112.

[JAC1] Chatfield, J. A., "Equivalence of integrals," *Proc. Amer. Math. Soc.* **38** (1973), 279–285.

[JAC2] Chatfield, J. A., "Continuous interval functions," *Tex. J. Sci.* **27** (1976), 27–31.

[JAC3] Chatfield, J. A., "Three Volterra integral equations," *Tex. J. Sci.* **27** (1976), 33–38.

[JAC4] Chatfield, J. A., "Product integration and eigenvalues," *Calcutta J. Math.* **69**, 1977.

[JAC5] Chatfield, J. A., "Eigenvalues of a Stieltjes-Volterra integral equation" (submitted for publication).

[JAC6] Chatfield, J. A., "Solution for an integral equation with continuous interval functions," *Pac. J. Math.* (to appear).

[JAC7] Chatfield, J. A., "A representation theorem for operators on a space of interval functions," *Int. J. Math. Math. Sci.* (to appear).

[PC] Chernoff, P., *Product Formulas, Nonlinear Semigroups, and Addition of Unbounded Operators*, Memoirs of the A.M.S. No. 140.

[EC] Coddington, E. A., and Levinson, N., *Theory of Ordinary Differential Equations*, McGraw-Hill, New York, Toronto, London, 1955.

[RC] Cox, R. H., "Integral equations in a Hilbert space," *Ill. J. Math.* **9** (1965), 256–260.

[MC1] Crandall, M., and Liggett, T., "Generation of semigroups of nonlinear transformations on general Banach spaces," *Amer. J. Math.* **93** (1971), 265–293.

[MC2] Crandall, M. G., and Pazy, A., "Nonlinear evolution equations in Banach spaces," *Israel J. Math.* **11** (1972), 57–94.

[WD] Davis, W. P., and Chatfield, J. A., "Concerning product integrals and exponentials," *Proc. Amer. Math. Soc.* **25** (1970), 743–747.

ISBN 0-201-13509-4

[JD1] Dollard, J., and Friedman, C., "Product integrals and the Schrödinger equation," *J. Math. Phys.* **18** (1977), 1598–1607.

[JD2] Dollard, J., and Friedman, C., Existence of the Moller wave operators for $V(r)=\lambda\sin(\mu r^{\alpha})/r^{\beta}$, *Ann. Phys.* **III** (1978), 251–266.

[JD3] Dollard, J., and Friedman, C., "On strong product integration," *J. Func. Anal.* **28** (1978), 309–354.

[JD4] Dollard, J., and Friedman, C., "Product integrals II: Contour integrals," *J. Func. Anal.* **28** (1978), 355–368.

[JD5] Dollard, J., and Friedman, C., "Asymptotic behavior of solutions of linear ordinary differential equations," *J. Math. Anal. Appl.* **66** (1978), 394–398.

[JD6] Dollard, J., and Friedman, C., "Product integration of measures and applications," *J. Diff. Eq.* **31** (1979), 418–464.

[AD] Dominguez, Alberto Gonzalez, "Propiedades en el contorno de funciones analiti cas," in *Cursos y seminarios de mathematics, Fas. 4*, Universidad Nacional de Buenos Aires, 1959.

[JRD1] Dorroh, J. R., "Integral equations in normed Abelian groups," *Pacific J. Math.* **13** (1963), 1143–1158.

[JRD2] Dorroh, J. R., "Semigroups of nonlinear transformations with decreasing domain," *J. Math. Anal. Appl.* **34** (1971), 396–411.

[WF1] Fitzgibbon, W., "Time-dependent nonlinear Cauchy problems in Banach spaces," *Proc. Amer. Math. Soc.* **36** (1972), 525–530.

[WF2] Fitzgibbon, W., "Abstract Volterra equations with infinite delay," preprint, Math. Dept., University of Houston, Houston, Texas.

[WF3] Fitzgibbon, W., "Approximations of nonlinear evolution equations," *J. Math. Soc. Japan* **25** (1973), 211–221.

[WF4] Fitzgibbon, W. E., "Weakly continuous nonlinear accretive operators in reflexive Banach spaces," *Proc. AMS* **41** (1973), 229–236.

[WF5] Fitzgibbon, W. E., "Product integration in reflexive Banach spaces," *Monat. Math.* **83** (1977), 113–119.

[JF1] Freeman, J. M., "Perturbation of the shift operator," *Trans. Amer. Math. Soc.* **114** (1965), 251–260.

[JF2] Freeman, J. M., "The tensor product of semigroups and the operator equation $SX-XT=A$," *J. Math. Mch.* **19** (1970), 819–828.

[CF1] Friedman, C. N., "Semigroup product formulas, compressions, and continual observations in quantum mechanics," *Indiana Univ. Math. J.* **21** (1972), 1001–1011.

[CF2] Friedman, C. N., "Continual measurements in the spacetime formulation of nonrelativistic quantum mechanics," *Ann. Phys.* **98** (1976), 87–97.

[FG] Gantmacher, F. R., *Matrix theory*, Chelsea Publishing Co., New York, 1960.

[VG] Georgescu, V., "Méthodes stationnaires pour des potentials à longue portée à symmétrie sphérique," These, Université de Génève, Département de Physique Théorique, 1974.

[WG1] Gibson, W. L., "Stieltjes and Stieltjes-Volterra integral equations," *Trans. Amer. Math. Soc.* **224** (1976).

[WG2] Gibson, W. L., "Embedding Stieltjes-Volterra integral equations in Stieltjes integral equations," *Trans. AMS* **227** (1977), 263–277.

[F-G] Gilbert, F., and Backus, G. E., "Propagator matrices in elastic wave and vibration problems," *Geophysics* **31** (1966), 326–332.

[JG1] Ginzburg, Ju. P., *Dokl. Akad. Nauk SSSR* **117** (1957), 171–173.

[JG2] Ginzburg, Ju. P., "The factorization of analytic matrix functions," *Soviet Math. Dokl.* **5** (1964), 1410–1515.

[JG3] Ginzburg, Ju. P., "Sobre la factorizacion de matrices de funciones analíticas," *Akad. Nauk. SSSR Comptes Rendus (Doklady), Nouv. Ser.* **159** (1964), 489–492.

ISBN 0-201-13509-4

[JG4] Ginzburg, Ju. P., "The factorization of analytic matrix functions," *Soviet Mathematics, AMS* **5** (1964), 1510–1514; translation of *Doklady TOM* **159**, #1–6.

[JG5] Ginzburg, Ju. P., "Multiplicative representations of bounded analytical operator-functions," *Dokl. Akad. Nauk SSSR* **170** (1966), 1125–1128.

[JG6] Ginzburg, Ju. P., *Uspekhi Mat. Nauk* **12** (1967), 163–167.

[JAG1] Goldstein, J. A., "Abstract evolution equations," *Trans. Amer. Math. Soc.* **141** (1969), 159–185.

[JAG2] Goldstein, J. A., "Approximation of nonlinear semigroups and evolution equations," *J. Math. Soc. Japan* **24** (1972), 558–573.

[CG] Groetsch, C. W., "A product integral representation of the generalized inverse," *Comm. Math. Univ. Coro.* **16** (1975), 13–20.

[JH1] Hamilton, J., "Multiplicative Riemann integration and logarithmic differentiation in Lie algebras and Lie groups," Doctoral dissertation, Indiana University, 1973.

[JH2] Hamilton, J. F., Jr., and Schulman, L. S., "Path integrals and product integrals," *J. Math. Phys.* **12** (1971), 160–164.

[BH1] Helton, B. W., "Integral equations and product integrals," *Pacific J. Math.* **16** (1966), 297–332.

[BH2] Helton, B. W., "A product integral representation for a Gronwall inequality," *Proc. Amer. Math. Soc.* **23** (1969), 493–500.

[BH3] Helton, B. W., "Solutions of $f(x)=f(a)+(RL)\int_a^x(fH+fG)$ for rings," *Proc. Amer. Math. Soc.* **25** (1970), 735–742.

[BH4] Helton, B. W., "The solution of a nonlinear Gronwall inequality," *Proc. Amer. Math. Soc.* **38** (1973), 337–342.

[BH5] Helton, B. W., "A product integral solution of a Riccati equation," *Pacific J. Math.* **56** (1975), 113–130.

[BH6] Helton, B. W., "The solution of a Stieltjes-Volterra integral equation for rings," *Pacific J. Math.* **64** (1976), 419–436.

[BH7] Helton, B. W., "A special integral and a Gronwall inequality," *Trans. AMS* **217** (1976), 163–181.

[JCH1] Helton, J. C., "Existence theorems for sum and product integrals," *Proc. Amer. Math. Soc.* **36** (1972), 407–413.

[JCH2] Helton, J. C., "An existence theorem for sum and product integrals," *Proc. Amer. Math. Soc.* **39** (1973), 149–154.

[JCH3] Helton, J. C., "Bounds for products of interval functions," *Pacific J. Math.* **49** (1973), 377–389.

{JCH4] Helton, J. C., "Product integrals and exponentials in commutative Banach algebras," *Proc. Amer. Math. Soc.* **39** (1973), 155–162.

[JCH5] Helton, J. C., "Existence of sum and product integrals," *Trans. Amer. Math. Soc.* **182** (1973), 165–174.

[JCH6] Helton, J. C., "Some interdependencies of sum and product integrals," *Proc. Amer. Math. Soc.* **37** (1973), 201–206.

[JCH7] Helton, J. C., "Product integrals and inverses in normed rings," *Pacific J. Math.* **51** (1974), 155–166.

[JCH8] Helton, J. C., "Mutual existence of product integrals," *Proc. Amer. Math. Soc.* **42** (1974), 96–103.

[JCH9] Helton, J. C., "Mutual existence of sum and product integrals," *Pacific J. Math.* **56** (1975), 495–516.

[JCH10] Helton, J. C., "Product integrals and the solution of integral equations," *Pacific J. Math.* **58** (1975), 87–103.

[JCH11] Helton, J. C., "Solution of integral equations by product integration," *Proc. Amer. Math. Soc.* **49** (1975), 401–406.

[JCH12] Helton, J. C., "Product integrals, bounds, and inverses," *Tex. J. Sci.* **26** (1975), 11–18.

ISBN 0-201-13509-4

[JCH13] Helton, J. C., "Mutual existence of product integrals in normed rings," *Trans. AMS* **211** (1975), 353–363.

[JCH14] Helton, J. C., "Existence of integrals and the solution of integral equations," *Trans. AMS* **229** (1977), 307–327.

[JCH15] Helton, J. C., "Two generalizations of the Gronwall inequality by product integration," *Rocky Mtn. J. Math.* **7** (1977), 733–749.

[JCH16] Helton, J. C., "Solution of two Volterra integral equations," *Rocky Mtn. J. Math.* (to appear).

[JCH17] Helton, J. C., "Nonlinear operations and the solution of integral equations," *Trans. AMS* **237** (1978), 373–390.

[JCH18] Helton, J., and Stuckwisch, S., "Numerical approximation of product integrals," *J. Math. Anal. Appl.* **56** (1976), 410–437.

[JCH19] Helton, J. C., and Stuckwisch, S., "An approximation technique for nonlinear integral operations," *J. Math. Anal. Appl.* (to appear).

[JVH1] Herod, J. V., "Solving integral equations by iteration," *Duke Math. J.* **34** (1967), 519–534.

[JVH2] Herod, J. V., "Multiplicative inverses of solutions for Volterra-Stieltjes integral equation," *Proc. Amer. Math. Soc.* **22** (1969), 650–656.

[JVH3] Herod, J. V., "Coalescence of solutions for nonlinear Volterra equations," *Not. AMS* **16** (1969), 834.

[JVH4] Herod, J. V., "A Gronwall inequality for linear Stieltjes integrals," *Proc. AMS* **23** (1969), 34–36.

[JVH5] Herod, J. V., "A pairing of a class of evolution systems with a class of generators," *Trans. Amer. Math. Soc.* **157** (1971), 247–260.

[JVH6] Herod, J. V., "A product integral representation for an evolution system," *Proc. Amer. Math. Soc.* **27** (1971), 549–556.

[JVH7] Herod, J. V., "Coalescence of solutions of nonlinear Stieltjes equations," *J. Reine Angew. Math.* **252** (1972), 187–194.

[JVH8] Herod, J. V., "The uniform limit of Lipschitz functions on a Banach space," *J. Austral. Math. Soc.* **15** (1973), 325–331.

[JVH9] Herod, J. V., "Generators for evolution systems with quasi-continuous trajectories," *Pacific J. Math.* **53** (1974), 153–161.

[DH] Herrero, D. A., "Valores limites de integrales multiplicativas de Stieltjes," *Rev. Union Matematica Argentina* **24** (1968), 59–64.

[TH1] Hildebrandt, T. H., "Integration in abstract spaces," *Bull. AMS* **59** (1953), 111–139.

[TH2] Hildebrandt, T. H., "On systems of linear differentio-Stieltjes-integral equation," *Ill. J. Math.* **3** (1959), 352–373.

[EH] Hille, E., and Phillips, R. S., *Functional Analysis and Semi-groups*, American Mathematical Society, Providence, 1957.

[DBH] Hinton, D., "A Stieltjes-Volterra integral equation theory," *Canad. J. Math.* **18** (1966), 314–331.

[TK1] Kato, T., "Integration of the equation of evolution in a Banach space," *J. Math. Soc. Japan* **5** (1953), 208–234.

[TK2] Kato, T., *Perturbation Theory for Linear Operators*, Springer-Verlag, New York, 1966.

[TK3] Kato, T., "Linear evolution equations of hyperbolic type," *J. Fac. Sci., Univ. Tokyo, Sec. 1, A Math.* **17** (1970), 241–258.

[TK4] Kato, T., "Linear evolution equations of hyperbolic type II," *J. Math. Soc. Japan* **25** (1973), 648–666.

[JK] Kay, J. A., "Nonlinear integral equations and product integrals," *Pacific J. Math.* **60** (1975), 203–222.

[CK] Kourkoumelis, C., and Nettel, S., "Operator functionals and path integrals," *Amer. J. Phys.* **45** (1977), 26–30.

[S1] Lang, S., *Algebra*, Addison-Wesley, Reading, Mass., 1965.

ISBN 0-201-13509-4

[ML1] Livsic, M. S., "On the theory of self-adjoint systems of differential equations," *Akad. Nauk. SSSR Comptes Rendus (Doklady), Nouv. Ser.* **72** (1959), 1013–1016.

[ML2] Livsic, M. S. (Russian), *Math. Sbornik* **34** (1954), 145–199.

[DL1] Lovelady, D. L., "A variation-of-parameters inequality," *Proc. AMS* **26** (1970), 598–602.

[DL2] Lovelady, D. L., "Bounded solutions of Stieltjes integral equations," *Proc. AMS* **28** (1971), 127–133.

[DL3] Lovelady, D. L., "Perturbations of solutions of Stieltjes integral equations," *Trans. Amer. Math. Soc.* **155** (1971), 175–187.

[DL4] Lovelady, D. L., "Addition in a class of nonlinear Stieltjes integrators," *Israel J. Math.* **10** (1971), 391–396.

[DL5] Lovelady, D. L., "Multiplicative integration of infinite products," *Canad. J. Math.* **23** (1971), 692–698.

[DL6] Lovelady, D. L., "Algebraic structure for a set of nonlinear integral operations," *Pacific J. Math.* **37** (1971), 421–427.

[DL7] Lovelady, D. L., "Product integrals for an ordinary differential equation in a Banach space," *Pacific J. Math.* **48** (1973), 163–168.

[GL] Lumer, G., and Rosenblum, M., "Linear operator equations," *Proc. Amer. Math. Soc.* **10** (1959), 32–41.

[JM1] MacNerney, J. S., "Stieltjes integrals in linear spaces," *Ann. Math.* **61** (1955), 354–367.

[JM2] MacNerney, J. S., "Continuous products in linear spaces," *J. Elisha Mitchell Sci. Soc.* **71** (1955), 185–200.

[JM3] MacNerney, J. S., "Determinants of harmonic matrices," *Proc. Amer. Math. Soc.* **7** (1956), 1044–1046.

[JM4] MacNerney, J. S., "Concerning quasi-harmonic operators," *J. Elisha Mitchell Sci. Soc.* **73** (1957), 257–261.

[JM5] MacNerney, J. S., "Integral equations and semigroups," *Ill. J. Math.* **7** (1963), 148–173.

[JM6] MacNerney, J. S., "A linear initial-value problem," *Bull. Amer. Math. Soc.* **69** (1963), 314–329.

[JM7] MacNerney, J. S., "An integration-by-parts formula," *Bull. AMS* **69** (1963), 803–805.

[JM8] MacNerney, J. S., "Note on succesive approximations," *Rend. Circ. Mat. Pal.* **12** (1963), 87–90.

[JM9] MacNerney, J. S., "A nonlinear integral operation," *Ill. J. Math.* **8** (1964), 621–638.

[GM] Marrah, G. W., and Proctor, T. G., "Solutions of some periodic Stieltjes integral equations," *Proc. AMS* **34** (1972), 121–127.

[JEM] Marsden, J., "On product formulas for nonlinear semigroups," *J. Func. Anal.* **13** (1973), 51–72.

[RM1] Martin, R. H., Jr., "A bound for solutions of Volterra-Stieltjes integral equations," *Proc. AMS* **23** (1969), 412–506.

[RM2] Martin, R. H., Jr., "Bounds for solutions to a class of nonlinear integral equations," *Trans. AMS* **160** (1971), 131–138.

[RM3] Martin, R. H., Jr., "Product integral approximations of solutions to linear operator equations," *Proc. Amer. Math. Soc.* **41** (1973), 506–512.

[PM1] Masani, P., "Multiplicative Riemann integration in normed rings," *Trans. Amer. Math. Soc.* **61** (1947), 147–192.

[PM2] Masani, P., "The rational approximation of operator-valued functions," *Proc. London Math. Soc.* **6** (1956), 43–58.

[PM3] Masani, P., "Sur les fonctions matricielles de la classe de Hardy H_2." *II, Academic Des Sciences, Paris Comptes Rendus Hebdomodoires de séances* **249** (1959), 906–907.

[PM4] Masani, P., "Une generalisation pour les fonctions matricielles de la classe de Hardy H_2 d'un theoreme de Nevanlinna," *C. R., Acad. Sci., Paris* **251** (1960), 318–320.

ISBN 0-201-13509-4

[HM] McKean, H. P., "Brownian motions on the 3-dimensional rotation group," *Memoirs of the College of Science, University of Kyoto, Series A* **33** (1960), 25.

[J-M] Mermin, J., "An exponential limit formula for nonlinear semi-groups," *Trans. Amer. Math. Soc.* **150** (1970), 469–476.

[IM] Miyadera, I., and Oharu, S., "Approximation of semi-groups of nonlinear operators," *Tohuku Math. J.* **22** (1970), 24–27.

[JN1] Neuberger, J. W., "Continuous products and nonlinear integral equations," *Pacific J. Math.* **8** (1958), 529–549.

[JN2] Neuberger, J. W., "Concerning boundary value problems," *Pacific J. Math.* **10** (1960), 1385–1392.

[JN3] Neuberger, J. W., "A generator for a set of functions," *Ill. J. Math.* **9** (1965), 31–39.

[JN4] Neuberger, J. W., "An exponential formula for one parameter semigroups of nonlinear transformations," *J. Math. Soc. Japan* **18** (1966), 154–157.

[JN5] Neuberger, J. W., "Product integral formulae for nonlinear expansive semigroups and non-expansive evolution systems," *J. Math. Mech.* **19** (1969), 403–410.

[SO] Oharu, S., "On the generation of semi-groups of nonlinear contractions," *J. Math. Soc. Japan* **22** (1970), 526–549.

[MP] Pease, M., *Methods of Matrix Algebra*, Academic Press, New York, 1965.

[DP] Pearson, D. B., "An example in potential scattering illustrating the breakdown of asymptotic completeness," preprint, Dept. of Theoretical Physics, University of Geneva, Geneva, Switzerland.

[AP1] Plant, A. J., "Hölder continuous nonlinear product integrals," in *Proc. Conf. Ordinary and Partial Diff. Equations*, Sleeman and Michael (eds.) Springer, Berlin, 1974.

[AP2] Plant, A. T., "Some aspects of non-linear semi-groups," in *Control theory and topics in functional analysis, V.* III, International Atomic Energy Agency, Vienna, 1976, pp. 291–304.

[AP3] Plant, A., "Nonlinear semigroups and evolution operators" (to appear).

[AP4] Plant, A., "Nonlinear Evolution equations with generators of irregular time dependence," University of Warwick, Report No. 14.

[AP5] Plant, A., "The product integral method for nonlinear evolution equations," Control Theory Centre, University of Warwick, Report No. 23.

[VP1] Potapov, V. P. (Russian), *Akad. Nauk. SSSR Comptes Rendus, Nouv. Ser.* **72** (1950), 849–852.

[VP2] Potapov, V. P., "The multiplicative structure of *J*-contractive matrix functions," *Trudi Mosk. Matem. Obsch.* **4** (1955), 125–236. (In Russian.) [*Amer. Math. Soc. Transl.* **15** (1960), 131–243.]

[GP] da Prato, G., *Applications Croissantes et Equations d'evolutions dans les espaces de Banach*, Academic Press, New York, 1976.

[GR1] Rasch, G., *Om matrixregning og dens anvendelse paa differens-og differentiallegninger*, Levin ana Munksgaards, Copenhagen, Denmark, 1931.

[GR2] Rasch, G., "Zur Theorie und Anwedung des Producktintegrals," *J. Reine Angew. Math.* **191** (1934), 65–119.

[SR] Rasmussen, S., *Nonlinear semi-groups, evolution equations and product integral representations*, Various Publication Series, No. 20, Aarhus Universitet, Denmark.

[MR1] Reed, M., and Simon, B., *Methods of Modern Mathematical Physics*, Vol. I, Academic Press, New York, San Francisco, London, 1972.

[MR2] Reed, M., and Simon, B., *Methods of Modern Mathematical Physics*, Vol. II, Academic Press, New York, San Francisco, London, 1975.

[JR1] Reneke, J. A., "A product integral solution of a Stieltjes-Volterra integral equation," *Proc. Amer. Math. Soc.* **24** (1970), 621–626.

[JR2] Reneke, J. A., "On the existence and representation of integrals," *Ill. J. Math.* **14** (1970), 99–112.

ISBN 0-201-13509-4

[JR3] Reneke, J. A., "Continuity for Stieltjes-Volterra integral equation," *Rev. Roum. Math. pures appl.* **17** (1972), 389–401.

[JR4] Reneke, J. A., *A variation of parameters formula*, Technical Report #87, Clemson University.

[JR5] Reneke, J. A., "Product integral solutions for hereditary systems," *Trans. Amer. Math. Soc.* **181** (1973), 483–493.

[FR] Riesz, F. and Sz.-Nagy, B., *Functional Analysis*, Frederick Ungar Publishing Co., New York, 1965.

[M-R] Rosenblum, M., "On the operator equation $BX - XA = Q$," *Duke Math. J.* **23** (1956), 263–269.

[AR] Roques, A. J., "Local evolution systems in general Banach spaces," *Pacific J. Math.* **62** (1976), 197–217.

[WR] Rudin, W., *Real and Complex Analysis*, McGraw-Hill, New York, 1966.

[DR] Rutledge, D., "A generator for a semi-group of nonlinear transformations," *Proc. AMS* **20** (1969), 491–498.

[LS1] Sahnovic, L. A. (Russian), *Uspekhi Mat. Nauk.* **XII** (1947), 211–220.

[LS2] Sahnovic, L. A. (Russian), *Akad. Nauk. SSSR, Izvestii, S IIA Mat.* **21** (1957), 235–262.

[LS3] Sahnovic, L. A., "On limit values of multiplicative integrals," *Usp. Mat. Nauk. Tom, XII ed.* **3** (1957), 205–210.

[LS4] Sahnovic, L. A. (Russian), *Mat. Sbornik N. S.* **76** (1968), 323–343.

[LS5] Sahnovic, L. A., "Dissipative Volterra operators," *Math. USSR Sbornik* **5** (1968), 311–331.

[LS6] Sahnovic, L. A., "Generalized wave operators," *Math. USSR Sbornik* **10** (1970), 197–216.

[L-S1] Schlesinger, L., *Vorlesungen über lineare differential-Gleichungen*, Leipzig, Teubner, 1908.

[L-S2] Schlesinger, L., *Einführung in die theorie der gewohnlichen differential-Gleischungen auf funktion-entheoretischer Grundlage*, Vereinigung Wissenschatflicher Verliger Watter De Gruyter and Co., Berlin, 1922.

[L-S3] Schlesinger, L., "Parrallelverschiebung und Krummungstensor," *Math. Annalen.* **99** (1928), 413–434.

[L-S4] Schlesinger, L., "Neue Grundlagen für einen Infinitesimalkalkul der Matrizen," *Math. Zeit.* **33** (1931), 33–61.

[L-S5] Schlesinger, L., "Weitere Beitragen zum Infinitesimalkalkul der Matrizen," *Math. Zeit.* **35** (1932), 485–501.

[GS] Schmidt, G., "On multiplicative Lebesque integration and families of evolution operators," *Math. Scand.* **29** (1971), 113–133.

[LIS] Schiff, L. I., *Quantum Mechanics*, 2nd ed., McGraw-Hill, New York, 1955.

[BS] Simon, B., "On positive engenvalues of one body Schroedinger operators," *Comm. Pure Appl. Math.* **22** (1969), 531–538.

[PS] Sobolevski, P. E., "On equations of parabolic type in a Banach space," *Trudy Moskov. Mat. Obsc.* **10** (1961), 297–350.

[JS1] Spellman, J. W., "Concerning the infinite diffenentiability of semi-group motion," *Pacific J. Math.* **30** (1969), 519–523.

[JS2] Spellman, J. W., "Product integral techniques for abstract hyperbolic partial differential equations," *Trans. AMS* **209** (1975), 353–365.

[FS] Stewart, F. M., "Integration in non-commutative spaces," *Trans. Amer. Math. Soc.* **68** (1950), 76–104.

[MS] Stone, M. H., *Linear transformations in Hilbert space and their applications to analysis*, Amer. Math. Soc. Colloquium Pub. Vol. 15, New York, 1932.

[JWS] Sullivan, J. W., "Product integral solutions of stochastic Volterra-Stieltjes integral equations with discontinuous integrators," Ph.D. dissertation, Georgia Inst. of Tech., 1976.

ISBN 0-201-13509-4

[HT] Trotter, H. F., "On the product of semigroups of operators," *Proc. Amer. Math. Soc.* **10** (1959), 545–551.

[VV1] Volterra, V., "Sulle equazioni differzeniali lineari," *Rendiconot Accademia dei Lincei* **3** (1887), 393–396.

[VV2] Volterra, V., "Sulle equazioni differenziali lineari," *Rendiconti del Circolo Mathematico di Palermo* **2** (1888), 69–75.

[VV3] Volterra, V., *Sui fondamenti della teoria delle equazioni differenziali lineari*, Memorie della Societa Itialiana delle Scienze, 3e série, t. VI, No. 8, parte prima (1887); 3e série, t. XII, parte seconda, p. 3–68 (1902).

[VV4] Volterra, V., and Hostinsky, B., *Opérations infinitesimales linéaires*, Gauthier-Villars, Paris, 1938.

[JV] Von Neumann, J., and Wigner, E. P., "Über Merkwurdige diskrete Eigenwerte," *Z. Phys.* **30** (1929), 465–467.

[AW] Wintner, A., "On Linear Asymptotic Equilibria," *Amer. J. Math.* **71** (1949), 853–858.

[HW] Wall, H. S., "Concerning harmonic matrices," *Arch. Math.* **5** (1954), 160–167.

[GW1] Webb, G. F., "Nonlinear evoluation equations and product integration in Banach spaces," *Trans. Amer. Math. Soc.* **148** (1970), 273–282.

[GW2] Webb, G. F., "Product integral representation of time-dependent nonlinear evolution equations in Banach spaces," *Pacific J. Math.* **32** (1970), 269–281.

[GW3] Webb, G. F., "Nonlinear evolution equations and product stable operators on Banach spaces," *Trans. Amer. Math. Soc.* **155** (1971), 1–18.

[GW4] Webb, G. F., "Representation of nonlinear non-expansive semi-groups of transformations in Banach space," *J. Math. Mech.* **19** (1969), 159–170.

[JW] Weidmann, J., "Zur Spectraltheorie von Sturm-Liouville-Operator," *Math. Zeitschr.* **98** (1967), 268–302.

[KY1] Yosida, K., "On the integration of the equation of evolution," *J. Fac. Sci. Univ. Tokyo, Sect. 1,* **9** (1963), 397–402.

[KY2] Yosida, K., *Functional Analysis*, 1st ed., Academic Press, New York, 1965.

[KY3] Yosida, K., *Functional Analysis*, 2nd ed., Academic Press, New York, 1968.

[KY4] Yosida, K., "Time dependent evolution equations in a locally convex space," *Math. Ann.* **162** (1965), 83–86.

ISBN 0-201-13509-4

APPENDIX II by P. R. Masani

APPENDIX II by P. R. Masani

The Place of Multiplicative Integration in Modern Analysis*

A.II.1 Introduction

Multiplicative integration was initiated by Volterra [33] in 1887 as a method of solving systems of linear differential equations. The early history of the subject bears the impress of this initiation. Schlesinger, who took up the further development of the theory and became its chief proponent, stressed the link with differential equations in his 1908 lectures on the subject [26]. But the historical growth of mathematics only roughly displays its authentic design as a logical edifice. From a logical or architectural standpoint multiplicative integration belongs to the area of Lie groups and Lie algebras. This was perceived by Birkhoff [2] in the late 1930s, and has been borne out by subsequent work. Accordingly, our first task here will be to lay bare the intrinsic Lie group-theoretic aspect of the subject. For this it is very convenient, as again Birkhoff indicated [2, Sec. 4], to consider the kinematics of fluid flows. It is with this that we shall begin.

Our exposition will not be rigorous in every detail. We shall not, for instance, define what "smooth" means in each case. Nor shall we tarry over questions concerning the existence or interchange of limits. We shall assume that the reader has the empathy to see that our assertions can be made correct by the imposition of reasonable restraints and that rigorization is feasible.

*Note to the reader: Some of the notation and terminology used in this Appendix differ from those of the text. In particular, Professor Masani refers to the product integral as a multiplicative integral. Some questions raised here are addressed in the text.—J. D. D. and C. N. F.

ENCYCLOPEDIA OF MATHEMATICS and Its Applications, Gian-Carlo Rota (ed.). Vol. 10: John D. Dollard and Charles N. Friedman, Product Integration with Applications to Differential Equations

ISBN 0-201-13509-4

The symbols $\mathbf{R}$, $\mathbf{C}$, and $\mathbf{N}$ will denote the real number field, the complex number field, and the ring of integers, respectively. The symbol $\mathbf{R}_{0+}$ will stand for the set of non-negative real numbers, and similar conventions will govern $\mathbf{R}_+$, $\mathbf{N}_{0+}$, etc.

The writer is grateful to Professors R. F. Arens, C. Y. Chao, G. W. Mackey, and A. H. Thompson for conversational remarks, which have been helpful in one way or another in preparing this paper.

A.II.2 Fluid Flows in Smooth Manifolds

Let $\mathcal{M}$ be a q-dimensional smooth (i.e., C^∞) connected manifold. Imagine $\mathcal{M}$ filled with a fluid in motion. Assume as usual that the flow is deterministic, cf. Ref. 19′, Sec. 3.4. Then it can be characterized kinematically by two functions whose values are operators on $\mathcal{M}$: (i) its *position-shift-operator function,* or briefly *propagator,* $T(\cdot,\cdot)$, and (ii) its *velocity-operator function* $V(\cdot)$.

The *propagator* $T(\cdot,\cdot)$ is an $\mathcal{M}$-to-$\mathcal{M}$ operator-valued function on $\mathbf{R}^2$ defined by: $\forall s, t \in \mathbf{R}$ and $\forall x \in \mathcal{M}$,

$$T(t, s)(x) = \text{the position at epoch } t \text{ of the fluid particle at } x \text{ at epoch } s. \tag{2.1}$$

We see at once that $\forall s, t, u \in \mathbf{R}$,

$$T(t, t) = I \underset{d}{=} \textit{identity operator on } \mathcal{M}, \tag{2.2a}$$

$$T(s, t) = T(t, s)^{-1}, \tag{2.2b}$$

$$T(u, s) = T(u, t)\cdot T(t, s), \textit{ for } t \textit{ between } s \textit{ and } u. \tag{2.2c}$$

On the topological side it is reasonable to suppose that all the shifts $T(t, s)$ are smooth functions on $\mathcal{M}$ onto $\mathcal{M}$, and that $T(\cdot,\cdot)$ is itself continuous. We shall for simplicity assume that $\forall s, t \in \mathbf{R}$,

$$\begin{aligned} T(t, s) &= \textit{an interior function on } \mathcal{M} \textit{ onto } \mathcal{M}, \\ T(t, s) &= \textit{a diffeomorphism on } \mathcal{M} \textit{ onto } \mathcal{M}; \end{aligned} \tag{2.3}$$

and that $T(\cdot,\cdot)$ is strongly continuous, i.e.,[1]

$$\operatorname*{slim}_{h\to 0} T(s + h, s) = I. \tag{2.4}$$

To define the velocity-operator $V(\cdot)$ and to understand its relationship to the propagator $T(\cdot,\cdot)$, it is convenient to deal first with flows in a linear manifold $\mathcal{M}$, and then return to flows in an arbitrary smooth manifold.

1. For a net of operators $(T_\lambda : \lambda \in \Lambda)$ on topological spaces X to Y, we write $\operatorname*{slim}_\lambda T_\lambda = T$ to mean that $\forall x \in X$, $\lim_\lambda T(x) = T(x)$ in Y.

ISBN 0-201-13509-4

A. Linear Manifolds

Let $\mathcal{M}$ be a q-dimensional Banach space over the field F (F = R or C). In this linear situation, the theory works just as well for $q = \infty$ as for $q < \infty$. Since $T(t, s)(x) - x$ is the displacement suffered by the particle which is at x at epoch s during the time-interval, $[s, t]$ or $[t, s]$, between the epochs s and t, it is natural to call

$$D(t, s) \underset{d}{=} T(t, s) - I \tag{2.5}$$

the *s-to-t displacement operator,* and to define the *velocity-operator function* $V(\cdot)$ on R by: $\forall t \in \mathrm{R}$,

$$V(t) \underset{d}{=} \operatorname*{slim}_{h \to 0} \frac{1}{h} D(t + h, t) = \operatorname*{slim}_{h \to 0} \frac{1}{h} \{T(t, s) - I\}. \tag{2.6}$$

It then follows of course that $\forall t \in \mathrm{R}$ and $\forall x \in \mathcal{M}$,

$$V(t)(x) = \textit{the fluid velocity at position } x \textit{ at epoch } t. \tag{2.7}$$

It is now easy to recover the propagator $T(\cdot,\cdot)$ from the velocity operator $V(\cdot)$. The restatement of (2.6) in the form

$$T(t + \Delta t, t) - I = V(t)\Delta t + o(\Delta t), \quad \text{as } \Delta t \to 0,$$

shows that for small Δt

$$T(t + \Delta t, t) = I + V(t)\, \Delta t \quad \text{(approx.).} \tag{2}$$

Now let $a, b \in \mathrm{R}$, say $a < b$. Let

$$a = t_0 < t_1 < \cdots < t_{n-1} < t_n = b.$$

Then assuming that this decomposition π of $[a, b]$ has a small mesh $|\pi| \underset{d}{=} \max_{1 \le k \le n} (t_k - t_{k-1})$, it follows from the multiplicative property (2.2c) and (2) that

$$T(b, a) = \prod_{k=n}^{1} T(t_k, t_{k-1}) = \prod_{k=n}^{1} \{I + V(t_{k-1})(t_k - t_{k-1})\},$$

approximately. For sufficiently smooth $V(\cdot)$ it is reasonable to expect the approximation to improve as $|\pi|$ decreases, and the error to vanish in

ISBN 0-201-13509-4

the limit as $|\pi| \to 0$. Thus

$$
\begin{aligned}
T(b, a) &= S \lim_{|\pi| \to 0} \prod_{k=n}^{1} [I + V(t_{k-1})(t_k - t_{k-1})] \\
&\underset{d}{=} \widehat{\int}_b^a (I + V(t)\, dt),
\end{aligned}
\tag{2.8}
$$

where the notation used in the last term is a natural extension of the familiar one for additive integrals. The symbol "$\widehat{\int}$" is meant to be a variant of P or Π for "product," much as "$\int$" is a variant of S or Σ for "sum." [2]

Our analysis shows that the differentiation and integration processes that emerge naturally in the study of fluid flows are not additive but *multiplicative*. This is self-evident for integration, cf. (2.8). But it is nearly so for differentiation as well. For if we introduce an initial epoch t_0, then by the multiplicative property (2.2) we may rewrite (2.6) in the form

$$
V(t) = S \lim_{h \to 0} \frac{1}{h} \{T(t + h, t_0) T(t, t_0)^{-1} - I\}.
$$

This "logarithmic" differentiation quotient shows up—not the familiar one, $\{T(t + h, t_0) - T(t, t_0)\}/h$, of the additive theory.

Before we return to nonlinear manifolds $\mathcal{M}$ it should be noted that in the present case even though the manifold $\mathcal{M}$ is linear, *the values of the functions* $T(\cdot,\cdot)$, $D(\cdot,\cdot)$, $V(\cdot)$ *are in general nonlinear operators on* $\mathcal{M}$ *to* $\mathcal{M}$. Thus our calculus is one for nonlinear operators. To see this consider the simplest (nonrectilinear) case, viz. $\mathcal{M} = \mathbb{R}^2$. In this case we can write $\forall x = (x_1, x_2) \in \mathbb{R}^2$,

$$
T(t + h, t)(x) = A_0(t, h) + A_1(t, h) \cdot \mathbf{x}^{(1)} + A_2(t, h) \cdot \mathbf{x}^{(2)} + \cdots, \tag{2.9}
$$

where $\mathbf{x}^{(1)}$, $\mathbf{x}^{(2)}$, $\mathbf{x}^{(3)}$, etc., are column vectors with components

$$
(x_1, x_2),\ (x_1^2, x_1 x_2, x_2^2),\ (x_1^3, x_1^2 x_2, x_1 x_2^2, x_2^3), \text{ etc.,}
$$

and $A_k(t, h)$ is a $2 \times (k + 1)$ matrix of real numbers. Thus $T(t + h, t)$ is in general a nonlinear operator on $\mathcal{M}$ onto $\mathcal{M}$. Next, our topological assumptions (2.4) and (2.3) require that

$$
S \lim_{h \to 0} A_k(t, h) = \delta_{k1} I, \qquad k \geq 0, \tag{3}
$$

2. We are writing $\widehat{\int}_b^a$ and not $\widehat{\int}_a^b$ in order to conform to the practice of using the latter for the limit of $\Pi_{k=1}^n$, cf., e.g., Ref. 18, p. 148, Eq. (1).

ISBN 0-201-13509-4

and that the entries of $A_k(t, h)$ be power series in t and h. From (1) it follows that for $k \neq 0$, these series, when rearranged as power series in h, will have no constant term. Hence, for each $k \geq 0$,

$$B_k(t) \underset{d}{=} S \lim_{h \to 0} \frac{1}{h} \{A_k(t, h) - \delta_{k1} I\} \quad \text{exists.}$$

Consequently, under mild restrictions, we would expect that

$$V(t)(x) = B_0(t) + B_1(t) \cdot \mathbf{x}^{(1)} + B_2(t) \cdot \mathbf{x}^{(2)} + \cdots.$$

Unless $B_0(t) = B_2(t) = \cdots = 0$, $V(t)$ will also be a nonlinear operator on $\mathcal{M}$ to $\mathcal{M}$.

Recall that the class $\mathbb{A}$ of all nonlinear operators on $\mathcal{M}$ to $\mathcal{M}$ is merely a *pseudolinear algebra*. More precisely, $\mathbb{A}$ is a vector space over $\mathbb{F}$ under functional addition, a (non-Abelian) semigroup with unit I under functional composition, and obeys the *right*-distributive and associative laws

$$(A + B)C = AC + BC, \qquad (ab)A = a(bA),$$

for $A, B, C \in \mathbb{A}$ and $a, b \in \mathbb{F}$. $\mathbb{A}$ is not a ring, and in general

$$C(A + B) \neq CA + CB, \qquad a(AB) \neq A \cdot aB.$$

Let A_{inv} be the multiplicative subgroup of invertible elements of the pseudolinear algebra $\mathbb{A}$. From (2.8) and (2.2b) we see that all multiplicative integrals belong to this group. Thus multiplicative integration carries functions from $\mathbb{R}$ to the pseudolinear algebra $\mathbb{A}$ into elements of the subgroup $\mathbb{A}_{\text{inv}}$.

B. Nonlinear Manifolds

In what form does the preceding analysis survive in the case where $\mathcal{M}$ is a smooth but nonlinear manifold? Since addition and subtraction are undefinable for operators on $\mathcal{M}$ to $\mathcal{M}$, the expression "$T(t, s) - I$" is now meaningless, and our definitions (2.5) for the displacement and (2.6) for the velocity break down. To proceed in the clean-cut, coordinate-free manner of Section 2A, we must find a suitable substitute for the formula (2.5) for $D(t, s)$.

By our topological assumptions (2.3) on the flow, the multiplicative group $\mathbb{G}$ generated by all the $T(t, s)$ for $t, s \in \mathbb{R}$ is a subgroup of the group $D(\mathcal{M}, \mathcal{M})$ of all diffeomorphisms on $\mathcal{M}$ onto $\mathcal{M}$. This latter group is an infinite dimensional "Lie group", cf. Birkhoff [3] and Lang [15]. We shall work under the plausible assumption that $\mathbb{G}$ *is a (finite-dimensional)*

ISBN 0-201-13509-4

Lie group.[3] G then possesses a Lie algebra A. The action of A on $\mathcal{M}$ is given by

$$\forall A \in \mathrm{A} \,\&\, \forall x \in \mathcal{M}, \qquad A(x) \in \mathcal{M}_x,$$

where $\mathcal{M}_x$ is the tangent space of $\mathcal{M}$ at x.[4]. Since the $V(t)$ we seek has also to carry x in $\mathcal{M}$ into a tangent vector in $\mathcal{M}_x$, it is apparent that $V(t)$ should lie in A. This suggests that we define the displacement operator $D(t, s)$ also as an element of A.

Now we know, cf. Chevalley [7, pp. 116–118], that there is a neighborhood U_0 of O in A and a neighborhood V_I of I in G such that the exponential function, exp, is a homeomorphism on U_0 onto V_I with $\exp O = I$. Consequently,

$$\log \underset{d}{=} \exp^{-1} = \text{a one–one function on } V_I \text{ onto } U_0.$$

Now from the strong continuity assumption (2.4) on the flow, we see that $\forall s \in \mathbf{R}$, $\exists \delta_s > 0$ such that

$$|t - s| < \delta_s \Rightarrow T(t, s) \in V_I \Rightarrow \log T(t, s) \in U_0.$$

This suggests that we define the displacement operator $D(t, s)$ only for $t \in (s - \delta_\epsilon, s + \delta_\epsilon)$ by

$$D(t, s) \underset{d}{=} \log T(t, s), \tag{2.10}$$

and then define the velocity operator $V(t)$, à la (2.6), by

$$V(t) \underset{d}{=} \lim_{h \to 0} \frac{1}{h} D(t + h, t) = \lim_{h \to 0} \frac{1}{h} \{\log T(t + h, t)\}. \tag{2.11}$$

Then, we can again assert, cf. (2.7), that $\forall t \in \mathbf{R}$ and $\forall x \in \mathcal{M}$,

$$V(t)(x) = \textit{the fluid velocity at position } x \textit{ at epoch } t.$$

The recovery of the propagator $T(t, s)$ from the velocity operator $V(t)$ can now be effected as in Section 2A. For this we restate (2.11) in the form

$$\log T(t + \Delta t, t) = V(t)\Delta t + o(\Delta t), \quad \text{as } \Delta t \to 0,$$

3. The finite-dimensionality requirement is imposed solely for simplicity, to avoid having to appeal to the as yet embryonic infinite-dimensional theory.

4. To see this, note that $\{\exp(At)x\colon t \in \mathbf{R}\}$ is a path in $\mathcal{M}$ passing through x at epoch 0, the tangent to which at x is $A(x)$.

ISBN 0-201-13509-4

and infer, cf. Chevalley [7, pp. 116—118], that

$$T(t + \Delta t, t) = \exp\{V(t)\Delta t\}, \quad \text{approx.}$$

Taking a decomposition π of $[a, b]$ of small mesh $|\pi|$, we conclude as in Section 2A that

$$T(b, a) = \prod_{k=n}^{1} \exp\{V(t_{k-1})(t_k - t_{k-1})\}, \quad \text{approx.}$$

For smooth $V(\cdot)$ on $[a, b]$ to A, it is again reasonable to expect the approximation to improve as $|\pi|$ diminishes, and the error to vanish in the limit as $|\pi| \to 0$. Thus

$$\begin{aligned} T(b, a) &= \lim_{|\pi| \to 0} \prod_{k=n}^{1} \exp\{V(t_{k-1})(t_k - t_{k-1})\} \\ &= \widehat{\int_b^a} \exp\{V(t)\, dt\} \in \mathrm{G}, \end{aligned} \tag{2.12}$$

where the symbol "$\widehat{\int}$" has the same multiplicative connotation it did in (2.8).

The formulas (2.6) and (2.8) are noticeably different from (2.11) and (2.12). For consistency we must show that in case $\mathcal{M}$ is linear, (2.11) and (2.12) reduce to (2.6) and (2.8). It will suffice to show that when the values of $V(\cdot)$ lie in a pseudolinear algebra A, we have

$$\begin{aligned} \widehat{\int_b^a} \exp\{V(t)\, dt\} &= \widehat{\int_b^a} (I + V(t)\, dt). \\ \lim_{h \to 0} \frac{1}{h} \{\log T(t + h, t)\} &= \lim_{h \to 0} \frac{1}{h} \{T(t + h, t) - I\}. \end{aligned} \tag{2.13}$$

These results were established by Schlesinger [29] for the algebra A of $q \times q$ matrices, and readily extend to Banach algebras A, cf. Ref. 18, Section 24.1. The proofs hinge on the estimate

$$\exp A - I - A = o(A), \quad \text{as } A \to 0, A \in \mathrm{A}. \tag{2.14}$$

If in our pseudolinear algebra A we define the exponential function by

$$\exp A \underset{d}{=} \lim_{n \to \infty} (I + A/n)^n, \quad A \in \mathrm{A},$$

and impose on A a suitable topological-algebraic axiom to compensate

ISBN 0-201-13509-4

for the loss of left-distributivity and ensure the estimate (2.14), then the connecting equations (2.13) will prevail, and both the linear and pseudo-linear theories will be subsumed in the general theory.

In the formulas (2.11) and (2.12) operator-addition does not appear, and as just noted its presence in formulas (2.6), (2.8) stems from the happy accident that the manifold $\mathcal{M}$ is linear. We may therefore conclude that *the natural differentiation and integration processes required for the analysis of the fluid flows in smooth manifolds are wholly multiplicative, and take us to and fro between a Lie group* $\mathbb{G}$ *and its Lie algebra* $\mathbb{A}$.

C. Steady Flows

If multiplicative integration and differentiation are that intrinsic to the study of flows, why is it that many mathematicians concerned with flows do not mention them? (See, for example, Mackey [17]). To understand this we must go back to a basic result on the exponential function from the Lie algebra $\mathbb{A}$ of a Lie group $\mathbb{G}$ to $\mathbb{G}$:

$$\begin{gathered} \forall A, B \in \mathbb{A}, [A, B] = 0, \quad \text{iff. } \forall s, t \in \mathbb{R}, \\ \exp(sA)\cdot\exp(tB) = \exp(sA + tB). \end{gathered} \tag{2.15}$$

Now assume that our flow has the grace to fulfill the miracle:

$$[V(s), V(t)] = 0, \quad \forall s, t \in [a, b]. \tag{2.16}$$

Then it easily follows from (2.15) that

$$\int_b^{\frown a} \exp\{V(t)\, dt\} = \exp\left\{\int_a^b V(t)\, dt\right\},^{5}$$

where the integral on the right-hand side is an ordinary additive Riemann integral. Thus, *for integrands on* $[a, b]$ *to* $\mathbb{A}$ *which fulfill (2.16) the multiplicative integral degenerates into the exponential of an additive integral.* Hence for a flow fulfilling (2.16) the formula (2.12) reduces to

$$T(b, a) = \exp\left\{\int_a^b V(t)\, dt\right\}, \tag{2.17}$$

in which the multiplicative integral is invisible.

The relation (2.16) is of course fulfilled in the trivial case of a *steady flow*, in which $V(\cdot)$ is constant-valued: $V(t) = V_0$. Then the fluid velocity

5. Since addition in $\mathbb{A}$ is commutative, we get $\int_a^b V(t)\, dt$ and not $\int_b^a V(t)\, dt$, which by definition is $-\int_a^b V(t)\, dt$.

ISBN 0-201-13509-4

at position x at epoch t is $V_0(x)$, and is therefore independent of t. Now $\int_a^b V(t)\,dt = V_0(b - a)$, and (2.17) further simplifies to

$$T(b, a) = \exp\{(b - a)V_0\}.$$

The propagator is now *translation-invariant*:

$$T(t + h, s + h) = T(t, s), \quad \forall h, s, t \in \mathbf{R}. \tag{2.18}$$

Letting

$$S(t) \underset{d}{=} T(a + t, a), \quad \text{any } a \in \mathbf{R}, \tag{2.19}$$

it follows that $\forall x \in \mathcal{M}$,

$S(t)(x) =$ *the position of the particle t seconds after*[6] *its position is x.*

Such a flow may be described as being *time-invariant*.

Reciprocally, if we start with a time-invariant flow, subject to the translational-invariance (2.18), and introduce $S(\cdot)$ as in (2.19), then clearly from (2.16) and (2.18)

$$V(t) = \operatorname*{slim}_{h\to 0} \frac{1}{h} \log\{T(t + h, t)\} \underset{d}{=} \operatorname*{slim}_{h\to 0} \frac{1}{h} \log S(h) = V_0. \tag{2.20}$$

Thus the flow is steady.

We see from (2.18), (2.19) that for time-invariant or steady flows the multiplicative property (2.2) of the propagator $T(\cdot,\cdot)$ can be expressed in the more familiar form:

$$S(0) = I, \quad S(-t) = S(t)^{-1}, \quad S(s + t) = S(s)\cdot S(t),$$

for all $s, t \in \mathbf{R}$. Thus, our multiplicative propagator $T(\cdot,\cdot)$ is replaceable by a one-parameter subgroup $(S(t)\colon t \in \mathbf{R})$ of the Lie group G, i.e., by a one-parameter group of operators on $\mathcal{M}$ onto $\mathcal{M}$. The constant-valued velocity V_0 of such a flow is just, cf. (2.20), the *infinitesimal generator* of this subgroup.

To sum up, from the standpoint of the theory of multiplicative integration, the steady flows fall in a rather small corner. Within this corner it is possible to forget such integration and deal exclusively with one-parameter groups and their infinitesimal generators. The fluid theorist will

6. "After" must become "before" when $t < 0$, to conform to English idiom.

ISBN 0-201-13509-4

encounter the multiplicative integral when he attends to unsteady flows, however. This is a universal situation. In any theory in which a time-domain is present, e.g., quantum mechanics or the theory of stochastic processes, multiplicative integration and differentiation in one form or another will confront the worker, no sooner he or she abandons underlying assumptions about stationarity or time invariance.

A.II.3 Abstract Formulation of the Theory

The preceding discussion of flows was meant to be an illustration of how multiplicative integration and differentiation, appear intrinsically in certain areas, and thereby meant to outline the objectives of a general study of these processes. Such a study would proceed at an abstract level, and hopefully admit several useful interpretations.

We see from Section 2B that a general theory must commence with an abstract Lie group G and its Lie algebra A. To meet the demands of contemporary analysis, we must allow G and A to be infinite dimensional, cf. Birkhoff [3] and Lang [15]. The general theory would demarcate a large class of "integrable" functions f from $\mathbb{R}$ to A, and define the assignment to each of a definite multiplicative integral in G. Membership in this class would not be restricted to continuous functions f, nor would the multiplicative integral be defined along the Riemannian lines of (2.12). The theory must be Lebesgue in spirit, i.e., allow measurable integrands f and countably additive complex-valued or even operator-valued measures μ over $\mathbb{R}$. It should include substitution rules for going from one μ to another. The case in which μ is a countably additive complex-valued measure of bounded variation, would then provide a theory of *multiplicative contour integration* of functions from $\mathbb{C}$ to A along rectifiable paths.

A differentiation theory to match this integration must of course be studied, and "fundamental theorems of the calculus," which assert the reciprocity of the two processes, must be established.

An important specialization of the theory would correspond to the case in which the Lie algebra A is pseudolinear and $[A, B] = AB - BA$. For this specialization we must establish the equivalent formulations of the integral and the derivative in which the exponential and logarithm are absent, cf. (2.6) and (2.8). A further important specialization would pertain to the case in which A is a linear algebra, for example, a Banach algebra.

It is not of course our intention to elaborate on such an abstract development. By way of history, the movement toward a Lebesgue theory was initiated by Schlesinger [29] for the case when A is the algebra of $q \times q$ matrices in 1931, and was continued by Birkhoff [2], Stewart [31], and others. A Riemann theory for Banach algebras was given by

ISBN 0-201-13509-4

the writer [18] in 1947, and more recently for Lie groups and Lie algebras by Hamilton [14]. In the sequel we shall explore the ways in which various facets of the general theory have been put to significant use in different parts of functional and probabilistic analysis. From some of these applications it will become clear that the abstract framework outlined in the preceding paragraphs does not always suffice. What is sometimes needed is a theory of multiplicative integration which covers the case in which the values of the integrand are *discontinuous* linear operators.

A.II.4 The Evolution Equation in a Pseudolinear Algebra

Let us revert to the theory of multiplicative integration of functions on $\mathbb{R}$ to a pseudolinear algebra $\mathbb{A}$, which we tentatively discussed in Section 2A. The ordinary additive derivative is latent in this theory as we shall now show, and its introduction yields a pair of (nonlinear) evolution equations for the multiplicative integral.

Let $T(\cdot,\cdot)$ be a multiplicative function on $\mathbb{R}^2$ to $\mathbb{A}_{\text{inv}}$ and let

$$A(t) \underset{d}{=} \operatorname*{slim}_{h\to 0} \frac{1}{h}\{T(t+h,t)-I\}. \tag{4.1}$$

Fix s and consider the function $F_s(\cdot) \underset{d}{=} T(\cdot, s)$ on $\mathbb{R}$. By (2.2),

$$T(t+h,t) = T(t+h,s)\cdot T(t,s)^{-1} = F_s(t+h)\cdot\{F_s(t)\}^{-1},$$

and (4.1) can be rewritten as

$$\begin{aligned} A(t) &= \operatorname*{slim}_{h\to 0} \frac{1}{h}\{F_s(t+h)-F_s(t)\}\{F_s(t)\}^{-1} \\ &= \dot{F}_s(t)\cdot\{F_s(t)\}^{-1}, \end{aligned}$$

where the (Newtonian) dot indicates ordinary differentiation. Thus

$$\dot{F}_s(t) = A(t)\cdot F_s(t), \quad t\in\mathbb{R}. \tag{1}$$

For each fixed s in $\mathbb{R}$, this an evolution equation in $\mathbb{A}$ in which the $\mathbb{A}$-valued function $A(\cdot)$ is given and the $\mathbb{A}$-valued $F_s(\cdot)$ is the unknown.

Next fix t and consider $G_t(\cdot) = T(t,\cdot)$ on $\mathbb{R}$. An easy calculation based on (2.2) shows that

$$I - T(s+h,s) = \{G_t(s)\}^{-1}\{G_t(s+h)-G_t(s)\}\cdot T(s,s+h)^{-1},$$

ISBN 0-201-13509-4

whence from (4.1), after division by h,

$$-A(s) = \{G_t(s)\}^{-1} \cdot G_t^{\cdot}(s) \cdot I.$$

Thus

$$G_t^{\cdot}(s) = -G_t(s)A(s), \quad s \in \mathbf{R}. \tag{2}$$

For each fixed t in $\mathbf{R}$, this is an evolution equation in A, with $A(\cdot)$ given and $G_t(\cdot)$ as the unknown.

We can combine (1) and (2) by writing the pair of evolution equations

$$\frac{\partial}{\partial t} T(t, s) = A(t)T(t, s), \tag{4.2a}$$

$$\frac{\partial}{\partial s} T(s, t) = -T(t, s)A(s). \tag{4.2b}$$

Their solution is given of course, cf. (2.8), by

$$T(t, s) = \prod_t^s (I + A(u)\, du), \quad t, s \in \mathbf{R}. \tag{4.3}$$

The interpretation in which s and t refer to the time and $t > s$ has suggested the nomenclature *forward equation* for (4.2a) and *backward equation* for (4.2b). In the concrete situation in which A is a class of nonlinear operators on $\mathbf{R}^q$ to $\mathbf{R}^q$, each of the equations (4.2a) and (4.2b) represents a system of q nonlinear differential equations in q unknowns. The possibility of solving such nonlinear systems by multiplicative integration was first pointed out by Birkhoff [2, Section 19]. We shall not pursue this interesting nonlinear phase of the subject any further, but refer the interested reader to Professor Browder's Foreword.

A.II.5 Linearization

In what types of applications may we regard our pseudolinear algebra A as being actually linear? Consider the case in which A is a class of operators on $\mathbf{R}^q$ into $\mathbf{R}^q$. In Section 2A, (2.9) *et seq.*, we saw how nonlinearities enter in the simple case $q = 2$. For $q \geq 2$ we associate with each $x = (x_1, \ldots, x_q)$ in $\mathbf{R}^q$ the column vectors $\mathbf{x}^{(1)}, \mathbf{x}^{(2)}, \ldots, \mathbf{x}^{(n)}$ whose components are

$$x_1, \ldots, x_q$$

$$x_1^2, x_1x_2, \ldots, x_1x_q; \quad x_2^2, \ldots, x_2x_q; \quad \ldots, x_q^2$$

$$x_1^{k_1} \cdots x_q^{k_q}, \quad \text{where } k_i \geq 0 \ \& \ \sum_1^q k_i = n,$$

ISBN 0-201-13509-4

respectively. Here we presume adherence to a definite order of enumeration of the components in each column vector $\mathbf{x}^{(n)}$.[7] Clearly $\mathbf{x}^{(n)}$ has

$$(q)_n \underset{d}{=} \binom{n+q-1}{n}$$

components. Our smoothness assumptions (2.3) now yield

$$T(t, s)(x) = A_0(t, s) + A_1(t, s)\cdot\mathbf{x}^{(1)} + \cdots + A_n(t, s)\cdot\mathbf{x}^{(n)} + \cdots, \qquad (5.1)$$

where $A_n(t, s)$ is a $q \times (q)_n$ matrix power series in s and t such that

$$\lim_{t\to s} A_n(t, s) = \delta_{n1} I.$$

Obviously the necessary and sufficient condition for $T(t, s)$ to be a linear operator on R^q to R^q is that

$$A_n(t, s) = 0, \quad \forall n \neq 1 \ \&\ \forall t, s \in \mathrm{R}. \qquad (5.2)$$

When is it reasonable to suppose that Eqs. (5.2) prevail? For an answer it is convenient to return to flows. If the flow has a fixed point, and we choose our coordinates with this point as origin O, then we see at once from (5.1) that $A_0(\cdot,\cdot) = O$ on R^2. Now assume that the fix-point O is one of *stable equilibrium,* i.e., that fluid particles sufficiently close to O remain, under the flow within a small neighborhood of O. The coordinates $(x_1, \ldots, x_q)$ of such a particle will then remain small for all times t, and we may neglect $x_i x_j$ and all higher-order products. Unless the entries in the matrices $A_n(t, s)$ grow very large with time, we may ignore the terms $A_n(t, s)\cdot\mathbf{x}^{(n)}$ for $n \geq 2$ in (5.1), and obtain the *linear approximation*

$$T(t, s)(x) = A_1(t, s)\cdot\mathbf{x}^{(1)}. \qquad (5.3)$$

In this $A_1(t, s)$ is a $q \times q$ matrix, since obviously $(q)_1 = q$.

Generally speaking, it is reasonable to assume that our operators are linear (or speaking at a more abstract level that A is a linear algebra) when our interest is confined primarily to a sufficiently small vicinity of a fix-point of stability of the system under consideration. In such situations (4.2) is a pair of *linear evolution equations*. For fixed s (or t) each represents a system of q ordinary linear differential equations in q unknowns. It was this case that engaged Volterra's pioneering efforts.

These remarks extend to the case in which $q = \infty$, i.e., to systems with

7. For example, one in which $x_1, \ldots, x_q$ appear in lexicographical order, as exhibited above in the case of the components of $\mathbf{x}^{(2)}$.

ISBN 0-201-13509-4

infinitely many degrees of freedom. We now have continuous linear operators on an infinite-dimensional Banach space $\mathcal{M}$. But at the abstract level, we *cannot* say that A is a Banach-algebra, for here as in (2.4) and (2.6) it is the strong operator topology which is natural. Only in special situations in which the uniform operator topology is an adequate substitute, may we say that A is a Banach algebra.

In linear situations, for example, when A is a Banach algebra, the multiplicative integral has more additive properties than in the general use. It is, for instance, expressible as a series of additive multiple integrals of higher and higher order:

$$\widehat{\int_b^a} (I + A(t)\, dt) = I + \int_a^b A(t)\, dt + \int_a^b \int_a^t A(t)A(s)\, ds\, dt + \cdots. \qquad (5.4)$$

This is called the *Peano series,* after its discoverer [22].

A.II.6 Discrete-State Markovian Processes with Continuous Time Domain

The discrete-state Markovian processes with continuous time domains constitute an important instance where linearity is instrinsic and not just an approximation to a basically nonlinear situation.

A stochastic process (briefly SP) $(x_t\colon t \geq 0)$, where x_t are random variates over a probability space $(\Omega, \mathcal{B}, P)$ with values in a Banach space $\mathfrak{X}$ is called *Markovian*, iff. for

$$0 \leq t_1 < \cdots < t_n < t \;\&\; S \in \mathrm{Bl.}(\mathfrak{X}) \Rightarrow$$
$$P\{x_t^{-1}(S) | x_{t_1}, \ldots, x_{t_n}\} = P\{x_t^{-1}(S) | x_{t_n}\}, \quad \text{a.e. } (P) \text{ on } \Omega.$$

In this $\mathrm{Bl.}(\mathfrak{X})$ is the family of Borel subsets of $\mathfrak{X}$ and $P(B | x_{t_1}, \ldots, x_{t_n}\}$ is shorthand for the conditional probability $P(B | \mathcal{B}\{x_{t_1}, \ldots, x_{t_n}\})$ for $B \in \mathcal{B}$, where $\mathcal{B}\{x_{t_1}, \ldots, x_{t_n}\}$ is the σ-algebra generated by the sets $x_{t_k}^{-1}(S)$ for $S \in \mathrm{Bl.}(\mathfrak{X})$ and $1 \leq k \leq n$. See Doob [8, Chap. I, Sect. 7] for the definition and properties of such (microscopic) conditional probabilities and expectations. The set $\mathcal{S} = \cup_{t \geq 0} x_t(\Omega)$ is called the *state space* of the process. Obviously $\mathcal{S} \overset{d}{\subseteq} \mathfrak{X}$.

A fundamental property of the Markovian SP $(x_t\colon t \geq 0)$ is given by the *general Chapman-Kolmogorov equations*: for $0 \leq s < t < u$ and $S \in \mathrm{Bl.}(\mathfrak{X})$,

$$P\{x_u^{-1}(S) | x_s\} = E[P\{x_u^{-1}(S) | x_t\} | x_s], \quad \text{a.e. } (P). \qquad (6.1)$$

ISBN 0-201-13509-4

Here $E(\phi \mid \psi)$ stands for the conditional expectation of the random variate ϕ with respect to the σ-algebra generated by $\psi^{-1}\{\text{Bl.}(\mathfrak{X})\} \subseteq \mathscr{B}$, ψ being another random variate, cf. Doob [8, p. 88, Eq. (6.16)].

Now let us drastically simplify matters by assuming that the state space $\mathscr{S}$ is finite: $\mathscr{S} = \{1, 2, ..., q\}$. Then

$$\sum_{j=1}^{q} P(x_t^{-1}\{j\}) = 1,$$

and it becomes possible to replace the (microscopic) conditional probabilities encountered above by the much simpler (macroscopic) conditional probabilities:

$$P(A \mid B) \underset{d}{=} P(A \cap B)/P(B) \quad \text{when } P(B) > 0.$$

To affect this replacement,[8] we define the functions $p_{ji}(\cdot,\cdot)$ for $1 \leq i, j \leq q$, on $\mathbf{R}_{0+}^2$ by

$$p_{ji}(t, s) = \begin{cases} P(x_j^{-1}\{t\} \mid x_i^{-1}\{s\}), & \text{if } P(x_i^{-1}\{s\}) > 0 \\ 0, & \text{if } P(x_i^{-1}\{s\}) = 0. \end{cases} \tag{6.2}$$

Then, cf. Doob [8, p. 235], Eq. (6.1) can be reformulated more simply as follows: for $0 \leq s < t < u$ and any i such that $P(x_s^{-1}\{i\}) > 0$,

$$p_{ki}(u, s) = \sum_{j=1}^{q} p_{kj}(u, t) \cdot p_{ji}(t, s), \quad 1 \leq k \leq q. \tag{1}$$

Since obviously $p_{ji}(s, s) = \delta_{ji}$, it is easy to see that Eqs. (1) hold for $s \leq t \leq u$. On defining the $q \times q$ matrix-valued function $P(\cdot,\cdot)$ on $\mathbf{R}_{0+}^2$ by

$$P(t, s) = [p_{ji}(t, s)], \quad s, t \geq 0,$$

Eqs. (1) can be stated in the form

$$P(u, s) = P(u, t) \cdot P(t, s), \quad P(s, s) = I, \quad 0 \leq s \leq t \leq u. \tag{6.3}$$

8. For this replacement, it is useful to recall that if f is a random variate whose range is $\{1, 2, ..., q\}$ such that each $P(f^{-1}\{k\}) > 0$, then for any $B \in \mathscr{B}$, the microscopic conditional probability $P(B \mid f)(\cdot)$ is a simple function on Ω, viz.,

$$P(B \mid f)(\omega) = \sum_{k=1}^{q} P(B \mid f^{-1}\{k\}) \chi_{f^{-1}\{k\}}(\omega).$$

ISBN 0-201-13509-4

This is the *simplified form of the Chapman-Kolmogorov equations* for Markovian processes with finite state space.

We have here a multiplicative linear problem, since the $P(t, s)$ belong to the linear algebra A of $q \times q$ matrices. And this linearity is intrinsic to the Markovian property, and not a matter of approximation. The question now arises as to whether

$$P(t, s) = \widehat{\int_t^s} (I + A(u)\, du) \tag{6.4}$$

for a suitable $q \times q$ matrix-valued function $A(\cdot)$ on $[0, \infty)$. The answer of course hinges on the smoothness of the function $P(\cdot,\cdot)$ on $\mathbf{R}^2_{0+}$. We proceed to show that with certain simple assumptions, which are empirically reasonable in many applications, $P(\cdot,\cdot)$ becomes smooth, and the answer is affirmative.

For Markovian processes in which $x_t(\omega)$ represents the size of a population at epoch t,[9] an important conditional probability, $c_i(t + h, t)$, is that of occurrence of a change in the value of $x_t(\omega)$ during the time-interval $[t, t + h]$, $h > 0$, given that $x_t(\omega) = i$. In symbols,

$$c_i(t + h, t) \underset{d}{=} P(x_{t+h}^{-1}(\mathscr{S} \setminus \{i\}) | x_t^{-1}\{i\}). \tag{2}$$

Another important conditional probability, $c_{ji}(t + h, t)$, is that of a transition from state i to a state $j \neq i$ during $[t, t + h]$, given that a transition has occurred during this interval. In symbols,

$$c_{ji}(t + h, t) = P[x_{t+h}^{-1}\{j\} | x_t^{-1}\{i\} \cap x_{t+h}^{-1}(\mathscr{S} \setminus \{i\})]. \tag{3}$$

A simple calculation shows that

$$c_{ji}(t + h, t) = p_{ji}(t + h, t)/c_i(t + h, t). \tag{4}$$

In many situations it is reasonable to assume that the first type of transition has at each epoch t a definite rate of change, i.e.,

$$d_i(t) \underset{d}{=} \lim_{h \to 0} \frac{1}{h} c_i(t + h, h) \quad \text{exists}, \quad 1 \leq i \leq q. \tag{5}$$

Equation (4) then suggests another reasonable assumption:

$$b_{ji}(t) \underset{d}{=} \lim_{h \to 0} k_{ji}(t + h, t) \quad \text{exists}, \quad j \neq i. \tag{6}$$

9. As, for instance, in bacterial growth or radioactive disintegration.

ISBN 0-201-13509-4

From (5) and (6) it follows by a simple calculation, omitted here, that

$$\lim_{h\to 0}\frac{1}{h}\{p_{ji}(t+h)-\delta_{ji}\}=\begin{cases}-d_i(t), & j=i,\\ b_{ji}(t)d_i(t), & j\neq i.\end{cases} \tag{6.5}$$

In matrix form Eq. (6.5) can be rendered

$$A(t)\underset{d}{=}\lim_{h\to 0}\frac{1}{h}\{P(t+h,t)-I\}=\{B(t)-I\}D(t), \tag{6.6}$$

where $D(t)$ is the diagonal matrix with diagonal entries $d_1(t), \ldots, d_q(t)$ given by (5), and $B(t) \underset{d}{=} [b_{ij}(t)]$, the entries being given by (6) for $j \neq i$, and $b_{ii}(t) \underset{d}{=} 0$. From (6.6) it follows of course, cf. Section 4, that we have the linear evolution equations

$$\frac{\partial}{\partial t}P(t,s)=A(t)\cdot P(t,s),$$

$$\frac{\partial}{\partial t}P(t,s)=-P(t,s)\cdot A(s),$$

and that (6.4) prevails with $A(\cdot)$ as in (6.6).

The finite state Markovian process with continuous time domain thus provides a very natural setting for the use of multiplicative integrals. The same holds when the state space $\mathcal{S}$ is countably infinite, but the corresponding theory is much harder since it involves $\infty \times \infty$ matrices. The product of two such matrices need not exist, and thus such matrices do not form an algebra. In effect they represent discontinuous linear operators.

Despite these difficulties, a theory of multiplicative integration for Markovian processes with countable state space was worked out by Arley [1] in 1943 in order to study cosmic-ray showers. With every $\infty \times \infty$ matrix-valued function $A(\cdot) = [a_{ij}(\cdot)]$, on $[\alpha, \beta]$ say, Arley associated the matrix

$$K=[k_{ij}],\quad k_{ij}\underset{d}{=}\sup_{\alpha\le t\le\beta}|a_{ij}(t)|.$$

He called $A(\cdot)$ *absolutely exponentiable*, iff. all the powers K^n, $n \geq 1$, exist and

$$\exp\{(\beta-\alpha)K\}\underset{d}{=}\sum_{n=0}^{\infty}\frac{1}{n!}(\beta-\alpha)^n K^n \quad \text{converges.}$$

ISBN 0-201-13509-4

He proved that if $A(\cdot)$ is continuous and absolutely exponentiable on $[\alpha, \beta]$, then its multiplicative integral on $[\alpha, \beta]$ exists and

$$\left| \widehat{\int_\alpha^\beta} (I + A(t)\, dt) \right| \leq \exp\{(\beta - \alpha)K\}, \quad [1, \text{p. } 47].$$

Here $|B| \leq C$ means that $|b_{ij}| \leq c_{ij}$, for $i, j = 1, 2, \ldots$. Since cosmic-ray showers involve a population with several species (electrons, protons, various mesons, etc.), Arley in fact proved a generalization of the last result adequate to cover r-variate Markovian processes, i.e., processes for which the state space is $\mathbf{N}_+^r$, r being the number of species.

A Markovian process with time domain $\mathbf{R}$ for which the transition probability matrix function is translation-invariant

$$P(t + h, s + h) = P(t, s), \quad h, s, t \in \mathbf{R}$$

is said to have a *stationary transition mechanism*. For the study of such processes, as for steady flows, cf. Section 2C, multiplicative integration is dispensable and we can fall back on the theory of one-parameter groups and semi-groups. But for many irreversible processes, for instance physical or biological birth and death processes, the stationarity assumption is untenable.[10] In such instances the time-varying evolution equations are often very hard to integrate. We know, however, that the solution does exist and is given by (6.4). This multiplicative integral formula sometimes provides a better way to "obtain some idea, if only a rough one" about the solution than the Peano series (5.4), cf. Arley [1, p. 122]. In Ref. 1, pp. 123–134 Arley uses the approximation $\Pi_i(I + A(t_i)\Delta_i)$ to compute the solution. His tables, though outdated,[11] suggest the efficacy of the multiplicative integral as a practical tool.

A.II.7 The Monodromy and Cousin Problems

Let f be a continuous function on a region D of $\mathbf{C}$ with values in a Banach algebra $\mathbf{A}$ over $\mathbf{C}$ with unit 1, and let C be a smooth path in D. Let $\zeta(\cdot)$ on $[a, b]$ be a parametrization of C. Then we can define the *multiplicative path integral of f along C* by

$$\widehat{\int_C} (1 + f(\zeta)\, d\zeta) \underset{d}{=} \widehat{\int_a^b} (1 + f\{\zeta(t)\}\zeta'(t)\, dt). \tag{7.1}$$

10. The cascade or multiple process caused by high-energy atomic interactions is a conspicuous example.

11. Arley's choice of $A(\cdot)$ was based on a modification of the now superceded Bhabha-Heitler theory of cascade showers.

ISBN 0-201-13509-4

The substitution rules for multiplicative integrals ensure that this definition is independent of the parametrization and depends on the path C alone.[12] For any circuit C in D,

$$M_f(C) \underset{d}{=} \widehat{\oint_C} (1 + f(\zeta)\, d\zeta) \in \mathrm{A}_{\text{inv}}$$

is called the *monodromic element of f for C*.

For any $a \in D$, the monodromic elements $M_f(C_a)$, for different circuits C_a in D with terminus[13] a constitute a multiplicative subgroup $\mathrm{M}_f(a)$ of A_{inv}, since

$$M_f(C_a)\cdot M_f(C'_a) = M_f(C_a\cdot C'_a), \quad \{M_f(C_a)\}^{-1} = M_f(C_a^{-1}),$$

where $C_a\cdot C'_a$ is the circuit obtained by traversing C_a and C'_a in succession, and C_a^{-1} is the circuit with the same range as C_a, but traversed in the opposite sense. $\mathrm{M}_f(a)$ is called the *monodromy group f at a*. Now let $b \in D$ and L be a path in D joining a to b. Then it is easy to check that the correspondence:

$$\phi_L\colon M_f(C_a) \to M_f(L^{-1}C_aL) \tag{7.2}$$

is an isomorphism on $\mathrm{M}_f(a)$ onto $\mathrm{M}_f(b)$. Speaking abstractly, we have just one group M_f, called the *monodromy group of f*.

Had we defined $M_f(C_a)$ to be the ordinary additive circuital integral of f along C, then in evaluating $M_f(L^{-1}C_aL)$ we could have cancelled the integrals along L^{-1} and L, and concluded that $M_f(L^{-1}C_aL) = M_f(C_a)$. The noncommutativity of multiplication in A precludes such cancellation, and we are stuck instead with the formula:

$$M_f(L^{-1}C_aL) = \widehat{\int_{L^{-1}}} (1 + f(\zeta)\, d\zeta)\cdot M_f(C_a)\cdot \widehat{\int_L} (1 + f(\zeta)\, d\zeta), \tag{7.3}$$

where the right-hand side is dependent on the path L in D joining a to b. It follows that *$M_f(C)$ depends on the terminus of the circuit C*. This severely complicates the study of multiplicative contour integration. For A, $B \in \mathrm{A}$ to write $B \equiv A$ to mean that B is *similar* to A, i.e., $B = C^{-1}AC$ for some $C \in \mathrm{A}_{\text{inv}}$. Then we see from (7.3) that

$$\phi_L\{M_f(C_a)\} \underset{d}{=} M_f(L^{-1}C_aL) \equiv M_f(C_a). \tag{7.4}$$

Next let the A-valued function f be holomorphic on the region D.

12. For $i = 1, 2$, we say that ζ_i on $[a_i, b_i]$ to C define the *same path*, iff. there is a strictly increasing continuous function ϕ on $[a_1, b_1]$ onto $[a_2, b_2]$ such that $\zeta_1 = \zeta_2\cdot\phi$.
13. That is, the starting point (or end point) of the circuit.

ISBN 0-201-13509-4

Then for any simple circuit C in D which encloses points of D alone, we have in analogy with Cauchy's classic theorem,

$$M_f(C) \underset{d}{=} \overset{\frown}{\oint_C} (1 + f(\zeta)\, d\zeta) = 1, \tag{7.5}$$

cf. Rasch [24, Secs. 11 and 28]. It follows readily that for any circuits C, C' in D,

$$C' \textit{ is homotopic to } C \textit{ in } D \Rightarrow M_f(C') \equiv M_f(C). \tag{7.6}$$

For circuits C, C' in D with the same terminus,

$$C' \textit{ is homotopic to } C \textit{ in } D \Rightarrow M_f(C') = M_f(C). \tag{7.7}$$

Another easy consequence of (7.5) is that $\phi_{L_1} = \phi_{L_2}$ for homotopic paths L_1, L_2 in D joining a to b. It can also be shown that the correspondence $C \to M_f(C)$ is a homomorphism on the fundamental group of D onto the monodromy group $\mathbb{M}_f$.

Let $A(\cdot)$ be an $\mathbb{A}$-valued function holomorphic on a region D of $\mathbb{C}$, let $a \in D$ and let $\forall z \in D$,

$$Y(z) = \overset{\frown}{\int_a^z} (1 + A(\zeta)\, d\zeta)$$

the integration being along some path Γ in D joining a to z. $Y(\cdot)$ will be in general a multivocal (many-valued) analytic function on D such that if $y \in \mathbb{A}$ is a value of Y at z, then after continuation of this branch along a circuit C_z in D with terminus z, the value becomes $\tilde{y} = M_a(C_z)\cdot y$. The values of Y at a fixed $z \in D$ are thus interrelated with the elements of the monodromy group $\mathbb{M}_A(z)$.

In the light of these preliminaries we can now formulate a celebrated problem in-the-large proposed by Riemann for the case in which $\mathbb{A}$ is the algebra of $q \times q$ matrices.

Riemann's Monodromy Problem. *Let (i) $c_1, \ldots, c_n$ be distinct points in $\mathbb{C}$, (ii) $M_1, \ldots, M_n \in \mathbb{A}_{\text{inv}}$ be such that $M_1 \cdots M_n = 1$. Find an $\mathbb{A}$-valued function $A(\cdot)$, holomorphic on the region $D \underset{d}{=} \mathbb{C} \setminus \{c_1, \ldots, c_n\}$ such that (a) for any simple circuit C_i in D which encloses only c_i among the points $c_1, \ldots, c_n$,*

$$M_A(C_i) \underset{d}{=} \overset{\frown}{\oint_{C_i}} (1 + A(\zeta)\, d\zeta) \equiv M_i,$$

ISBN 0-201-13509-4

and (b) for each branch of the indefinite integral

$$Y(z) \underset{d}{=} \widehat{\int_{\Gamma^a}^{z}} (1 + A(\zeta)\, d\zeta), \qquad a, z \in D, \qquad \Gamma \subset D,$$

we have $|Y(z)| = O(|z - c_i|^{-r_i})$, *as* $z \to c_i$, *for some positive integer* r_i, $i = 1, \ldots, n$.

For matrices this problem was solved by Hilbert and Plemelj, and a more advanced version, with additional demands on $A(\cdot)$, by Birkhoff [5]. What light does the theory of multiplicative integration shed on this problem? To find out it is convenient to first see its role in another global problem, that due to Cousin. For its enunciation it will be convenient to denote by $\widehat{\text{Hol}}.(D)$ the class of holomorphic functions f on *a* region D, and by $\widehat{\text{Hol}}.(D)$ the class of such functions for which both $f(\cdot)$ and its reciprocal $f(\cdot)^{-1}$ are holomorphic on D.

COUSIN'S MULTIPLICATIVE PROBLEM. *Let (i) R be a bounded region in* $\mathbb{C}$ *and* $D_1, \ldots, D_n$ *be open disks in* $\mathbb{C}$ *such that* $R \subseteq D_1 \cup \cdots \cup D_n$, *(ii)* $f_1, \ldots, f_n$ *be* $\mathbb{A}$*-valued functions such that* $f_k \in \text{Hol}.(D_k)$, *and in case* $D_j \cap D_k$ *is non-void,*

$$f_k = f_j g_{jk}, \quad \textit{where } g_{jk} \in \widehat{\text{Hol}}.(D_j \cap D_k).$$

Find an $\mathbb{A}$*-valued function* $F \in \text{Hol}.(R)$ *such that for each* $j = 1, \ldots, n$,

$$F = f_j G_j, \quad \textit{where } G_j \in \widehat{\text{Hol}}.(D_j).$$

The solution of this problem rests on the following difficult lemma on the Laurent-type factorization of holomorphic functions in lens-shaped regions due to Cartan [6]:

CARTAN'S FACTORIZATION LEMMA. *Let (i)* D_1, D_2 *be open disks in* $\mathbb{C}$ *for which* $D_1 \cap D_2$ *is non-void, (ii) R be a region such that* $R \supset$ clos.$(D_1 \cap D_2)$, *(iii)* $F(\cdot)$ *be an* $\mathbb{A}$*-valued function in* $\widehat{\text{Hol}}.(R)$. *Then* $\exists$ $\mathbb{A}$*-valued functions* F_1, F_2 *such that* $F_1 \in \widehat{\text{Hol}}.(D_1)$, $F_2 \in \widehat{\text{Hol}}.(D_2)$ *and*

$$F(\cdot) = F_1(\cdot) \cdot F_2(\cdot) \textit{ on } D_1 \cap D_2.$$

A more elegant version of this result in which R is eliminated in favor of $D_1 \cap D_2$ is deducible from the above version, cf. [21]. Cartan's proof rested on an interesting and ingenious process of successive approximations.

To see how the Cartan lemma could be handled more directly by multiplicative integration, grant for a moment that we have at our disposal

ISBN 0-201-13509-4

a multiplicative analogue of Cauchy's additive formula:

$$f(z) = \frac{1}{2\pi i} \oint_{\partial D} \frac{f(\zeta)}{\zeta - z}\, d\zeta, \quad z \in D,$$

for an $\mathbb{A}$-valued function f, holomorphic on a region D with a smooth boundary ∂D and continuous on ∂D. For $f(\cdot)$ with commuting values on ∂D, the analogue we seek would be, cf. (2.16) *et seq.*,

$$\exp\{f(z)\} = \widehat{\oint_{\partial D}} (1 + K(z, \zeta)\, f(\zeta)\, d\zeta), \quad z \in D, \tag{$*$}$$

where $K(z, \zeta) = (1/2\pi)(\zeta - z)$. Let us pretend that there is a suitable complex-valued kernel $K(\cdot,\cdot)$ for which the formula $(*)$ survives even when the values of f do not commute. Now apply this $(*)$ taking $D \underset{d}{=} D_1 \cap D_2$ and $f(\cdot) \underset{d}{=} \log F(\cdot)$ on D, where D_1, D_2 and F are as in Cartan's lemma. Then since $\partial D = C_1 \cup C_2$,

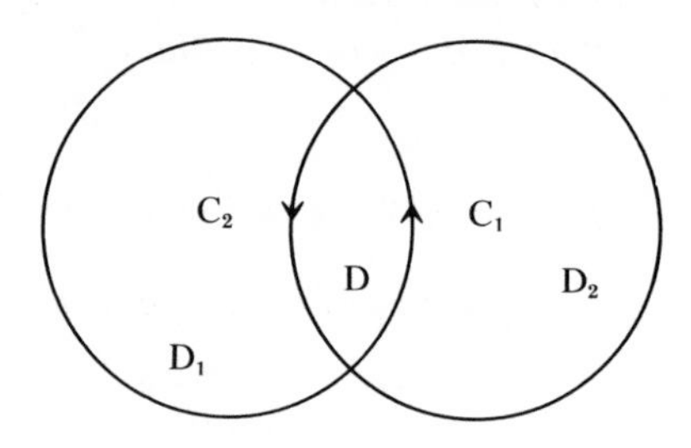

where C_1, C_2 are the two circular arcs, we get after orienting these arcs so that $C_1 \cdot C_2$ becomes a simple circuit:

$$\begin{aligned} F(z) &= \widehat{\int_{C_1}} (1 + K(z, \zeta)\, f(\zeta)\, d\zeta) \cdot \widehat{\int_{C_2}} (1 + K(z, \zeta)\, f(\zeta)\, d\zeta) \\ &= F_1(z) \cdot F_2(z), \quad \text{say}. \end{aligned}$$

For a large class of kernels $K(\cdot,\cdot)$, it would readily follow that $F_1 \in \widehat{\text{Hol}}.(D_1)$, $F_2 \in \widehat{\text{Hol}}.(D_2)$, and we would have an elegant proof of Cartan's lemma.

All formulas of the form $(*)$ are untenable, however. For one thing, the right-hand side will, cf. (7.3) *et seq.*, depend on the terminus of the circuit over D. This writer has obtained several (unpublished) correct formulations, which take account of (7.3), for instance

$$\exp\{f(z)\} = \widehat{\oint_{C_a}} \left[1 + \widehat{D_\zeta} \left\{ \exp \frac{1}{2\pi i} \int_{C_a}^{\zeta} \frac{f(w)}{w - z}\, dw \right\} d\zeta \right],$$

ISBN 0-201-13509-4

where $a \in \partial D$ is arbitrary, C_a is any circuit over ∂D with terminus a, and $\widehat{D}_\zeta$ stands for logarithmic differentiation. But none of these formulas is deep enough to yield a proof of Cartan's lemma or do the jobs one might expect of a Cauchy integral formula.

To revert to Riemann's monodromy problem, we should emphasize that its solution like that of Cousin's problem rests on a factorization lemma for matrix-valued functions. The best version of this was given by Birkhoff [4]. It is the much deeper analogue of Cartan's lemma for the (doubly-connected) annulus:

BIRKHOFF'S FACTORIZATION LEMMA. *Let D be any open disk with center O in $\mathbb{C}$, and let F be a $q \times q$ matrix-valued function in $\widehat{\mathrm{Hol}}.(\mathbb{C} \setminus D)$. Then*

$$F(z) = F_+(z) \cdot F_-(z) \cdot D(z), \quad z \in \mathbb{C} \setminus D,$$

where $F_+ \in \widehat{\mathrm{Hol}}.(\mathbb{C})$, $F_- \in \widehat{\mathrm{Hol}}.(\mathbb{C} \cup \{\infty\} \setminus D)$ with $F_-(\infty) = I$, and $D(z)$ is a diagonal matrix with entries $z^{k_1}, \ldots, z^{k_q}$, where $k_1, \ldots, k_q$ are integers.

Birkhoff's proof rested on solving a system of Fredholm integral equations, and then modifying the solutions obtained by heavy determinental manipulations. When we contrast it with the easy proof of the trivial additive version of this theorem by Cauchy's formula, we again feel the need for a suitable Cauchy formula for multiplicative integrals. Briefly, circuital multiplicative integrals are basic ingredients of the theory of analytic matrix-valued functions, and the absence of a good Cauchy formula for them is a serious handicap in building both the local and global aspects of the theory. The discovery of such a formula would be a major contribution.

A.II.8 The Matricial Hardy and Nevanlinna Classes

Factorization theorems of the type considered by Birkhoff and Cartan (Sec. 7) play a crucial role in several areas of analysis. An interesting theorem of this sort was encountered by Wiener for Hermitian matrix-valued functions in the theory of forcasting of q-variate stochastic processes [35] in 1950. Its complete solution resulted in the following theorem, cf. [36, I, 7.13]:

8.1 THEOREM. *Let F be a $q \times q$ non-negative Hermitian matrix-valued function on the unit circle C in $\mathbb{C}$ such that both F and $\log \det F$ are in $L_1(C)$. Then*

$$F(e^{i\theta}) = \Phi(e^{i\theta}) \cdot \Phi(e^{i\theta})^*, \quad \text{a.e. (Leb)},$$

ISBN 0-201-13509-4

where Φ *is the radial limit of an optimal function* Φ_+ *in the matricial Hardy class* H_2 *on the inner disk* D_+.

For the purposes of this survey we may define a function Φ_+ in H_2 on D_+ with radial limit Φ to be *optimal*[14] iff.

$$\log\{\det) \Phi_+(0)\} = \frac{1}{2\pi}\int_0^{2\pi} \log\{\det \Phi(e^{i\theta})\}\, d\theta. \tag{8.2}$$

This result, though purely analytical, was proved by interpreting $F(\cdot)$ as the spectral density matrix of a purely nondeterministic stationary sequence in Hilbert space and employing the so-called Wold decomposition, cf. [36, I, p. 148]. In the corresponding classical factorization (with $q = 1$), the exponential of Cauchy's expression $\int_C [F(\zeta)/(\zeta - z)]\, d\zeta$, where the last is now a Lebesgue integral, plays the central role, and yields a direct proof. This suggests, once again (cf. Sec. 7), that new light would be shed on the problem by the discovery of a good Cauchy formula for multiplicative integrals which with a Lebesgue-interpretation would continue to be valid on the boundary of the region of holomorphy.

The difficulties in obtaining such a formula are enormous, and it is nowhere in sight. It is therefore all the more remarkable that in the matricial extension of a factorization problem closely related to Wiener's, a multiplicative integral formula, close to Cauchy's, was discovered and put to significant use by Potapov [23] and his colleagues of the Odessa school. To state Potapov's result let us recall the solution of the corresponding scalar factorization problem due to Riesz and Nevanlinna. If $f_+(\cdot)$ is a (nonzero) complex-valued function of the Hardy class H_2 on the unit disk D_+, then

$$f_+(z) = B(z)\cdot\exp\left\{\frac{1}{2\pi}\int_0^{2\pi} S(z, e^{i\theta})\mu(d\theta)\right\}, \quad z \in D_+, \tag{8.3}$$

where $B_+(\cdot)$ is a Blaschke product, $\mu(\cdot)$ is a (bounded) countably-additive real-valued measure on the Borel σ-algebra Bl.$[0, 2\pi]$, and $S(z, \zeta) \underset{d}{=} (\zeta - z)/(\zeta + z)$ is the Schwarz kernel.

It is easy to check that in the special case in which $|f_+(\cdot)| \leq 1$ on D_+, and therefore f_+ falls in the Hardy class H_∞, we have $\mu(\cdot) \leq 0$ on Bl.$[0, 2\pi]$. Potapov [23] obtained a full-fledged matricial extension of this special case of (8.3):

8.4 Potapov's Theorem. *Let* $F_+(\cdot)$ *be a* $q \times q$ *matrix-valued function in the Hardy class* H_∞ *on the unit disk* D_+ *in* C *such that* det $F_+(\cdot)$

14. *Outer* in Beurling's terminology.

ISBN 0-201-13509-4

does not vanish everywhere on D_+, *and*[15]

$$0 \leq F_+(z) \cdot F_+(z)^* \leq I, \qquad \forall z \in D_+.$$

Then

$$F_+(z) = B_+(z) \cdot \widehat{\int_0^l} \exp\{-S(z, e^{i\theta(t)})H(dt)\}, \quad z \in D_+$$

where $B_+(\cdot)$ *is a matrix-valued Blaschke product corresponding to the countable set of zeros of* det $F_+(\cdot)$, $\theta(\cdot)$ *is a monotone increasing function on* $[0, l]$ *to* $[0, 2\pi]$, *and* $H(\cdot)$ *is a non-negative* $q \times q$ *matrix-valued measure on* Bl.$[0, l]$ *such that trace* $H(0, t] = t, 0 \leq t \leq l$, *and* $S(\cdot,\cdot)$ *is the Schwarz kernel as before.*

In the Potapov product integral, it is the integrand which is scalar and it is the measure which takes values in a non-Abelian algebra. Unless such a measure is absolutely continuous with respect to a scalar measure μ, the integral is not of the type discussed above, and has to be defined *ab initio*, cf. Potapov [23, App., Sec. 1]. But the measure $H(\cdot)$ occurring in Potapov's theorem is non-negative Hermitian, and by a result of Rosenberg [25, Sec. 2], $H(\cdot) \ll \tau(\cdot)$, where $\tau(\cdot) \underset{d}{=}$ trace $H(\cdot)$. By Potapov's choice of parametrization, $\tau(\cdot)$ is the Lebesgue measure over $[0, l]$. It would be worthwhile to find out if the integral in his theorem, for any scalar integrand $\phi(\cdot)$, could be defined by

$$\widehat{\int_0^l} \exp\{\phi(t)H(dt)\} = \widehat{\int_0^l} \exp\left\{\phi(t)\frac{dH}{dL}(t)\, dt\right\},$$

where dH/dL is the Radon-Nikodym derivative of $H(\cdot)$ with respect to the Lebesgue measure.

In Ref. 23, Potapov actually proves the matricial analogue of the more general classical result of Nevanlinna in which $f_+(\cdot)$ is merely a bounded meromorphic function on D_+. He and his colleagues, notably Ginsburgh [11, 12], have developed a far reaching theory, which plays a significant ancillary role in the spectral theory of non-self-adjoint operators on a Hilbert space due to Krein, Brodsky, Livsic, and others. This occurs through the application of Potapov-type results to the so-called "matrix characteristic-function" of the operator.

It is well known that the second factor on the right-hand side of (8.3) can in turn be factored into a product of two factors of the same type, involving measures μ_1, μ_2, and that $\mu_1 \ll L \underset{d}{=}$ Lebesgue measure, with

15. For matrices A, B we write $A \leq B$ to mean that $A - B$ is non-negative Hermitian.

ISBN 0-201-13509-4

$d\mu_1/dL = \log|f(\cdot)|$, where f is the radial limit of f_+, and μ_2 is singular with respect to L. The first factor with the measure μ_1 is called the *optimal* (or *outer*) factor of f_+. It is the scalar counterpart of the factor Φ_+ met with in 8.1 and 8.2. In [19] the writer gave an argument to show that the second factor on the right-hand side in Potapov's theorem 8.4 also admits a factorization into a product of two factors, the first of which is optimal in the sense of (8.2). A flaw in this argument was pointed out by Ginsburgh [11]. But his much harder analysis [12, 13] has shown that the conclusion is essentially correct.

These matricial results fall short of the classical, however, in that no integral representation of the optimal factor of F_+ in terms of its radial limit F has been given. We are confronted with the following open problem:

Problem. Let $F_+(\cdot)$ be as in Theorem 8.3, *and let $\Phi_+(\cdot)$ be its optimal factor (so that* (8.2) *holds). Express $\Phi_+(z)$, $z \in D_+$, by a multiplicative integral of the type occurring in Theorem* 8.4. *Show that the measure $H(\cdot)$ therein is absolutely continuous with respect to the Lebesgue measure L, and express the Radon-Nikodym derivative dH/dL in terms of the radial limit $F(\cdot)$ of $F_+(\cdot)$.*

A satisfactory solution of this problem would revolutionize the prediction theory of q-variate stochastic processes.

A.II.9 Holonomy

Let $\mathcal{M}$ be a smooth q-dimensional manifold with affine connection Γ, let x_0, x_1 be points in $\mathcal{M}$, and let C be a path in $\mathcal{M}$ from x_0 to x_1 with the parametrization $x(\cdot)$ on $[a, b]$, so that $x(a) = x_0$, $x(b) = x_1$. Let $u_0 \in \mathcal{M}_{x_0} \underset{d}{=}$ the tangent space at x_0, and for $t \in [a, b]$ let $u(t) \in \mathcal{M}_{x(t)}$ be obtained by *parallel transport* of u_0 along C. The parallelism is of course in relation to the affine connection Γ.

In terms of a Gaussian coordinate system, if $x \in \mathcal{M}$ is represented by the (row-) vector $\vec{x} = (x_1, \ldots, x_q) \in \mathbb{R}^q$ and $u(t)$ by the (row-) vector $\vec{u}(t) = (u_1(t), \ldots, u_q(t))$, then as is well known[16] the functions $u_1(\cdot), \ldots, u_q(\cdot)$ must satisfy the system of n linear differential equations

$$u_j'(t) = -\sum_{i=1}^{q}\left[u_i(t)\sum_{k=1}^{q}\Gamma_{ik}^{j}\{\vec{x}(t)\}x_k'(t)\right], \quad 1 \leq j \leq q, \tag{1}$$

in which the coefficients depend of course on the components of the affine connection as well as the path C. In terms of the $q + 1$ $q \times q$

16. Cf., e.g., Levi-Civita [16, p. 138, Eq. (5.2)], where the notation is different.

ISBN 0-201-13509-4

matrix-valued functions defined by

$$\Gamma_k(\cdot) \underset{d}{=} [\Gamma^j_{ik}(\cdot)],\ 1 \le k \le q, \quad \& \quad A_C(\cdot) \underset{d}{=} \sum_{k=1}^{q} \Gamma_k\{\vec{x}(\cdot)\}x'_k(\cdot), \tag{9.1}$$

we can rewrite (1) in the form

$$\vec{u}\,'(t) = \vec{u}(t)\cdot A_C(t). \tag{9.2}$$

It follows, cf. Section 4, that

$$\vec{u}(t) = \vec{u}_0 \cdot \widehat{\int_a^t} [I + A_C(\tau)\, d\tau\}, \quad a \le t \le b. \tag{9.3}$$

This formula gives us the vector $\vec{u}(t) \in \mathcal{M}_{x(t)}$ obtained by the parallel transport of $u_0 \in \mathcal{M}_{x_0}$ along the path C from x_0 to $x(t)$. Had we started with q linearly independent vectors, $u_1, \ldots, u_q \in \mathcal{M}_{x_0}$instead of just u_0, and formed their component-matrix $U_0 = [u_{ij}]$, where u_{ij} is the jth component of u_i, then the q formulas corresponding to (9.3) could have been written

$$U(t) = U_0 \cdot \widehat{\int_a^t} \{I + A_C(\tau)\, d\tau\}, \quad a \le t \le b. \tag{9.3$'$}$$

In this the rows of $U(t)$ constitute the basis in $\mathcal{M}_{x(t)}$ obtained by the parallel transport of the basis $\{u_1, \ldots, u_q\}$ of $\sum_{x_0}$ along C. If the path C is a circuit from x_0 to x_0, then

$$M(C) \underset{d}{=} \widehat{\int_C} \{I + A_C(t)\, dt\} \tag{9.4}$$

is called the *holonomy matrix at* x_0 *for the circuit* C. If $\tilde{u}_0 \in \mathcal{M}_{x_0}$ is the vector obtained by the parallel transport of u_0 around C, then by (9.3)

$$\tilde{\vec{u}}_0 = \vec{u}_0 \cdot M(C). \tag{9.5}$$

Considerations similar to those engaging us in Section 7 show that the matrices $M(C_x)$ for circuits in $\mathcal{M}$ with the same terminus x constitute a group M_x. This matricial group is relative to the coordinate system adopted. It can be shown, however, that for each such circuit C_x there is a one-to-one linear *holonomy transformation* $H(C_x)$ on $\mathcal{M}_x$ such that for all $u \in \mathcal{M}_x$, $H(C_x)(u)$ is the vector in $\mathcal{M}_x$ obtained by the parallel transport of u along the circuit C_x. Our earlier $M(C_x)$ are the matrices

ISBN 0-201-13509-4

of $H(C_x)$ in different coordinate systems. The transformations $H(C_x)$ for different circuits C_x in $\mathcal{M}$ with the same terminus x also form a group $\mathrm{H}(C_x)$. This is called the *holonomy group at x for the connection* Γ.

The above somewhat sketchy discussion shows how naturally the multiplicative integral intrudes into the study of the holonomy of affine-connected manifolds. Is this intrusion deep enough to yield formulas connecting the holonomy of $\mathcal{M}$ to its Riemannian curvature? The answer is affirmative as Schlesinger showed in an interesting paper [28] he wrote in 1928.

Schlesinger's point of departure lay in defining for a smooth $q \times q$ matrix-valued function $A(\cdot,\cdot)$ on a region $R \subset \mathrm{R}^2$ of the type

$$R = \{(x, y): c \leq y \leq d \quad \& \quad \phi(y) \leq x \leq \psi(y)\},$$

with ϕ, ψ given, the *mixed iterated integral*

$$\widehat{\iint_R} \{I + A(x, y)\, dx\, dy\} \underset{d}{=} \widehat{\int_c^d} \left[I + \left\{ \int_{\phi(y)}^{\psi(y)} A(x, y)\, dx \right\} dy \right], \qquad (9.6)$$

and establishing for it a "Green's Theorem."[17] The enunciation of this theorem contains a *multiplicative circuital integral* involving two smooth $q \times q$ matrix-valued functions F_1, F_2 on a smooth circuit C in R^2, which is defined by

$$\widehat{\oint_C} (I + F_1\, dx_1 + F_2\, dx_2)$$

$$\underset{d}{=} \widehat{\int_a^b} (I + [F_1(\vec{x}(t)\}x_1'(t) + F_2\{\vec{x}(t)\}x_2'(t)]\, dt), \qquad (9.7)$$

where $x(\cdot) = (x_1(\cdot), x_2(\cdot))$ is a parametrization of C. The theorem then reads as follows, cf. Ref. 28, Eq. (22),

THEOREM (GREEN'S THEOREM). *For a rectangle $R = [a, b] \times [c, d]$ in R^2 and smooth $q \times q$ matrix-valued functions $F_1(\cdot,\cdot)$, $F_2(\cdot,\cdot)$ on $R \cup \partial R$,*

$$\widehat{\oint_{\partial R}} (I + F_1\, dx_1 + F_2\, dx_2)$$

$$= \widehat{\iint_R} [I + L_F(x, y)\cdot[F](x, y)\cdot L_F(x, y)^{-1}\, dx\cdot dy], \qquad (9.8)$$

17. A "Green's theorem" is also available for the multiplicative iterated integral $\widehat{\int\int}$, cf. [34, p. 92, Theorem I], but this, more full-fledged, version does not seem to have a bearing on the problem at hand.

ISBN 0-201-13509-4

where for $(x, y) \in R$,

$$L_F(x, y) \underset{d}{=} \widehat{\int_{b.}^{y}} [I + F_2(a, \eta)\, d\eta] \widehat{\int_{a_1}^{x}} [I + F_1(\xi, y)\, d\xi]$$

and

$$[F](x, y) \underset{d}{=} \frac{\partial F_2}{\partial x_1} - \frac{\partial F_1}{\partial x_2} - \{F_2(\cdot)F_1(\cdot) - F_1(\cdot)F_2(\cdot)\}.$$

Note that when the values of F_1 and F_2 commute, the cumbersome right-hand side of (9.8) collapses to

$$\exp \iint_R \left(\frac{\partial F_2}{\partial x_1} - \frac{\partial F_1}{\partial x_2}\right) dx_1\, dx_2,$$

and the left-hand side to $\exp \int_{\partial R} (F_1\, dx_1 + F_2\, dx_2)$, and so (9.8) reduces to the familiar (additive) Green's formula.

Our next step consists in applying the formula (9.8) to well-chosen functions F_1, F_2 defined with the aid of the affine connection Γ of our q-dimensional smooth manifold $\mathcal{M}$. Let $\phi(\cdot)$ be a smooth one–one function on the closed rectangle $R \cup \partial R$ to $\mathcal{M}$. Then the two-dimensional surface $\phi(R)$ in $\mathcal{M}$ (a warped parallelogram) has the boundary $C = \phi(\partial R)$. Denoting the coordinates in the plane of R by (ξ_1, ξ_2) or (ξ, η) instead of (x, y), and the Gaussian coordinates in $\mathcal{M}$ by $\vec{x} = (x_1, \ldots, x_q)$, we define F_1, F_2 on $R \cup \partial R$ by

$$F_j(\xi_1, \xi_2) \underset{d}{=} \sum_{k=1}^{q} \Gamma_k\{\vec{\phi}(\xi_1, \xi_2)\} \frac{\partial \phi_k}{\partial \xi_j}(\xi_1, \xi_2), \quad j = 1, 2. \tag{9.9}$$

Computation based on (9.1) show that with this choice of F_1, F_2, and C,

$$\widehat{\int_{\partial R}} (I + F_1\, d\xi_1 + F_2\, d\xi_2) = \widehat{\oint_C} \{I + A_C(\tau)\, d\tau\} = M(C).$$

Thus, under the present setting the left-hand side of (9.8) reduces to $M(C)$. To make a corresponding reduction of the right-hand side of (9.8) we must introduce the *Riemann-Christoffel curvature tensor field*:[18]

$$R^j_{irs} \underset{d}{=} \frac{\partial \Gamma^j_{is}}{\partial x_r} - \frac{\partial \Gamma^j_{ir}}{\partial x_s} + \sum_{k=1}^{q} (\Gamma^k_{is}\Gamma^j_{kr} - \Gamma^k_{ir}\Gamma^j_{ks}), \tag{2}$$

18. Our index conventions are those of Schlesinger.

ISBN 0-201-13509-4

and, in the spirit of (9.1), the q^2 associated $q \times q$ matrix-valued functions

$$R_{rs} = [R^i_{irs}]. \tag{9.10}$$

Equations (2) then boil down to

$$R_{rs} = \frac{\partial \Gamma_s}{\partial x_r} - \frac{\partial \Gamma_r}{\partial x_s} + \Gamma_s \Gamma_r - \Gamma_r \Gamma_s, \quad 1 \le r < s \le q. \tag{9.11}$$

Calculations then show that the right-hand side of (9.8) (with the dummy variables ξ, η replacing x, y) reduces to

$$\widehat{\iint_R} \left[I + \sum_{r<s} L_F(\xi, \eta) \cdot R_{sr}\{\vec{\phi}(\xi, \eta)\} \cdot L_F(\xi, \eta)^{-1} \cdot \frac{\partial(\phi_r, \phi_s)}{\partial(\xi, \eta)} \, d\xi \, d\eta \right].$$

We thus wind up with Schlesinger's important formula [28, Eq. (34a)]:

$$M(C) = \widehat{\iint_R} \left[I + \sum_{r<s} L_F(\xi, \eta) \cdot R_{sr}\{\vec{\phi}(\xi, \eta)\} \cdot L_F(\xi, \eta)^{-1} \cdot \frac{\partial(\phi_r, \phi_s)}{\partial(\xi . \eta)} \, d\xi \, d\eta \right], \tag{9.12}$$

connecting the holonomy matrix for a (parallelogram-like) circuit C in $\mathcal{M}$ with the values of the Riemannian curvature at points in any two-dimensional manifold stretched over C and lying in $\mathcal{M}$.

The result (9.12) is powerful enough to yield several corollaries. It shows, for instance, that if $R_{sr} = 0$ on the patch $\phi(R) \subset \mathcal{M}$, then $M(C) = I$, and that if $R_{sr} = 0$ on $\mathcal{M}$, then the holonomy group H_x at any $x \in \mathcal{M}$ collapses to $\{I\}$. In this case at any point $y \in \mathcal{M}$, the vector $v \in \mathcal{M}_y$ obtained by the parallel transport of a vector $u \in \mathcal{M}_x$ along a path C in $\mathcal{M}$ from x to y will not depend on C. Consequently, the tangent vectors obtained at different points y of $\mathcal{M}$ by the parallel transport of a given tangent vector elsewhere will be the values $u(y)$ of a (single-valued) *vector field* $u(\cdot)$ on $\mathcal{M}$. In this case we have Euclidicity. As another corollary we have Schlesinger's result that for a two-dimensional Riemannian manifold, (9.12) yields the Gauss-Bonnet formula [28, Eq. (69)].

How would a differential geometer unacquainted with multiplicative integration formulate the relationship between holonomy and curvature? The answer may be found in the section in Levi-Civita's classic [16, Chap. VII, Sec. 2] entitled "Cyclic displacement over an *elementary* parallelogram" (my emphasis). For sufficiently small $\epsilon > 0$, let C_ϵ be the "infinitesimal parallelogram" with vertex at $x \in \mathcal{M}$ and sides $\epsilon\alpha$, $\epsilon\beta$, where $\alpha, \beta \in \mathcal{M}_x$ are tangent vectors. Levi-Civita shows that the change Δu_j suffered by the component u_j of a vector $u \in \mathcal{M}_x$ by its parallel

ISBN 0-201-13509-4

transport round C_ϵ is given (in our notation) by

$$\Delta u_j = \sum_{i=1}^{q} u_j \left\{ \sum_{h,k=1}^{q} R^j_{ihk} \cdot \epsilon\alpha_h \cdot \epsilon\beta_k \right\}, \quad \text{approx.} \tag{3}$$

If we introduce the $q \times q$ matrix

$$A \underset{d}{=} [a_{ij}], \quad a_{ij} \underset{d}{=} \sum_{h,k=1}^{q} R^j_{ihk}\alpha_h\beta_k,$$

we can rephrase (3) in the form

$$\Delta\vec{u} = \epsilon^2 \cdot \vec{u} \cdot A, \quad \text{approx.} \tag{4}$$

Since, cf. (9.5), $\Delta\vec{u} \underset{d}{=} \vec{\bar{u}} - \vec{u} = \vec{u}\cdot\{M(C_\epsilon) - I\}$, we see from (4) that

$$M(C_\epsilon) = I + \epsilon^2 A + o(\epsilon^2), \quad \text{as } \epsilon \to 0. \tag{9.13}$$

This establishes a nexus between small circuital holonomy at x and the curvature at x.

The local result (9.13) is available without the use of multiplicative integration. It is somewhat akin to the local version

$$\int_{\partial(\Delta S)} (V \cdot ds) = \text{curl } V \cdot |\Delta S| \tag{5}$$

of Stokes theorem for an infinitesimal surface element ΔS. But while ordinary surface integration suffices to go from (5) to Stokes theorem, we need Schlesinger's mixed iterated integral to recover a global result from (9.13). The utility of the multiplicative integral here is about the same as its utility in the study of monodromy (Sec. 7).

A.II.10 Perturbation and Partial Integration

Just as the flow of a fluid is describable by its propagator $T(\cdot,\cdot)$ or its velocity $V(\cdot)$, with the connection (2.12), so the behavior of many systems evolving in time can be described by an operator-valued function $T(\cdot,\cdot)$ on $\mathbb{R}^2$ or an operator-valued function $A(\cdot)$ on $\mathbb{R}$, bearing the interrelation:

$$T(t, s) = \widehat{\int_t^s} \exp\{A(t)\, dt\}. \tag{1}$$

The function $A(\cdot)$ is the counterpart in the general case of the infinitesimal generator appearing in the steady state.

ISBN 0-201-13509-4

For systems governed by differential laws $A(\cdot)$ is the primary function. In many situations our knowledge of this function will be subject to error, and it is a germane question as to how $T(\cdot,\cdot)$ is affected by a perturbation of the function $A(\cdot)$. To emphasize the dependence of $T(\cdot,\cdot)$ on $A(\cdot)$, let us write $T_A(\cdot,\cdot)$ in place of $T(\cdot,\cdot)$, so that (1) reads

$$T_A(t, s) = \widehat{\int_t^s} \exp\{A(t)\, dt\}. \tag{10.1}$$

Then the general perturbation problem is to determine the deviation of $T_{A+B}(\cdot,\cdot)$ from $T_A(\cdot,\cdot)$, where $B(\cdot)$ is another operator-valued function akin to $A(\cdot)$.

In the linear case the relationship between $T_{A+B}(\cdot,\cdot)$ and $T_A(\cdot,\cdot)$ is given by the analogue of the law of *integration-by-parts*, cf. Volterra and Hostinsky [34, Sec. 4, p. 64], which in the light of (2.13) may be given the following (exponential) formulation:

$$T_{A+B}(b, a) = T_A(b, a)\cdot \widehat{\int_b^a} \exp[T_A(t, a)^{-1}\cdot B(t)\cdot T_A(t, a)\, dt]. \tag{10.2}$$

This in the *general perturbation formula* for the linear case, i.e., the case in which the functions $A(\cdot)$, $B(\cdot)$ have values in a Banach algebra A and $T(\cdot,\cdot)$ in A_{inv}.

A generalization of (10.2) covering the case in which $T(\cdot,\cdot)$ takes values in a Lie group G and $A(\cdot)$, $B(\cdot)$ take values in its Lie algebra A has been given by Hamilton in his unpublished dissertation [14, p. 54]. (It is reproduced in Ref. 20, Sec. 3.1.) This result, which involves the so-called adjoint representation, shows that (10.2) will survive when the values of $A(\cdot)$ and $B(\cdot)$ lie in a pseudolinear algebra, cf. Section 4.

The general formula (10.2) simplifies appreciably when the values of $A(\cdot)$ and $B(\cdot)$ commute. In the steady-state case in which these functions are constant-valued, we can deduce from (10.2) the *product-formula*:

$$\exp\{(A + B)t\} = \lim_{n\to\infty} [\exp(At/n)\cdot\exp(Bt/n)]^n, \tag{10.3}$$

the matricial version of which was known to Lie.

Lie's result has been extended by Trotter [32] to the case in which the A, B in (10.3) are *discontinuous linear operators* on a Banach space, and the limit is taken in the strong-operator topology. Trotter's result is of course much deeper than (10.3), and its derivation from a more general perturbation formula, akin to (10.2), will require an extension of the theory of multiplicative integration to the realm of discontinuous operators. As these issues are discussed more fully in this book and also in the writer's recent paper [20], we shall say no more about them.

ISBN 0-201-13509-4

A.II.11 Concluding Remarks

There are many facets of the subject of multiplicative integration which we have not discussed in this survey. But what we have said should suffice to show how the subject ramifies in far-flung areas in mathematics, and the way in which it brings to the forefront certain underlying unities.

All mathematicians are acquainted with the additive integral, but very few with the multiplicative. Of course the latter, by virtue of its confinement to totally ordered domains, is much narrower in scope than the former. But this factor alone cannot account for the disparity in the knowledge people have of the two integrals. One reason for it is the dearth of expository literature. The very fine treatises of Schlesinger [26, 27] have been out of circulation for decades, and their only follow-up has been the 1938 monograph of Volterra and Hostinsky [34] with its unfortunate notation, and an isolated chapter in Gantmacher's book [9].[19] The latter is perhaps the only connected account of the subject available in English, and while it is very useful, especially for its revelation of errors in the early work of Volterra and Birkhoff, its scope is limited to just one application of the subject.

The new book by Professors Dollard and Friedman will therefore be a significant addition to the literature. Those who have read this Appendix will, we hope, be encouraged to pursue what precedes it in this book. As for those who are led to the book by a specific interest or problem, it is our hope that this Appendix may convince them that they are dealing with a subject of wide scope and elegance—indeed with one of Volterra's lasting contributions—in which there are interesting open questions.

References

1. N. Arley, *On the Theory of Stochastic Processes and their Applications to the Theory of Cosmic Radiation*. Wiley, New York, 1943.
2. Garrett Birkhoff, On product integration, *J. Math. Phys.* **16** (1937), 105–132.
3. Garrett Birkhoff, Analytic groups, *Trans. Amer. Math. Soc.* **43** (1938), 61–101.
4. G. D. Birkhoff, A theorem on matrices of analytic functions, *Math. Ann.* **74** (1913), 122–133.
5. G. D. Birkhoff, The generalized Riemann problem for linear differential equations and the allied problems for linear difference and q-difference equations, *Proc. Amer. Acad. Arts Sci.* **49** (1913), 521–568.
6. H. Cartan, Sur les matrices holomorphes de n variables complexes, *J. Math. Pure Appl.* **19** (1940), 1–26.
7. C. Chevalley, *Theory of Lie Groups*. Princeton Univ. Press, N.J., 1946.
8. J. L. Doob, *Stochastic Processes*. Wiley, New York, 1953.
9. F. R. Gantmacher, *Applications of the Theory of Matrices*. Interscience Publishers, New York, 1959.

19. Possibly there are books in Russian that we do not know of.

ISBN 0-201-13509-4

10. Ju. P. Ginzburg, On J-contractive operator functions, *Dokl. Akad. Nauk SSSR* **117** (1957), 171–173.
11. Ju. P. Ginzburg, The factorization of analytic matrix functions, *Sov. Math., AMS* (Nov.-Dec. 1964), **5**(6); translation of *Doklady TOM* **159**(1-6), 1510–1514.
12. Ju. P. Ginzburg, Multiplicative representations of bounded analytical operator-functions, *Dokl. Akad. Nauk SSSR* **170** (1966), 1125–1128.
13. Ju. P. Ginzburg, *Usp. Mat. Nauk* **12** (1967) 163–167.
14. J. F. Hamilton, Multiplicative Riemann integration and logarithmic differentiation in Lie algebras and Lie groups, Indiana University Dissertation, 1973 (unpublished).
15. S. Lang, *Introduction to Differentiable Manifolds*. Interscience Publishers, New York, 1962.
16. T. Levi-Civita, *The Absolute Differential Calculus*, Blackie, London, 1946.
17. G. W. Mackey, *Unitary Group Representations in Physics, Probability and Number Theory*. Benjamin, Advanced Book Program, Reading, Mass., 1978.
18. P. R. Masani, Multiplicative Riemann integration in normed rings, *Trans. Amer. Math. Soc.* **61** (1947), 147–192.
19. P. Masani, Une généralisation pour les fonctions matricielles de la classe de Hardy H_2 d'un théorème de Nevanlinna, C.R., Acad. Sci., Paris **251** (1960), 318–320.
19′. P. Masani, Propagators and dilations, in *Probability Theory on Vector Spaces*, (A. Weron, ed.) Springer Lecture Notes No. 656, pp. 95–117, 1978.
20. P. Masani, Multiplicative partial integration and the Trotter product formula, *Advs. in Math.* (to appear).
21. P. Masani and T. Vijayaraghavan, An analogue of Laurent's theorem for a simply connected region, *J. Ind. Math. Soc.* **16** (1952), 25–30.
22. G. Peano, Intégration par séries des équations différentielles linéaires, *Math. Ann.* **32** (1888), 450–456.
23. V. P. Potapov, The multiplicative structure of J-contractive matrix functions, *Trudi Mosk. Matem. Obsch* **4** (1955), 125–236. (in Russian) (*Amer. Math. Soc. Transl.* **15** (1960), 131–243).
24. G. Rasch, Zur Theorie und Anwendung des Productintegrals, *J. Reine Angew. Math.* **171** (1934), 65–119.
25. M. Rosenberg, The square integrability of matrix-valued functions with respect to a non-negative measure, *Duke Math. J*, **31** (1964), 291–298.
26. L. Schlesinger, *Vorlesungen über Lineare Differentialgleichungen*. Berlin, 1908.
27. L. Schlesinger, *Einführung in die Theorie der gewöhnlichen Differentialgleichungen auf funktiontheoretischer Grundlage*. Berlin, 1922.
28. L. Schlesinger, Parallelverschiebung und Krummungstensor, *Math. Ann.* **99** (1928), 413–434.
29. L. Schlesinger, Neue Grundlagen für einen Infinitesimalkalkul der Matrizen, *Math. Zeit.* **33** (1931), 33–61.
30. L. Schlesinger, Weitere Beiträge zum Infinitesimalkalkül der Matrizen, *Math. Zeit.* **35** (1932), 485–501.
31. F. M. Stewart, Integration in non-commutative systems, *Trans. Amer. Math. Soc.* **68** (1950), 76–104.
32. H. F. Trotter, On the product of semigroups of operators, *Proc. Amer. Math. Soc.* **10** (1959), 545–551.
33. V. Volterra, Sulle equazioni differzeniali lineari, *Rendiconti Accademia dei Lincei* **3** (1887), 393–396.
34. V. Volterra and B. Hostinsky, *Operations Infinitesimales Lineaires*. Paris, 1938.
35. N. Wiener, *Comprehensive View of Prediction Theory*, Proc. Int. Cong. Math., Vol. 2, pp. 308–321, Cambridge, Mass., 1950.
36. N. Wiener and P. Masani, The prediction theory of multivariate stochastic processes, Part 1, *Acta Math.* **98** (1957), 111–150; Part II, *Acta Math.* **99** (1958), 93–137.

ISBN 0-201-13509-4

Index

Index